Grundlehren der mathematischen Wissenschaften 253

A Series of Comprehensive Studies in Mathematics

Bernard Dwork

Lectures on *p*-adic Differential Equations

Springer-Verlag
New York Heidelberg Berlin

Bernard Dwork
Department of Mathematics
Princeton University
Princeton, NJ 08540
U.S.A.

AMS Subject Classifications (1980): 10D30, 14G20, 33A30

Library of Congress Cataloging in Publication Data

Dwork, Bernard M.
Lectures on p-adic differential equations.

(Grundlehren der mathematischen Wissenschaften; 253)
Bibliography: p.
1. Differential equations. 2. p-adic numbers.
I. Title. II. Series.
QA372.D86 515.3′5 82-5764
AACR2

With 5 Illustrations

Softcover reprint of the hardcover 1st edition 1982

Typeset by Composition House Ltd., Salisbury, England.
Printed by Malloy Lithographing, Inc., Ann Arbor, MI.

9 8 7 6 5 4 3 2 1

ISBN-13:978-1-4613-8195-2 e-ISBN-13:978-1-4613-8193-8
DOI: 10.1007/978-1-4613-8193-8

Preface

The present work treats p-adic properties of solutions of the hypergeometric differential equation

$$\left(x(1 - x)\frac{d^2}{dx^2} + (c(1 - x) + (c - 1 - a - b)x)\frac{d}{dx} - ab\right)y = 0,$$

with a, b, c in $\mathbb{Q} \cap \mathbb{Z}_p$, by constructing the associated Frobenius structure. For this construction we draw upon the methods of Alan Adolphson [1] in his 1976 work on Hecke polynomials. We are also indebted to him for the account (appearing as an appendix) of the relation between this differential equation and certain L-functions. We are indebted to G. Washnitzer for the method used in the construction of our dual theory (Chapter 2).

These notes represent an expanded form of lectures given at the U. L. P. in Strasbourg during the fall term of 1980. We take this opportunity to thank Professor R. Girard and IRMA for their hospitality.

Our subject—p-adic analysis—was founded by Marc Krasner. We take pleasure in dedicating this work to him.

Contents

Introduction

The p-adic theory of an ordinary differential equation seems to have begun with the 1937 result of Lutz [25]: A formal power series solution of a first-order system without singularity converges p-adically in a nontrivial disk.

The p-adic theory of linear differential equations was started by John Tate in 1958 [9] with a result equivalent to the following: Let λ lie in an extension field of $\mathbb{Q}$ with p-adic valuation. Suppose that the image in the residue class field of the elliptic curve $y^2 = x(x-1)(x-\lambda)$ is not supersingular. Then, in the completion of the maximally unramified extension of $\mathbb{Q}_p(\lambda)$ there exists a constant C such that the equation

$$(0.1) \qquad C\frac{du}{dx} = \frac{u}{\sqrt{x(x-1)(x-\lambda)}}$$

has for generic choice of x_0 a power series solution in $\mathbb{Q}_p(\lambda, C, x_0)[[x - x_0]]$ which converges for all x in some universal domain satisfying the condition $|x - x_0| < 1$. Furthermore, if the residue class field of $\mathbb{Q}_p(\lambda)$ is finite then C may be used to compute the unit root of the zeta-function of the reduced elliptic curve.

In testing the work of Tate we found that the hypergeometric series

$$F(x) = F(\tfrac{1}{2}, \tfrac{1}{2}, 1, x) = \sum_{s=0}^{\infty} \left(\frac{(\frac{1}{2})_s}{s!}\right)^2 x^s, \qquad \left((\tfrac{1}{2})_s = \prod_{j=0}^{s-1}(\tfrac{1}{2} + j)\right),$$

which obviously converges for $|x| < 1$ ($p \neq 2$), has the remarkable property that

(0.2) $F(x)/F(x^p)$ can be continued analytically to the complement in Ω (our universal p-adic domain) of $(p-1)/2$ residue classes.

(0.3) F'/F can be continued analytically in the same region.

The reader may find a proof for general (but not for all) a, b, c in Theorem 9.6. The above result means that the differential equation

$$x(1-x)\frac{d^2}{dx^2} + (1-2x)\frac{d}{dx} - \frac{1}{4} \tag{0.4}$$

satisfied by F has almost everywhere a two-dimensional solution space with a canonically defined one-dimensional subspace. This is completely different from the classical theory.

It has been known since 1958 that the $(p-1)/2$ missing disks are precisely the supersingular values of the λ invariant. The impossibility of extending F'/F into the supersingular domain (see Theorem 16.11) was not understood until 1969 at which time it was also found that if we replace $x \mapsto x^p$ by the canonical lifting, ϕ_1 (cf. Chapter 13) then $F(x)/F(\phi_1(x))$ may be extended into the supersingular domain. Furthermore, this property characterizes the canonical lifting.

As known to Euler, $F(a, b, c; \lambda)$ is a period of a differential on a certain curve. If we reduce to characteristic p then the Frobenius mapping $x \to x^p$, $\lambda \to \lambda^p$ induces a mapping of this differential into a differential on a related curve. By lifting this Frobenius mapping to characteristic zero (and here we need p-adic limits) we obtain a mapping between periods. This general idea is worked out explicitly in this volume. We obtain a matrix $B(\lambda)$ defined everywhere, except in a small disk about $\lambda = 1$, such that solutions (u_1, u_2) of $(2.4.5.1)_{a',b',c'}$ at λ_0^p are mapped into solutions of $(2.4.5.1)_{a,b,c}$ at λ_0 by

$$(u_1, u_2) \mapsto (u_1, u_2)(\lambda^p)B(\lambda). \tag{0.5}$$

By applying an automorphism to the coefficients of u we obtain a semilinear mapping of solutions at λ_0 of the first equation into solutions at the same point of the second equation. By l-fold iteration (l being the period of p modulo d, the common denominator of a, b, c) we obtain a semilinear transformation of solutions of $(2.4.5.1)_{a,b,c}$ at λ_0 into solutions of the same equation at the same point. The "eigenvectors" of this transformation are the central objects of our study. In the situation indicated here we say that our differential equation has a strong Frobenius structure. All this has been generalized in terms of cohomology of curves and of varieties of higher dimension. As we shall see shortly the general p-adic theory of differential equations, i.e., of equations without known Frobenius structure, has not had a strong development. Even for the second-order case little is known. The object of this work is to use the Frobenius structure of $F(a, b, c, x)$ to obtain information about a class of second-order differential equations which we hope may be sufficiently large to give useful hints for the general case.

We now try to summarize the little that is known without the hypothesis of strong Frobenius structures.

Let K be an extension of $\mathbb{Q}$, let $D = d/dx$ and let $\mathbf{D}$ be the ordinary linear differential operator

$$\mathbf{D} = D^n + g_{n-1}D^{n-1} + \cdots + g_0,$$

with coefficients $g_1, \ldots, g_n \in K(x)$. We imbed K in some universal domain Ω which is algebraically closed, complete under a rank one valuation extending the p-adic valuation of $\mathbb{Q}$, and whose residue class field contains an element $\bar{t}$ transcendental over the residue class field of K.

For each $x_0 \in \Omega$ we consider formal power series solutions $y \in \Omega[[x - x_0]]$ of the equation $\mathbf{D}y = 0$. Naturally we are also interested in solutions at infinity, i.e., in solutions in $\Omega[[1/x]]$. We may also consider the situation in which x_0 is a singularity of $\mathbf{D}$. We consider the system

$$\text{(0.6)} \qquad \frac{d}{dx}\begin{pmatrix} y_1 \\ y_2 \\ \vdots \\ y_n \end{pmatrix} = \begin{pmatrix} 0 & 1 & 0 & \cdots & 0 \\ 0 & 0 & 1 & \cdots & 0 \\ \cdots & \cdots & \cdots & \cdots & \cdots \\ 0 & 0 & 0 & \cdots & 1 \\ -g_0 & -g_1 & -g_2 & \cdots & -g_{n-1} \end{pmatrix} \begin{pmatrix} y_1 \\ \vdots \\ y_n \end{pmatrix}$$

and for each $x_0 \in \Omega$ associate an $n \times n$ matrix U with coefficients in $\Omega[[x - x_0]]$ (resp. $\Omega[[\xi]]$ where $\xi = (x - x_0)^{1/n!}$) if x_0 is not (resp. is) an irregular singularity of $\mathbf{D}$ by the condition that

$$\text{(0.7)} \qquad U(x - x_0)\exp[C \log(x - x_0)]$$

$$\left(\text{resp. } U(\xi)\exp\left[C \log \xi + \Delta\left(\frac{1}{\xi}\right)\right]\right),$$

be a formal solution matrix, it being understood that C is an $n \times n$ constant matrix whose eigenvalues are roots of the indicial polynomial of $\mathbf{D}$ at x_0 and in the irregular case that Δ is a diagonal polynomial matrix.

Subject to the hypothesis that none of the roots of the indicial polynomial at x_0 are p-adic Liouville numbers, the matrix U has a nonzero radius convergence (Clark [6], Baldassarri [4a]). This condition is certainly satisfied if the roots in question are algebraic over $\mathbb{Q}$. Unless otherwise indicated we shall use the expression "solutions of $\mathbf{D}$ at x_0" to designate the entries in the first row of U.

We view the solutions of $\mathbf{D}$ as functions on a disk in Ω. Since each triangle in Ω is isosceles there is no Weierstrass theory of analytic continuation. In particular, solutions at x_0 need not converge up to the nearest singularity. An example is given by $\exp x = \sum x^s/s!$ which converges only for $\operatorname{ord} x > 1/(p - 1)$ and not on all of Ω as should be expected.

Among the questions to be posed we may list

(0.8) What is the radius of convergence of the solutions at x_0?

(0.9) What is the order of growth of solutions as $(x - x_0)$ approaches the radius of convergence?

(0.10) How does the filtration by growth conditions vary from disk to disk?

We have mentioned "the" generic residue class $\bar{t}$. We recall that $D(0, 1^+)$ coincides with the ring of integers of Ω and $D(0, 1^-)$ with the maximal ideal. We decompose $D(0, 1^+)$ into translates of $D(0, 1^-)$. These are the finite residue classes and we define the infinite residue class to be the image of $D(0, 1^-)$ under $x \mapsto 1/x$. Let t be a representative of $\bar{t}$. We define the gauss norm $|g|_{\text{gauss}} = |g(t)|$ for elements $g \in \Omega(x)$.

Subject to the mild hypothesis that $|g_j|_{\text{gauss}} \leq 1$ $(j = 0, 1, \ldots, n-1)$ we may conclude that the solutions of $\mathbf{D}$ at t converge in the disk

$$\operatorname{ord}(x - t) > \frac{1}{p-1}.$$

This is the best that can be said as Deligne has observed using (0.1) with $C = 1$ to construct a Lamé differential equation with rational exponents which has exactly this radius of convergence.

In general if the solutions of $\mathbf{D}$ at t all converge in $D(t, r^-)$ $(r \leq 1)$ then the solutions at x_0 converge in $D(x_0, r^-)$ provided x_0 does not lie in a *singular* residue class, i.e., a residue class containing singularities of $\mathbf{D}$. The filtration of solutions at t according to radius of convergence leads to a factorization of $\mathbf{D}$ over a ring of superadmissible functions, i.e., analytic elements whose support meets every residue class and contains almost every class. This type of filtration will not play a role in the present work and will appear in the theory of differential equations associated with geometry only in the treatment of exceptional primes (e.g., the divisor of numerator or denominators of a, b, c).

We are unaware of any general way to determine the radius of convergence of solutions at t. If, however, the solutions of $\mathbf{D}$ all converge in a residue class $D(x_0, 1^-)$ which contains at most just one regular singular point with exponents in $\mathbb{Z}_p$ then we may conclude that the solutions at t converge in $D(t, 1^-)$.

In particular if there is a power series solution of $\mathbf{D}$ converging in $D(0, 1^-)$ then there must be at least one solution converging and bounded in $D(t, 1^-)$. For second-order differential equations such as in the present volume we can settle the radius of convergence by checking one power series solution at zero and checking that the wronskian has a branch at t which converges in $D(t, 1^-)$. This method is effective for the differential equation treated in this volume (we give a more direct treatment in Corollary 4.6.3) but we repeat that no general results are known for the second-order differential equation with four singular points, all regular and all exponents rational.

Somewhat more subtle are questions (0.9) and (0.10). Let $|x_0| = 1$. The function

$$\log \frac{x}{x_0} = \log\left(1 + \frac{x - x_0}{x_0}\right) = -\sum_{m=1}^{\infty} \left(\frac{x_0 - x}{x_0}\right)^m \frac{1}{m}$$

is, of course, unbounded on the disk $D(x_0, 1^-)$. If this disk contains no singularities of **D** and if the solutions of **D** at x_0 all converge in the disk then they are bounded (as $|x - x_0| \to 1$) by the order of growth of $(\log(x/x_0))^{n-1}$. This estimate can be made effective and an *ad hoc* proof of the result for differential equation of the second order is given in Corollary 11.4. The question of growth (as well as that of radius of convergence) is only partially understood in the singular disks. For the second-order case the noneffective result remains valid in the case of a singular disk with just one regular singularity whose exponents lie in $\mathbb{Z}_p$.

Suppose now that the solutions of **D** at t all converge in $D(t, 1^-)$. If these solutions are all bounded on that disk then the solutions on any nonsingular residue class are also bounded and the filtration by order of growth is trivial.

Suppose now that not all solutions are bounded on $D(t, 1^-)$. It is known that at least one solution must be bounded and it is known that there exists a nontrivial factor $\mathbf{D}_0$ of **D** which annihilates precisely the solutions of **D** bounded in $D(t, 1^-)$. The coefficients of $\mathbf{D}_0$ lie in $K(x)^{\text{comp}}$, the completion of $K(x)$ under the gauss norm. (For an *ad hoc* proof in the second-order case see §11.10.) For each real number $\alpha > 0$ we may construct $\mathbf{D}_\alpha$ the factor of **D** which annihilates the solutions of **D** at t whose growth in $D(t, 1^-)$ are dominated by $\log^\alpha(x/t)$. The coefficients of each $\mathbf{D}_\alpha$ also lie in $K(x)^{\text{comp}}$.

We say that a nonsingular residue class $D(x_0, 1^-)$ is G-supersingular if the growth filtration of solutions on that disk is different from that on the generic disk. Roughly speaking (precisely if $n = 2$) a residue class is G-ordinary (i.e., the opposite of G-supersingular) if the coefficients of the $\mathbf{D}_\alpha$ are analytic on the residue class. This definition of supersingularity may also be applied to singular residue classes containing at worst one regular singularity with exponents in $\mathbb{Z}_p$. (For $F(\frac{1}{2}, \frac{1}{2}, 1, x)$ compare this with our starting point (0.3).)

At this moment we cannot show (in general) even for $n = 2$ that there are any ordinary residue classes which are *algebraic* over the residue class field of K. Of course all the generic residue classes are ordinary. We conjecture that almost all residue classes algebraic over the residue class field of K are ordinary.

In the present work we use a different criterion for supersingularity, namely the more conventional one involving the magnitudes of the eigenvalues of Frobenius. However in Theorem 9.10.5 we show how the G-supersingular locus may be determined explicitly from a finite number of the

X_s (cf. equation (11.2)) associated with the differential equations of $F(a, b, c, x)$, $F(a', b', c', x)$ etc. without explicitly determining the Frobenius matrix B. Since the X_s exist for all second-order equations even for those without known Frobenius structure, this may be a useful illustration. In the following, unless otherwise indicated, "ordinary" and "supersingular" will be used in the slope sense of Definition 13.5.

We determine the orders of growth of solutions in ordinary disks (Corollary 9.7) and in supersingular disks (Lemma 16.9) subject to the restriction that $(p - 1)(a, b, c) \varepsilon \mathbb{Z}^3$, i.e., that $l = 1$. This restriction could be removed by further study of the slopes (in the sense of Dieudonné theory) of products of matrices such as the matrix A in Proposition 12.2. The estimates for singular disks are not given explicitly but could be worked out by the estimates of Lemma 21.5.

To clarify the relation of the two definitions of supersingularity we observe that for (2.4.5.1) subject to (9.0.1)–(9.0.5), each disk ordinary in the slope sense is also G-ordinary. The converse holds for $l = 1$.

Much of the present work involves an analysis of how the Frobenius structure may be put in a simpler form at the expense of replacing λ^p in (0.5) by a more complicated lifting of Frobenius. By means of this canonical lifting the matrix B is made triangular. This plays a critical role in the determination of growth in supersingular disks. The canonical lifting is constructed without hypothesis concerning l.

Perhaps the most interesting result of the present work is the determination of the support of the canonical lifting under the hypothesis $l = 1$. The treatment involves two parts, Chapter 15 where the support is shown to contain a certain part of supersingular disks and Chapter 18 where it is shown that this part is indeed maximal. For the case $(a, b, c) = (\frac{1}{2}, \frac{1}{2}, 1)$ this had been established by Deligne (unpublished) in 1968 on the basis of the modular equation, i.e., the algebraic relation between λ and λ' where $\tau(\lambda') = p\tau(\lambda)$, τ being the ratio of two branches of $F(\frac{1}{2}, \frac{1}{2}, 1, \lambda)$. (Indeed Deligne did more since he treated elliptic modular functions of all levels, not just level two.) The treatment in this volume makes no use of the modular equation. Lubin's theory [24] of canonical subgroups may give an alternate approach to this problem.

Despite the length of the present work we have not treated such topics as:

(1) Relation to crystalline cohomology.
(2) Application to the unit root zeta-function.
(3) Variation of Frobenius structure with a, b, c, in particular the degeneration when $a, b, c - a, c - b$ assume integral values.

We now say a few words about previous p-adic studies of hypergeometric functions. Our 1969 work (Dwork [11]) was largely restricted to $F(\frac{1}{2}, \frac{1}{2}, 1, \lambda)$. Adolphson's 1976 work [1] on Hecke polynomials was largely devoted to symmetric powers of that differential equation. These articles treated the Frobenius structure.

The papers by Dwork [14] and Robba [31] were based to a large extent on an analysis of the simple nature of the solutions at the singular point. This method is particularly limited if the singular disk is also supersingular. Robba avoided this question since he restricted his attention to the case $c = 1$. Honda [20] also limited himself to the case in which $c = 1$. The analogue of Gauss's evaluation of $F(a, b, c, 1)$ has been treated by Koblitz [22] and Diamond [8]. They are concerned with the case in which $D(0, 1^-)$ and $D(1, 1^-)$ are ordinary disks. This restriction is avoided in our treatment in Chapters 24–26.

For further information on the p-adic theory of hypergeometric functions, the reader may consult the 1981 Princeton thesis of M. Heiligman.

Chapter 1

The Space $\mathbf{L}$ (Algebraic Theory)

Let a, b, c be elements of $\mathbb{Z}_p \cap \mathbb{Q}$, such that

$$(1.1) \qquad c - a,\ c - b,\ b,\ a \quad \text{all lie outside } \mathbb{Z}.$$

Let λ be transcendental over $\mathbb{Q}$ and let $\mathbf{L}$ be the $\mathbb{Q}(\lambda)$ space of elements of $\mathbb{Q}(\lambda, x)$ which as functions of x have poles only at $x = 0, 1, \infty, 1/\lambda$.

Let

$$E = x \frac{\partial}{\partial x}.$$

Let $f = x^{b-1}(1 - x)^{c-b}(1 - x\lambda)^{-a}$, an algebraic function. Then

$$D_f = \frac{1}{xf} \circ E \circ xf = E + \frac{E(xf)}{xf} = E + c - a + \frac{b - c}{1 - x} + \frac{a}{1 - \lambda x}$$

$$= E + b + \frac{(b - c)x}{1 - x} + \frac{a\lambda x}{1 - x\lambda}$$

is stable on $\mathbf{L}$.

1.2 **Lemma.**

(i) $\mathbf{W}_f = \mathbf{L}/D_f\mathbf{L}$ *is a two-dimensional* $\mathbb{Q}(\lambda)$ *space.*

(ii) $$\sigma_f = \frac{\partial}{\partial \lambda} + \frac{1}{xf}\frac{\partial}{\partial \lambda}(xf) = \frac{\partial}{\partial \lambda} + \frac{ax}{1 - \lambda x}$$

induces a mapping of $\mathbf{W}_f$ *into itself;*

$$\sigma_f(u\xi) = u\sigma_f\xi + \frac{\partial u}{\partial \lambda}\cdot \xi, \qquad \forall u \in \mathbb{Q}(\lambda), \quad \xi \in \mathbf{W}_f.$$

(iii) $$\sigma_f\begin{pmatrix} 1 \\ \dfrac{1}{1 - x}\end{pmatrix} = \begin{pmatrix} -\dfrac{c}{\lambda} & \dfrac{c - b}{\lambda} \\ \dfrac{c - a}{1 - \lambda} & \dfrac{a + b - c}{1 - \lambda}\end{pmatrix}\begin{pmatrix} 1 \\ \dfrac{1}{1 - x}\end{pmatrix} \bmod D_f\mathbf{L}.$$

Proof. (i) $\{x^{s-1}, 1/x^s, 1/(1-x)^s, 1/(1-x\lambda)^s\}_{s=1,2,\ldots}$ is a basis of **L** as an infinite dimensional $\mathbb{Q}(\lambda)$ space. Our object is to show that each of these elements may be reduced mod $D_f\mathbf{L}$ to a linear combination of 1, $1/(1-x)$. For $s \geq 1$

$$D_f x^{s-1} = (s-1+c-a)x^{s-1} + (b-c)\frac{x^{s-1}}{1-x} + \frac{ax^{s-1}}{1-\lambda x}$$

$$= (s-1+c-a)x^{s-1} + \frac{b-c}{1-x} + \frac{a}{\lambda^{s-1}}\frac{1}{1-\lambda x}$$

$$+ (c-b)\frac{1-x^{s-1}}{1-x} - \frac{a}{\lambda^{s-1}}\frac{(x\lambda)^{s-1}-1}{x\lambda-1}.$$

$$D_f x^{-s} = (-s+c-a)\frac{1}{x^s} + \frac{(b-c)}{(1-x)x^s} + \frac{a}{x^s(1-\lambda x)}$$

$$= -(s+a-c)\frac{1}{x^s} + \frac{b-c}{1-x} + \frac{a\lambda^s}{1-\lambda x}$$

$$+ \frac{(b-c)}{x^s}\frac{1-x^s}{1-x} + \frac{a\lambda^s}{\lambda^s x^s}\frac{(1-\lambda^s x^s)}{1-\lambda x}$$

$$= -(s-b)\frac{1}{x^s} + \frac{b-c}{1-x} + \frac{a\lambda^s}{1-\lambda x}$$

$$+ (b-c)\sum_{\substack{i+j=s-1\\ i>0}} \frac{x^i}{x^s} + a\lambda^s \sum_{\substack{i+j=s-1\\ i>0}} \frac{(\lambda x)^i}{(\lambda x)^s}.$$

$$D_f \frac{1}{(1-x)^s} = \frac{sx}{(1-x)^{s+1}} + \frac{c-a}{(1-x)^s} + \frac{b-c}{(1-x)^{s+1}} + \frac{a}{(1-\lambda x)(1-x)^s}$$

$$= \frac{s+b-c}{(1-x)^{s+1}} + \frac{c-a-s}{(1-x)^s} + \frac{a\lambda^s}{(\lambda-1)^s}\frac{1}{1-x\lambda}$$

$$+ \frac{a}{1-\lambda x}\left(\frac{1}{1-x} - \frac{\lambda}{\lambda-1}\right)\sum_{i+j=s-1}\left(\frac{\lambda}{\lambda-1}\right)^i \frac{1}{(1-x)^j}$$

$$= \frac{s+b-c}{(1-x)^{s+1}} + \frac{c-a-s}{(1-x)^s} + \frac{a\lambda^s}{(\lambda-1)^s}\frac{1}{1-x\lambda}$$

$$+ \frac{a}{1-\lambda}\frac{1}{1-x}\sum_{i+j=s-1}\left(\frac{\lambda}{\lambda-1}\right)^i \frac{1}{(1-x)^j}.$$

$$D_f \frac{1}{(1-x\lambda)^s} = \frac{sx\lambda}{(1-x\lambda)^{s+1}} + \frac{c-a}{(1-x\lambda)^s} + \frac{b-c}{(1-x)(1-x\lambda)^s} + \frac{a}{(1-x\lambda)^{s+1}}$$

$$= \frac{s+a}{(1-x\lambda)^{s+1}} + \frac{c-a-s}{(1-x\lambda)^s} + \frac{b-c}{(1-\lambda)^s}\frac{1}{1-x}$$

$$+ \frac{b-c}{1-x}\left(\frac{1}{1-x\lambda} - \frac{1}{1-\lambda}\right) \sum_{i+j=s-1} \frac{1}{(1-x\lambda)^i}\frac{1}{(1-\lambda)^j}$$

$$= \frac{s+a}{(1-x\lambda)^{s+1}} + \frac{c-a-s}{(1-x\lambda)^s} + \frac{b-c}{(1-\lambda)^s}\frac{1}{1-x}$$

$$- \frac{\lambda(b-c)}{1-\lambda}\frac{1}{1-x\lambda} \sum_{i+j=s-1} \frac{1}{(1-\lambda)^j}\frac{1}{(1-x\lambda)^i}.$$

These reduction formulae show that

$$\mathbf{L} = V + D\mathbf{L},$$

where

$$V = \mathbb{Q}(\lambda)\cdot 1 + \mathbb{Q}(\lambda)\frac{1}{1-x}.$$

To complete the proof we must show that this sum is direct. It is simpler for this investigation to consider $\xi \in \mathbf{L}$ such that

$$D_f \xi = \alpha \frac{1}{1-\lambda x} + \beta \frac{1}{1-x}.$$

Let

$$V_1 = \mathbb{Q}(\lambda)\frac{1}{1-\lambda x} + \mathbb{Q}(\lambda)\frac{1}{1-x}.$$

We write ξ uniquely in the form

$$\xi = P_0\left(\frac{1}{x}\right) + P_1\left(\frac{1}{1-x}\right) + P_{1/\lambda}\left(\frac{1}{1-x\lambda}\right) + P_\infty(x),$$

where $P_0, P_1, P_{1/\lambda}$ are polynomials in one variable with coefficients in $\mathbb{Q}(\lambda)$ with no constant term and $P_\infty(x) \in \mathbb{Q}(\lambda)[x]$. Our reduction formulas show that

$$D_f P_0\left(\frac{1}{x}\right) \in V_1 + \text{polynomial in } \frac{1}{x} \text{ of degree equal to that of } P_0,$$

$$D_f P_1\left(\frac{1}{1-x}\right) \in V_1 + \text{polynomial in } \frac{1}{1-x} \text{ of degree } 1 + \deg P_1,$$

$$D_f P_{1/\lambda}\left(\frac{1}{1-x\lambda}\right) \in V_1 + \text{polynomial in } \frac{1}{1-\lambda x} \text{ of degree } 1 + \deg P_{1/\lambda},$$

$$D_f P_\infty(x) \in V_1 + \text{polynomial in } x \text{ of degree equal to that of } P_\infty.$$

Remark. $P_0, P_1, P_{1/\lambda}$ cannot be of degree zero but can be zero, i.e., of degree $-\infty$. It follows that $1 + \deg P_1 \leq 1$, $1 + \deg P_{1/\lambda} \leq 1$, $\deg P_0 \leq 0$. Hence $P_0, P_1, P_{1/\lambda}$ are all zero. This then shows that a polynomial of degree P_∞ lies in V_1 and hence $P_\infty = 0$. This shows $\xi = 0$. This completes the proof of the first part of the lemma.

(ii) We observe that $\sigma_f = \partial/\partial\lambda$ twisted by xf and D_f is $x(\partial/\partial x)$ twisted by xf. Since $\partial/\partial\lambda$ commutes with $\partial/\partial x$ on $xf\mathbf{L}$, it follows that

$$D_f \circ \sigma_f = \sigma_f \circ D_f$$

on $\mathbf{L}$ and hence σ_f operates on $\mathbf{W}_f$ by passage to quotients.

(iii)
$$\sigma_f 1 = \frac{ax}{1-x\lambda} = -\frac{1}{\lambda}\left(b + \frac{(b-c)x}{1-x}\right) = -\frac{1}{\lambda}\left(c + \frac{b-c}{1-x}\right).$$

$$\sigma_f \frac{1}{1-x} = \frac{ax}{(1-x)(1-x\lambda)} = \frac{a}{1-\lambda}\frac{1}{1-x} + \frac{a}{\lambda-1}\frac{1}{1-x\lambda}$$
$$= \frac{a}{1-\lambda}\frac{1}{1-x} + \frac{1}{1-\lambda}\left(c - a + \frac{b-c}{1-x}\right)$$
$$= \frac{1}{1-\lambda}\left(c - a + \frac{a+b-c}{1-x}\right). \qquad \square$$

Remark. Condition (1.1) is well known as a condition which implies the irreducibility of the equation

$$l = \frac{d^2}{d\lambda^2} + \left(\frac{c}{\lambda} - \frac{a+b-c+1}{1-\lambda}\right)\frac{d}{d\lambda} - \frac{ab}{\lambda(1-\lambda)}$$

satisfied by $F(a, b; c, \lambda)$.

In fact if l is reducible over $\mathbb{C}(\lambda)$ then there exists a solution u such that u'/u is rational, say

$$\frac{u'}{u} = R(\lambda)$$

($R \in \mathbb{C}(\lambda)$). Let $\alpha_1, \ldots, \alpha_m$ be the finite singularities of R, then

$$R(\lambda) = \sum \frac{A_i}{\lambda - \alpha_i} + P_\infty(\lambda) + \sum_{i=1}^{m} \frac{1}{(\lambda - \alpha_i)^2} P_i\left(\frac{1}{\lambda - \alpha_i}\right),$$

where A_i is the residue of R at α_i ($i = 1, \ldots, m$) and $P_\infty, P_1, \ldots, P_m$ are polynomials in one variable with coefficients in $\mathbb{C}$. Hence

$$\log u = \sum_{i=1}^{m} A_i \log(\lambda - \alpha_i) + R_1,$$

where $R_1 \in \mathbb{C}(\lambda)$.

Thus

$$u = \prod_{i=1}^{m} (\lambda - \alpha_i)^{A_i} \cdot \exp R_1.$$

Since l has only regular singularities, we conclude that R_1 must be a constant and so

$$u = \prod_{i=1}^{m} (\lambda - \alpha_i)^{A_i}.$$

The only singularities may be at 0, 1, ∞ and hence

$$u = \lambda^{A_0}(1 - \lambda)^{A_1} P(\lambda),$$

where $P \in \mathbb{C}[\lambda]$. Thus the equation l has exponents e_α at α ($\alpha = 0, 1, \infty$) such that

$$\begin{aligned} e_0 &= A_0, \\ e_1 &= A_1, \\ -e_\infty &= A_0 + A_1 + \deg P. \end{aligned}$$

Thus

$$-\deg P = e_0 + e_1 + e_\infty.$$

There are eight possibilities for the right side:

$$0 + 0 + a = a,$$

$$0 + 0 + b = b,$$

$$0 + c - a - b + a = c - b,$$

$$0 + c - a - b + b = c - a,$$

$$1 - c + 0 + a = a - c + 1,$$

$$1 - c + 0 + b = b - c + 1,$$

$$1 - c + c - a - b + a = 1 - b,$$

$$1 - c + c - a - b + b = 1 - a.$$

Thus for reducibility one of these eight quantities must be a negative integer. This is excluded by (1.1).

Note. It is well known that the negation of (1.1) implies reducibility of l (cf. Goursat [18], §§21–23).

Chapter 2

Dual Theory (Algebraic)

Our object is to define a two-dimensional $\mathbb{Q}(\lambda)$ space, $\mathbf{K}_f$, dual to $\mathbf{W}_f$ together with a natural map of $\mathbf{K}_f$ into $\mathbf{W}_{1/f}$. Let $\Omega_2 = \mathbb{Q}(\lambda)$.

2.1

We introduce the notation

$$S = \{0, 1, \quad 1/\lambda, \infty\},$$

$$T_0 = x,$$

$$T_1 = 1 - x,$$

$$T_{1/\lambda} = 1 - \lambda x,$$

$$T_\infty = 1/x.$$

For $v = 0, 1, 1/\lambda$, we define R_v, $\mathbf{R}'_v$

$$R_v = \Omega_2[[T_v]] \subset R'_v = \Omega_2((T_v))$$

and for $v = \infty$

$$R_\infty = T_\infty \Omega_2[[T_\infty]] \subset R'_\infty = \Omega_2((T_\infty)).$$

We set

$$R = R_0 \oplus R_1 \oplus R_{1/\lambda} \oplus R_\infty \hookrightarrow R'_\infty \oplus R'_1 \oplus R'_{1/\lambda} \oplus R'_\infty = R'.$$

Thus we think of each element ξ of R as a 4-tuple

$$\xi = (\xi_0, \xi_1, \xi_{1/\lambda}, \xi_\infty),$$

where each $\xi_v \in R_v$ and there is no relation among the components. Similarly for R'.

We imbed $\mathbf{L}$ into R' diagonally.

$$\mathbf{L} \ni \eta \to (\eta, \eta, \eta, \eta) \in R'.$$

We define the operation of taking principal parts in each R'_v $(v \in S)$

$$P_v: R'_v \to \mathbf{L}$$

by

$$\sum_{j=-m}^{\infty} a_j T_v^j \mapsto \begin{cases} \displaystyle\sum_{j=-m}^{-1} a_j T_v^j & \text{if } v \neq \infty, \\ \displaystyle\sum_{j=-m}^{0} a_j T_v^j & \text{if } v = \infty. \end{cases}$$

We thus have a natural projection of R' into $\mathbf{L}$

$$\gamma_+ : R' \to \mathbf{L}$$

by

$$(\xi_0, \xi_1, \xi_{1/\lambda}, \xi_\infty) \mapsto \sum_{v \in S} P_v(\xi_v) \in \mathbf{L}.$$

Since $\mathbf{L}$ has been imbedded in R', it makes sense to define

(2.1.1) $$\gamma_- = I - \gamma_+$$

as a map of R' into itself. In fact γ_- is a map of R' into R.

2.1.2 *Remark.* γ_- annihilates $\mathbf{L}$; indeed in a natural way,

$$R' = R \oplus \mathbf{L}$$

and γ_- (resp. γ_+) is the projection onto the first (resp. second) summand.

2.2

We define the residue map for R'_v $(v \in S)$

(2.2.0) $$\operatorname{Res}_v : R'_v \to \Omega_2.$$

$$\sum_{j=-m}^{\infty} a_j T_v^j \to \begin{cases} -a_1 & \text{if } v = \infty, \\ a_{-1} k_v & \text{if } v \neq \infty, v \in S, \end{cases}$$

where $k_0 = 1$, $k_1 = -1$, $k_{1/\lambda} = -1/\lambda$.

$$\text{(2.2.1)} \qquad \text{For } v \in S,\ v \neq \infty, \quad \eta \in R_v \quad \text{implies } \operatorname{Res}_v \eta = 0,$$

$$v = \infty, \quad \eta \in T_\infty R_v \quad \text{implies } \operatorname{Res}_\infty \eta = 0.$$

2.2.2 **Lemma.** $(\xi, \eta) \mapsto \operatorname{Res}_v \xi\eta$ *gives a pairing of* R'_v *with itself which induces a perfect pairing of* R_v *with* R'_v/R_v.

Proof. The mapping is obviously a pairing of R'_v with itself. That it induces a pairing of R_v with R'_v/R_v follows from the fact that

$$R_v \cdot R_v = \begin{cases} R_v & \text{if } v \neq \infty, \\ T_\infty^2 \Omega_2[[T_\infty]] & \text{if } v = \infty, \end{cases}$$

and hence $R_v \cdot R_v$ lies in the kernel of Res_v.

To show that the pairing is perfect we must show:

(a) If $\xi \in R_v$ and $\operatorname{Res}_v \xi\eta = 0$ for all $\eta \in R'_v$ then $\xi = 0$.
(b) If $\eta \in R'_v$ and if $\operatorname{Res} \xi\eta = 0$ for all $\xi \in R_v$ then $\eta \in R_v$.

For (a) let a_m be the coefficient of T_v^m in ξ (so $m \geq 0$ (resp. $m \geq 1$) if $v \neq \infty$ (resp. $v = \infty$)). Let

$$\eta = \begin{cases} 1/T_v^{m+1} & \text{if } v \neq \infty, \\ 1/T_\infty^{m-1}, & \end{cases}$$

so by hypothesis

$$0 = a_m \begin{cases} \operatorname{Res}_v 1/T_v & \text{if } v \neq \infty, \\ \operatorname{Res}_\infty T_\infty, & \end{cases}$$

which shows that $a_m = 0$.

For (b) let b_m be the coefficient of $1/T_v^m$ in η with $m \geq 1$ (resp. $m \geq 0$) for $v \neq \infty$ (resp. $v = \infty$). Then let $\xi = T_v^{m-1}$ (resp. T_∞^{m+1}) for $v \neq \infty$ (resp. $v = \infty$) so in either case $\xi \in R_v$ and so by hypothesis

$$0 = \operatorname{Res}_v \xi\eta = b_m \begin{cases} \operatorname{Res}_v 1/T_v, \\ \operatorname{Res}_\infty T_\infty, \end{cases}$$

so that $b_m = 0$, as asserted. $\square$

2.2.3 **Definition.** We pair R' with itself by

$$\langle \xi, \eta \rangle = \sum_{v \in S} \operatorname{Res}_v \xi_v \eta_v.$$

2.2.4 **Lemma.**

$$\text{(2.2.4.1)} \qquad \textit{For } \xi \in \mathbf{L}, \qquad \langle 1, \xi \rangle = 0,$$

$$\text{(2.2.4.2)} \qquad \xi, \eta \in \mathbf{L}, \qquad \langle \eta, \xi \rangle = 0,$$

$$\text{(2.2.4.3)} \qquad \xi, \eta \in R, \qquad \langle \eta, \xi \rangle = 0.$$

Proof. Clearly the kernel of Res_v (in R'_v) contains the image of R'_v under $\partial/\partial x$. By the decomposition of elements of **L** into principal parts,

$$\mathbf{L} = \Omega_2 \frac{1}{x} + \Omega_2 \frac{1}{1-x} + \Omega_2 \frac{1}{1-\lambda x} + \frac{d}{dx}\mathbf{L}.$$

As elements of R_∞,

$$\frac{1}{1-x} = -\left(\frac{1}{x} + \frac{1}{x^2} + \cdots\right),$$

$$\frac{1}{1-\lambda x} = -\left(\frac{1}{\lambda x} + \frac{1}{\lambda^2 x^2}\right).$$

Thus

$$\operatorname{Res}_\infty \frac{1}{1-x} = +1,$$

$$\operatorname{Res}_\infty \frac{1}{1-\lambda x} = \frac{1}{\lambda},$$

and so we may check that $\langle 1, \xi \rangle = 0$ for $\xi = 1/x, 1/(1-x), 1/(1-\lambda x)$. This completes the proof of (2.2.4.1). Assertion (2.2.4.2) is a direct consequence of (2.2.4.1). Assertion (2.2.4.3) follows from the fact that $R_v \cdot R_v$ ($=R_v$ if $v \neq \infty$) lies in the kernel of Res_v for all $v \in S$. □

2.2.5 **Lemma.** *Under the pairing of Definition* 2.2.3, R *may be identified with* $\operatorname{Hom}(\mathbf{L}, \Omega_0)$.

Proof. Clearly for $\eta \in R$, we may define $\theta_\eta \in \mathrm{Hom}(\mathbf{L}, \Omega_0)$ by setting

$$\theta_\eta(\xi) = \langle \eta, \xi \rangle, \qquad \forall \xi \in \mathbf{L}.$$

If $\theta_\eta = 0$ then let a_m be the coefficient of T_ν^m in η_ν, $m \geq 0$ (resp. $m \geq 1$) if $\nu \neq \infty$ (resp. $\nu = \infty$) and so

$$0 = \begin{cases} \theta(1/T_\nu^{m+1}) = a_m \operatorname{Res}_\nu 1/T_\nu & \text{if } \nu \neq \infty, \\ \theta(1/T_\infty^{m-1}) = a_m \operatorname{Res}_\infty T_\infty & \text{if } \nu = \infty. \end{cases}$$

This shows $a_m = 0$ and hence $\eta = 0$. Thus the mapping $\eta \mapsto \theta_\eta$ is injective. We assert that it is also surjective. Thus given $\theta \in \mathrm{Hom}(\mathbf{L}, \Omega_0)$ put

$$\eta_\nu = \begin{cases} \displaystyle\sum_{m=1}^{\infty} \frac{T_\nu^{m-1}\theta(1/T_\nu^m)}{\operatorname{Res}_\nu 1/T_\nu} & \text{if } \nu \neq \infty, \\ \displaystyle\sum_{m=0}^{\infty} \frac{T_\infty^{m+1}\theta(1/T_\infty^m)}{\operatorname{Res}_\infty T_\infty} & \text{if } \nu = \infty. \end{cases}$$

Then $\theta = \theta_\eta$. This completes the proof. □

2.2.6 **Lemma.** *The pairing $\langle\ ,\ \rangle$ of R' with itself restricts to a perfect pairing of R with* $\mathbf{L}$.

Proof. We have shown (Lemma 2.2.5) that if $\eta \in R$ annihilates $\mathbf{L}$ then $\eta = 0$. Thus suppose $\xi \in \mathbf{L}$ annihilates R, then let a_m, $m \geq 1$ (resp. $m \geq 0$) be the coefficient of $1/T_\nu^m$ for $\nu \neq \infty$ (resp. $\nu = \infty$) in the Mittag–Leffler decomposition of ξ. Let η be the element of R whose component in R_μ is zero except for $\mu = \nu$, and let $\eta_\nu = T_\nu^{m-1}$ (resp. T_∞^{m+1}) if $\nu \neq \infty$ (resp. $\nu = \infty$). We conclude that $a_m = 0$. This shows that $\xi = 0$ which completes the proof. □

2.2.7 **Proposition.** *If $\xi \in R'$, $\eta \in \mathbf{L}$ then*

$$\langle \xi, \eta \rangle = \langle \gamma_- \xi, \eta \rangle.$$

This follows trivially from (2.1.2) and (2.2.4.2).

2.2.8 **Lemma.** *Let T be an endomorphism of R' as Ω_2 space which is stable on* $\mathbf{L}$. *Let T^* be an endomorphism of R' dual to T under $\langle\ ,\ \rangle$. Then*

$$(\gamma_- \circ T^*)|R$$

is the unique endomorphism of R adjoint to $T|L$ under the induced pairing of R with $\mathbf{L}$.

Proof. If θ_1, θ_2 are endomorphisms of R such that

$$\langle \theta_i \eta, \xi \rangle = \langle \eta, T\xi \rangle, \qquad \forall \eta \in R, \xi \in \mathbf{L}, i = 1, 2,$$

then

$$\langle (\theta_1 - \theta_2)\eta, \xi \rangle = 0, \qquad \forall \eta \in R, \xi \in \mathbf{L}$$

and hence $\theta_1(\eta) = \theta_2(\eta)$ by Lemma 2.2.6. This shows uniqueness. For $\xi \in \mathbf{L}$, $\eta \in R$ we have by Proposition 2.2.7.

$$\langle \gamma_-(T^*\eta), \xi \rangle = \langle T^*\eta, \xi \rangle = \langle \eta, T\xi \rangle.$$

It only remains to observe that for $\eta \in R$, $T^*\eta$ lies in R' but $\gamma_-(T^*\eta)$ lies in R. This completes the proof. □

2.3

We apply the above result to the mapping D_f of Chapter 1.

Let $H = (H_0, H_1, H_{1/\lambda}, H_\infty) \in R'$. Then H operates on R' by componentwise multiplication. Clearly H is self dual relative to $\langle \;,\; \rangle$. In particular if $H \in \mathbf{L}$ then multiplication by H is stable on $\mathbf{L}$. Hence $\gamma_- \circ H$ is the endormorphism of R adjoint to multiplication by H as operator on $\mathbf{L}$.

For $\xi, \eta \in R'$,

$$\begin{aligned}\left\langle \xi, \frac{\partial}{\partial x}\eta \right\rangle &= \sum_{v \in S} \mathrm{Res}_v\, \xi_v \frac{\partial}{\partial x}\eta_v \\ &= -\sum_{v \in S} \mathrm{Res}_v\, \eta_v \frac{\partial \xi_v}{\partial x} \\ &= \left\langle -\frac{\partial}{\partial x}\xi, \eta \right\rangle.\end{aligned}$$

Thus for $H \in \mathbf{L}$, $E + H$ is endomorphism of $\mathbf{L}$ which has $\gamma_- \circ (-\partial/\partial x \circ x + H)$ as adjoint operator. In particular then

$$D_f = E + \frac{E(xf)}{xf} \tag{2.3.1}$$

as operator on $\mathbf{L}$ has

$$D_f^* = \gamma_-\left(-E - 1 + \frac{E(xf)}{xf}\right) = -\gamma_- D_{1/xf} \tag{2.3.2}$$

as an adjoint operator on R.

2.3.3 **Definition.** $\mathbf{K}_f$ is the kernel of D_f^* in R.

2.3.4 **Lemma.** *The pairing* $\langle\ ,\ \rangle$ *of* R *with* $\mathbf{L}$ *induces a perfect pairing of* $\mathbf{K}_f$ *with* $\mathbf{W}_f = \mathbf{L}/D_f\mathbf{L}$.

Proof. Clearly each element of $\mathbf{K}_f$ acts via $\langle\ ,\ \rangle$ on $\mathbf{L}/D_f\mathbf{L}$ as linear map into Ω_2. If $\xi \in \mathbf{K}_f$ annihilates $\mathbf{W}_f$ then it annihilates $\mathbf{L}$ and hence is zero by Lemma 2.2.5. Hence $\mathbf{K}_f$ may be identified with a subspace of the dual space $\mathrm{Hom}(\mathbf{W}_f, \Omega_2)$ of $\mathbf{W}_f$. Conversely if θ is linear function on $\mathbf{W}_f$ then θ may be viewed as linear function on $\mathbf{L}$ and hence by Lemma 2.2.5 there exists, ξ, unique element of R such that

$$\theta(\eta) = \langle \xi, \eta \rangle$$

for all $\eta \in \mathbf{L}$. Hence ξ annihilates $D_f\mathbf{L}$ and so $D_f^*\xi$ annihilates $\mathbf{L}$ which shows that ξ lies in $\mathbf{K}_f$. Thus $\mathbf{K}_f \simeq \mathrm{Hom}(\mathbf{W}_f, \Omega_2)$. This completes the proof. $\square$

2.3.4.1 **Corollary.**

$$\dim_{\Omega_2} \mathbf{K}_f = 2.$$

2.3.5

2.3.5.1 **Notation.** For $\theta \notin -\mathbb{N}$, $m \in \mathbb{N}$ let

$$U_{\theta,m} = \sum_{j=0}^{m} \mathbb{Z}_p \frac{1}{\theta + j}.$$

2.3.5.2 We define t_v, θ_v, and recall T_v

v	0	1	$1/\lambda$	∞
t_v	λ	$\lambda/(\lambda-1)$	$1/(1-\lambda)$	$1/\lambda$
T_v	x	$1-x$	$1-\lambda x$	$1/x$
θ_v	$1-b$	$b-c$	a	$c-a$

We denote by $\mathbf{P}_{m+1}(U_{\theta,m}; t)$ the set of all polynomials of degree bounded by $m+1$ in t with coefficients in $U_{\theta,m}$.

Theorem. *If* $\xi \in \mathbf{K}_f$ *then the coefficient of* T_v^m *in the component,* ξ_v (resp. $(1/\lambda)\xi_{1/\lambda}$ *if* $v = 1/\lambda$) *is an element of* $\mathbb{Q}(\lambda)$ *which lies in*

$$\langle \xi, 1 \rangle \mathbf{P}_{m+1}(U_{\theta_v,m}; t_v) + \left\langle \xi, \frac{1}{1-x} \right\rangle \mathbf{P}_{m+1}(U_{\theta_v,m}; t_v)$$

and the $\mathbf{P}_{m+1}$ *may be replaced by* $\mathbf{P}_m$ *unless* $v = 0$. *If* $v = \infty$ *then* $U_{\theta_v,m}$ *may be replaced by* $U_{\theta_v,m-1}$.

Proof. The proof involves a lengthy calculation. We write

$$\xi = (\xi_0, \xi_1, \xi_{1/\lambda}, \xi_\infty) \in \mathbf{K}_f,$$

so by definition

$$D_{1/xf}\xi = \gamma_+ D_{1/xf}\xi = \sum_{v \in S} P_v(D_{1/xf}\xi_v). \tag{2.3.5.3}$$

Thus for $\beta \in S$

$$D_{1/xf}\xi_\beta = \sum_{v \in S} P_v\left(E\xi_v + \frac{E(1/f)}{1/f}\,\xi_v\right). \tag{2.3.5.4}$$

The right side is independent of β. We first assert that

$$P_v E\xi_v = 0. \tag{2.3.5.5}$$

For all $v \in S$, R_v is closed under $\partial/\partial x$ while for $v \neq \infty$, $x \in R_v$. Thus (2.3.5.5) is clear for $v \neq \infty$, while for $v = \infty$, we recall that $R_\infty = (1/x)\Omega_2[[1/x]]$ which is clearly closed under E.

The calculation of the right side of (2.3.5.4) is reduced to that of $P_v(\xi_v[E(1/f)/1/f])$.

$$\frac{E(1/f)}{1/f} = 1 - c + a - \frac{b-c}{1-x} - \frac{a}{1-\lambda x}.$$

Now

$$\left.\begin{aligned} &1 - c + a \\ &-\frac{b-c}{1-x} \\ &-\frac{a}{1-\lambda x} \end{aligned}\right\} \in R_v \quad \text{unless} \left\{\begin{aligned} &v = \infty, \\ &v = 1, \\ &v = \frac{1}{\lambda}. \end{aligned}\right.$$

Hence

$$\sum_{v} P_v\left(\xi_v \frac{E(1/f)}{1/f}\right) = P_\infty(\xi_\infty \cdot (1 - c + a) + P_1\left(-\frac{b-c}{1-x}\xi_1\right) + P_{1/\lambda}\left(-\frac{a}{1-\lambda x}\xi_{1/\lambda}\right)$$

$$= P_1\left(-\frac{b-c}{1-x}\xi_1\right) + P_{1/\lambda}\left(-\frac{a}{1-\lambda x}\xi_{1/\lambda}\right)$$

$$= -\frac{b-c}{1-x}e_1 - \frac{a}{1-\lambda x}e_2,$$

where e_1 (resp. e_2) is the coefficient of T_1^0 (resp. $T_{1/\lambda}^0$) in ξ_1 (resp. $\xi_{1/\lambda}$). On the other hand

$$\left\langle\frac{1}{1-x}, \xi\right\rangle = \sum_{v} \operatorname{Res}_v \xi_v \cdot \frac{1}{1-x} = \operatorname{Res}_1 \xi_1 \cdot \frac{1}{1-x} = -e_1,$$

$$\left\langle\frac{1}{1-\lambda x}, \xi\right\rangle = \operatorname{Res}_{1/\lambda} \xi_{1/\lambda} \frac{1}{1-\lambda x} = -\frac{e_2}{\lambda}.$$

Thus

$$\sum_{v\in S} P_v\left(\xi_v \frac{E(1/f)}{1/f}\right) = \frac{b-c}{1-x}\left\langle\frac{1}{1-x}, \xi\right\rangle + \frac{a\lambda}{1-\lambda x}\left\langle\frac{1}{1-\lambda x}, \xi\right\rangle. \tag{2.3.5.6}$$

Since

$$c - a + \frac{b-c}{1-x} + \frac{a}{1-\lambda x} = D_f 1,$$

the right side of (2.3.5.6) may be calculated in terms of $\langle 1/(1-x), \xi\rangle$ and $\langle 1, \xi\rangle$.

Putting

$$A = (b-c)\left\langle\frac{1}{1-x}, \xi\right\rangle,$$

$$B = a\lambda\left\langle\frac{1}{1-\lambda x}, \xi\right\rangle,$$

equation (2.3.5.4) takes the form

$$\left(f \circ E \circ \frac{1}{f}\right)\xi_v = \frac{A}{1-x} + \frac{B}{1-\lambda x},$$

which we write

$$\text{(2.3.5.7)} \qquad \frac{\partial}{\partial x}\left(\frac{1}{f}\xi_v\right) = \frac{1}{xf}\left(\frac{A}{1-x} + \frac{B}{1-\lambda x}\right).$$

We rewrite in terms of T_v

$$x\frac{\partial}{\partial x} = T_0\frac{\partial}{\partial T_0} = -(1-T_1)\frac{\partial}{\partial T_1} = -(1-T_{1/\lambda})\frac{\partial}{\partial T_{1/\lambda}} = -T_\infty\frac{\partial}{\partial T_\infty}.$$

x	T_0	$1-T_1$	$(1/\lambda)(1-T_{1/\lambda})$	$1/T_\infty$
$1-x$	$1-T_0$	T_1	$((\lambda-1)/\lambda)(1-t_{1/\lambda}T_{1/\lambda})$	$-(1-T_\infty)/T_\infty$
$1-\lambda x$	$1-t_0T_0$	$(1-\lambda)(1-t_1T_1)$	$T_{1/\lambda}$	$-\lambda(1-t_\infty T_\infty)/T_\infty$

$$\text{(2.3.5.8)} \qquad \frac{1}{f} = T_0^{1-b}(1-T_0)^{b-c}(1-t_0T_0)^a$$

$$= (1-T_1)^{1-b}T_1^{b-c}(1-t_1T_1)^a(1-\lambda)^a$$

$$= (1-T_{1/\lambda})^{1-b}(1-t_{1/\lambda}T_{1/\lambda})^{b-c}T_{1/\lambda}^a\frac{(\lambda-1)^{b-c}}{\lambda^{1-c}}$$

$$= T_\infty^{c-a-1}(1-T_\infty)^{b-c}(1-t_\infty T_\infty)^a\lambda^a(-1)^{a+b-c}.$$

$$\text{(2.3.5.9)} \qquad \frac{A}{1-x} + \frac{B}{1-x\lambda} = \frac{A}{1-T_0} + \frac{B}{1-t_0T_0}$$

$$= \frac{A}{T_1} + \frac{B}{(1-\lambda)(1-t_1T_1)}$$

$$= \frac{A\lambda}{\lambda-1}\frac{1}{1-t_{1/\lambda}T_{1/\lambda}} + \frac{B}{T_{1/\lambda}}$$

$$= -\frac{AT_\infty}{1-T_\infty} - \frac{BT_\infty}{\lambda(1-t_\infty T_\infty)}.$$

Thus

(2.3.5.10)

$$T_0 \frac{\partial}{\partial T_0}\left[\xi_0 T_0^{1-b}(1-T_0)^{b-c}(1-t_0T_0)^a\right]$$

$$= T_0^{1-b}(1-T_0)^{b-c}(1-t_0T_0)^a\left[\frac{A}{1-T_0}+\frac{B}{1-t_0T_0}\right],$$

$$-(1-T_1)\frac{\partial}{\partial T_1}\left[\xi_1(1-T_1)^{1-b}T_1^{b-c}(1-t_1T_1)^a\right]$$

$$= (1-T_1)^{1-b}T_1^{b-c}(1-t_1T_1)^a\left[\frac{A}{T_1}+\frac{B}{(1-\lambda)(1-t_1T_1)}\right],$$

$$-(1-T_{1/\lambda})\frac{\partial}{\partial T_{1/\lambda}}\left[\xi_{1/\lambda}(1-T_{1/\lambda})^{1-b}(1-t_{1/\lambda}T_{1/\lambda})^{b-c}T_{1/\lambda}^a\right]$$

$$= (1-T_{1/\lambda})^{1-b}(1-t_{1/\lambda}T_{1/\lambda})^{b-c}T_{1/\lambda}^a\left[\frac{A\lambda}{\lambda-1}\frac{1}{1-t_{1/\lambda}T_{1/\lambda}}+\frac{B}{T_{1/\lambda}}\right],$$

$$-T_\infty\frac{\partial}{\partial T_{1/\infty}}\left[\xi_\infty T_\infty^{c-a-1}(1-T_\infty)^{b-c}(1-t_\infty T_\infty)^a\right]$$

$$= T_\infty^{c-a-1}(1-T_\infty)^{b-c}(1-t_\infty T_\infty)^a\left[-\frac{AT_\infty}{1-T_\infty}-\frac{BT_\infty}{\lambda(1-t_\infty T_\infty)}\right].$$

Thus

$$\xi_0 = T_0^{b-1}(1-T_0)^{c-b}(1-t_0T_0)^a\int T_0^{-b}(1-T_0)^{b-c}(1-t_0T_0)^a$$

$$\times\left[\frac{A}{1-T_0}+\frac{B}{1-t_0T_0}\right]dT_0$$

$$= T_0^{b-1}\sum_{m=0}^{\infty}(-T_0)^m\sum_{i+j=m}\binom{c-b}{i}\binom{a}{j}t_0^j\int dT_0\,T_0^{-b}$$

$$\times\sum_{m=0}^{\infty}(-T_0)^m\sum_{i+j=m}\left[A\binom{b-c-1}{i}\binom{a}{j}+B\binom{b-c}{i}\binom{a-1}{j}\right]t_0^j$$

$$= \sum_{m=0}^{\infty}(-T_0)^m\sum_{i+j=m}\binom{c-b}{i}\binom{a}{j}t_0^j\cdot\sum_{m=0}^{\infty}(-1)^m\frac{T_0^m}{m+1-b}\sum_{i+j=m}t_0^j$$

$$\times\left[A\binom{b-c-1}{i}\binom{a}{j}+B\binom{b-c}{i}\binom{a-1}{j}\right].$$

Thus

$$\xi_0 = \sum_{m=0}^{\infty}(-T_0)^m \sum_{r=0}^{m} t_0^r \sum_{\substack{k+l+i+j=m \\ l+j=r}} \binom{c-b}{k}\binom{a}{l}$$

$$\times\left[A\binom{b-c-1}{i}\binom{a}{j} + B\binom{b-c}{i}\binom{a-1}{j}\right]\frac{1}{i+j+1-b}.$$

$$-\xi_1 = (1-T_1)^{b-1}T_1^{c-b}(1-t_1T_1)^{-a}\int dT_1(1-T_1)^{-b}T_1^{b-c}(1-t_1T_1)^a$$

$$\times\left[\frac{A}{T_1} + \frac{B}{1-\lambda}\frac{1}{1-t_1T_1}\right]$$

$$= T_1^{c-b}\sum_{m=0}^{\infty}(-T_1)^m \sum_{k+l=m}\binom{b-1}{k}\binom{-a}{l}t_1^l \int dT_1$$

$$\cdot\left[AT_1^{b-c-1}\sum_{m=0}^{\infty}(-T_1)^m \sum_{i+j=m}\binom{-b}{i}\binom{a}{j}t_1^j\right.$$

$$\left.+\frac{B}{1-\lambda}T_1^{b-c}\sum_{m=0}^{\infty}(-T_1)^m \sum_{i+j=m}\binom{-b}{i}\binom{a-1}{j}t_1^j\right]$$

$$= \sum_{m=0}^{\infty}(-T_1)^m \sum_{k+l=m}\binom{b-1}{k}\binom{-a}{l}t_1^l$$

$$\times\left[A\sum_{m=0}^{\infty}\frac{T_1^m(-1)^m}{m+b-c}\sum_{i+j=m}\binom{-b}{i}\binom{a}{j}t_1^j\right.$$

$$\left.+\frac{B}{1-\lambda}\sum_{m=0}^{\infty}\frac{T_1^{m+1}(-1)^m}{m+b-c+1}\sum_{i+j=m}\binom{-b}{i}\binom{a-1}{j}t_1^j\right]$$

$$= A\sum_{m=0}^{\infty}(-T_1)^m \sum_{r=0} t_1^r \sum_{\substack{k+l+i+j=m \\ l+j=r}}\binom{b-1}{k}\binom{-a}{l}\binom{-b}{i}\binom{a}{j}\frac{1}{i+j+b-c}$$

$$+ B\frac{1}{1-\lambda}\sum_{m=0} T_1(-T_1)^m \sum_{r=0} t_1^r \sum_{\substack{k+l+i+j=m \\ l+j=r}}$$

$$\times\binom{b-1}{k}\binom{-a}{l}\binom{-b}{i}\binom{a-1}{j}\frac{1}{i+j+b-c+1}.$$

$$-\xi_{1/\lambda} = (1-T_{1/\lambda})^{b-1}(1-t_{1/\lambda}T_{1/\lambda})^{c-b}T_{1/\lambda}^{-a}\int dT_{1/\lambda}$$

$$\times(1-T_{1/\lambda})^{-b}(1-t_{1/\lambda}T_{1/\lambda})^{b-c}$$

$$\times T_{1/\lambda}^{a}\left[\frac{A\lambda}{\lambda-1}\frac{1}{1-t_{1/\lambda}T_{1/\lambda}} + B\frac{1}{T_{1/\lambda}}\right]$$

$$= T_{1/\lambda}^{-a} \sum_{m=0}^{\infty} (-T_{1/\lambda})^m \sum_{k+l=m} \binom{b-1}{k} \binom{c-b}{l} t_{1/\lambda}^l \int dT_{1/\lambda}$$

$$\times \left[\frac{A\lambda}{\lambda - 1} T_{1/\lambda}^a \sum_{m=0}^{\infty} (-T_{1/\lambda})^m \sum_{i+j=m} \binom{-b}{i} \binom{b-c-1}{j} t_{1/\lambda}^j \right.$$

$$\left. + BT_{1/\lambda}^{a-1} \sum_{m=0}^{\infty} (-T_{1/\lambda})^m \sum_{i+j=m} \binom{-b}{i} \binom{b-c}{j} t_{1/\lambda}^j \right]$$

$$= \sum_{m=0}^{\infty} (-T_{1/\lambda})^m \sum_{k+l=m} \binom{b-1}{k} \binom{c-b}{l} t_{1/\lambda}^l$$

$$\times \left\{ \frac{A\lambda}{\lambda - 1} \sum_{m=0}^{\infty} (-1)^m \frac{T_{1/\lambda}^{m+1}}{a+m+1} \sum_{i+j=m} \binom{-b}{i} \binom{b-c-1}{j} t_{1/\lambda}^j \right.$$

$$\left. + B \sum_{m=0}^{\infty} (-1)^m \frac{T_{1/\lambda}^m}{a+m} \sum_{i+j=m} \binom{-b}{i} \binom{b-c}{j} t_{1/\lambda}^j \right\}$$

$$= \frac{A\lambda}{\lambda - 1} \sum_{m=0}^{\infty} T_{1/\lambda} (-T_{1/\lambda})^m \sum_{r=0} t_{1/\lambda}^r \sum_{\substack{k+l+i+j=m \\ l+j=r}}$$

$$\times \binom{b-1}{k} \binom{c-b}{l} \binom{-b}{i} \binom{b-c-1}{j} \frac{1}{a+i+j+1}$$

$$+ B \sum_{m=0}^{\infty} (-T_{1/\lambda})^m \sum_{r=0}^{\infty} t_{1/\lambda}^r \sum_{\substack{k+l+i+j=m \\ l+j=r}}$$

$$\times \binom{b-1}{k} \binom{c-b}{l} \binom{-b}{i} \binom{b-c}{j} \frac{1}{a+i+j}.$$

$$+\xi_\infty = T_\infty^{1+a-c} (1 - T_\infty)^{c-b} (1 - t_\infty T_\infty)^{-a} \int dT_\infty T_\infty^{c-a-2}$$

$$\times (1 - T_\infty)^{b-c} (1 - t_\infty T_\infty)^a \left[\frac{AT_\infty}{1 - T_\infty} + \frac{B}{\lambda} \frac{T_\infty}{1 - t_\infty T_\infty} \right]$$

$$= T_\infty^{1+a-c} \sum_{m=0}^{\infty} (-T_\infty)^m \sum_{k+l=m} \binom{c-b}{k} \binom{-a}{l} t_\infty^l \int dT_\infty \cdot T_\infty^{c-a-1}$$

$$\times \left[A \sum_{m=0} (-T_\infty)^m \sum_{i+j=m} \binom{b-c-1}{i} \binom{a}{j} t_\infty^j \right.$$

$$\left. + \frac{B}{\lambda} \sum_{m=0}^{\infty} (-T_\infty)^m \sum_{i+j=m} \binom{b-c}{i} \binom{a-1}{j} t_\infty^j \right]$$

$$= \sum_{m=0}^{\infty} (-T_\infty)^m \sum_{k+l=m} \binom{c-b}{k} \binom{-a}{l} t_\infty^l \cdot \sum_{m=0}^{\infty} (-1)^m \frac{T_\infty^{m+1}}{m+c-a} \sum_{i+j=m}$$

$$\times \left[A \binom{b-c-1}{i} \binom{a}{j} + \frac{B}{\lambda} \binom{b-c}{i} \binom{a-1}{j} \right] t_\infty^j,$$

so

$$\xi_\infty = \sum_{m=0}^{\infty} T_\infty(-T_\infty)^m \sum_r t_\infty^r \sum_{\substack{k+l+i+j=m\\ l+j=r}} \binom{c-b}{k}\binom{-a}{l}$$
$$\times \left[A\binom{b-c-1}{i}\binom{a}{j} + \frac{B}{\lambda}\binom{b-c}{i}\binom{a-1}{j}\right]\frac{1}{c-a+i+j}.$$

The theorem now follows using the relation

$$A = (b-c)\left\langle \frac{1}{1-x}, \xi \right\rangle,$$

$$B = a\lambda\left\langle \frac{1}{1-\lambda x}, \xi \right\rangle,$$

$$(c-a)\langle 1, \xi\rangle + A + \frac{B}{\lambda} = 0,$$

and the definition of the t_v. □

2.3.6 **Corollary.** *Let $\xi \in \mathbf{K}_f$, let Ω be an extension of $\mathbb{Q}_p$ with an extended valuation. If we specialize λ to an element λ_0 of Ω, $\lambda_0 \neq 0, 1, \infty$, λ_0 not a pole of either $\langle 1, \xi\rangle$ or $\langle 1/(1-x), \xi\rangle$, then for $v \in S$, the component ξ_v of ξ specializes to an element of $\Omega[[T_v]]$ which converges for*

$$|T_v| < \mathrm{Min}\left(1, \left|\frac{1}{t_v}\right|\right).$$

Proof. This follows from the theorem since

$$\mathrm{ord}(\theta_v + j) = 0(\log j). \quad \square$$

2.4 Dual of σ_f

We define

$$\sigma_f^* = -\gamma_-\left(\frac{\partial}{\partial\lambda} - \frac{1}{f}\frac{\partial}{\partial\lambda}f\right) = -\gamma_-\sigma_{1/f}$$

as a mapping of R' into itself (and hence into R).

2.4.1 **Lemma.** *For $\xi \in R'$, $\eta \in \mathbf{L}$,*

$$\langle \xi, \sigma_f \eta\rangle = \langle \sigma_f^* \xi, \eta\rangle + \frac{\partial}{\partial\lambda}\langle \xi, \eta\rangle.$$

Proof. We assert that for $\xi_\nu \in R'_\nu$

$$\xi_\nu = \sum_{j=-m}^{\infty} A_j T_\nu^j,$$

(2.4.1.1)
$$\mathrm{Res}_\nu \frac{\partial}{\partial \lambda} \xi_\nu = \frac{\partial}{\partial \lambda} \mathrm{Res}_\nu\, \xi_\nu.$$

Indeed

$$\frac{\partial}{\partial \lambda} \xi_\nu = \sum_{j=-m}^{\infty} \frac{\partial A_j}{\partial \lambda} T_\nu^j + \sum_{j=-m}^{\infty} A_\nu j T_\nu^{j-1} \frac{\partial T_\nu}{\partial \lambda}.$$

For $\nu = 0, 1, \infty$, $\partial T_\nu / \partial \lambda = 0$ and so the assertion is clear while for $\nu = 1/\lambda$, it is simpler to write

$$\xi_{1/\lambda} = \sum_{j=-m}^{\infty} A_j (x - \lambda^{-1})^j,$$

so that

$$\frac{\partial \xi_{1/\lambda}}{\partial \lambda} = \sum_{j=-m} \frac{\partial A_j}{\partial \lambda} \left(x - \frac{1}{\lambda}\right)^j + \sum_{j=-m} A_j j (x - \lambda^{-1})^{j-1} \frac{1}{\lambda^2},$$

so that

$$\mathrm{Res}_{1/\lambda} \frac{\partial \xi_{1/\lambda}}{\partial \lambda} = \frac{\partial A_{-1}}{\partial \lambda} = \frac{\partial}{\partial \lambda} \mathrm{Res}_{1/\lambda}\, \xi_{1/\lambda},$$

which completes the verification of (2.4.1.1). Thus

(2.4.1.2)
$$\begin{aligned}
\left\langle \xi, \frac{\partial \eta}{\partial \lambda} \right\rangle &= \sum_{\nu \in S} \mathrm{Res}_\nu \left(\xi_\nu \frac{\partial \eta}{\partial \lambda} \right) \\
&= \sum_{\nu \in S} \mathrm{Res}_\nu \left(\frac{\partial}{\partial \lambda} (\xi_\nu \eta) - \eta \frac{\partial \xi_\nu}{\partial \lambda} \right) \\
&= \frac{\partial}{\partial \lambda} \sum_{\nu \in S} \mathrm{Res}_\nu (\xi_\nu \eta) - \sum_{\nu \in S} \mathrm{Res}_\nu \left(\eta \frac{\partial \xi_\nu}{\partial \lambda} \right) \\
&= \frac{\partial}{\partial \lambda} \langle \xi, \eta \rangle + \left\langle -\frac{\partial \xi}{\partial \lambda}, \eta \right\rangle,
\end{aligned}$$

where in this last equation $\partial/\partial\lambda$ acts on R' componentwise. Since $\sigma_f = \partial/\partial\lambda + (1/f)(\partial f/\partial\lambda)$, we compute by (2.4.1.2),

$$\begin{aligned}\langle \xi, \sigma_f \eta \rangle &= \left\langle \xi, \frac{\partial \eta}{\partial \lambda} \right\rangle + \left\langle \xi, \frac{\eta}{f} \frac{\partial f}{\partial \lambda} \right\rangle \\ &= \frac{\partial}{\partial \lambda} \langle \xi, \eta \rangle + \left\langle -\frac{\partial \xi}{\partial \lambda}, \eta \right\rangle + \left\langle \xi \frac{1}{f} \frac{\partial f}{\partial \lambda}, \eta \right\rangle \\ &= \frac{\partial}{\partial \lambda} \langle \xi, \eta \rangle + \langle -\sigma_{1/f} \xi, \eta \rangle.\end{aligned}$$

The lemma now follows from Proposition 2.2.7. □

2.4.2 **Lemma.** *D_f^* and σ_f^* commute as mappings of R' into itself.*

Proof. By definition (2.3.2),

$$\begin{aligned}D_f^* \circ \sigma_f^* &= \gamma_- \circ D_{1/xf} \circ \gamma_- \circ \sigma_{1/f} \\ &= \gamma_- \circ D_{1/xf} \circ (I - \gamma_+) \circ \sigma_{1/f} \\ &= \gamma_- \circ D_{1/xf} \circ \sigma_{1/f} - \gamma_- \circ D_{1/xf} \circ \gamma_+ \circ \sigma_{1/f}.\end{aligned}$$

Now $\sigma_{1/f}$ maps R' into itself, while by 2.1.2, γ_+ maps R' into **L** which is stable under $D_{1/xf}$ while **L** is annihilated by γ_-. Thus

(2.4.2.1) $$D_f^* \circ \sigma_f^* = \gamma_- \circ D_{1/xf} \circ \sigma_{1/f}.$$

By a similar argument

(2.4.2.2) $$\sigma_f^* \circ D_f^* = \gamma_- \circ \sigma_{1/f} \circ D_{1/xf}.$$

The lemma now follows from the commutativity of $\sigma_{1/f}$ and $D_{1/xf}$ as mappings of R' into itself. □

2.4.2.3 *Note.* That commutativity may be verified most easily by writing

$$\sigma_{1/f} = f \circ \frac{\partial}{\partial \lambda} \circ \frac{1}{f},$$

$$D_{1/xf} = f \circ E \circ \frac{1}{f}.$$

2.4.3 **Corollary.** *σ_f^* and D_f^* commute as mappings of R into itself.*

This follows from the lemma since σ_f^* and D_f^* are stable on R.

2.4.3.1 **Corollary.** *σ_f^* is stable on $\mathbf{K}_f$ and for $u \in \mathbb{Q}(\lambda)$, $\xi \in \mathbf{K}_f$,*

$$\sigma_f^*(u\xi) = -\frac{\partial u}{\partial \lambda} \cdot \xi + u\sigma_f^* \xi.$$

2.4.4 **Corollary.** *Let 1^*, $(1/(1-x))^*$ be the basis of $\mathbf{K}_f$ dual to the basis of $\mathbf{W}_f$ represented by* 1, $1/(1-x)$. *Then*

$$\sigma_f^* \begin{pmatrix} 1^* \\ \left(\dfrac{1}{1-x}\right)^* \end{pmatrix} = \begin{pmatrix} -\dfrac{c}{\lambda} & \dfrac{c-a}{1-\lambda} \\ \dfrac{c-b}{\lambda} & \dfrac{a+b-c}{1-\lambda} \end{pmatrix} \begin{pmatrix} 1^* \\ \left(\dfrac{1}{1-x}\right)^* \end{pmatrix}.$$

Proof. If $\xi \in \mathbf{K}_f$, $\eta \in \mathbf{W}_f$ and $\langle \xi, \eta \rangle$ is independent of λ then by Lemma 2.4.1

$$\langle \xi, \sigma_f \eta \rangle = \langle \sigma_f^* \xi, \eta \rangle.$$

This is precisely the situation if ξ and η are elements of dual bases and hence the lemma follows from Lemma 1.2(iii). □

2.4.5 We say that $u_1 1^* + u_2(1/(1-x))^*$ is a *horizontal element* of $\mathbf{K}_f$ if

$$\sigma_f^*\left(u_1 1^* + u_2\left(\frac{1}{1-x}\right)^*\right) = 0.$$

With the aid of Corollary 2.4.4 this reduces to the system of differential equations

$$(2.4.5.1)_{a,b,c} \qquad \frac{\partial}{\partial \lambda}(u_1, u_2) = (u_1, u_2)\begin{pmatrix} -\dfrac{c}{\lambda} & \dfrac{c-a}{1-\lambda} \\ \dfrac{c-b}{\lambda} & \dfrac{a+b-c}{1-\lambda} \end{pmatrix}.$$

By an easy calculation, this means that u_1 satisfies the equation for $F(a, b; 1+c; \lambda)$:

$$(2.4.5.2)_{a,b,c} \qquad \left[\frac{d^2}{d\lambda^2} + \left(\frac{1+c}{\lambda} + \frac{c-(a+b)}{1-\lambda}\right)\frac{d}{d\lambda} - \frac{ab}{\lambda(1-\lambda)}\right]u_1 = 0$$

and that u_2 satisfies the equations for $F(a, b; c; \lambda)$

$$(2.4.5.3)_{a,b,c} \qquad \left[\frac{d^2}{d\lambda^2} + \left(\frac{c}{\lambda} + \frac{c-1-a-b}{1-\lambda}\right)\frac{d}{d\lambda} - \frac{ab}{\lambda(1-\lambda)}\right]u_2 = 0.$$

Of course there are no horizontal elements in $\mathbf{K}_f$ but there are in the differential module obtained from $\mathbf{K}_f$ by extension of base field.

2.5 Symplectic Structure

We recall the definitions

$$\begin{aligned} D_f &= \frac{1}{xf} \circ E \circ xf, \\ \mathbf{W}_f &= \mathbf{L}/D_f\mathbf{L}, \\ D_f^* &= -\gamma_- \circ D_{1/xf}, \\ \mathbf{K}_f &= \text{kernel of } D_f^*. \end{aligned}$$

Let

$$\tilde{f} = \frac{1}{x(1 - x\lambda)f},$$

the image of f under the transformation

$$(a, b, c) \mapsto (1 - a, 1 - b, 1 - c).$$

We define

$$\hat{D}_f = (1 - x\lambda)D_{1/xf} \tag{2.5.1}$$

as componentwise mapping of R' into itself.

Lemma. *$\hat{D}_f$ maps $\mathbf{K}_f$ into $\mathbf{L}$ and by composition with the natural map of $\mathbf{L}$ onto $\mathbf{W}_{\tilde{f}}$ we obtain a mapping whose matrix is given by*

$$\hat{D}_f\begin{pmatrix} 1^* \\ \left(\dfrac{1}{1-x}\right)^* \end{pmatrix} \equiv \begin{pmatrix} (a-c)\lambda & 0 \\ 0 & (b-c)(1-\lambda) \end{pmatrix}\begin{pmatrix} 1 \\ 1 \\ \dfrac{1}{1-x} \end{pmatrix} \bmod D_{\tilde{f}}\mathbf{L}. \tag{2.5.2}$$

Proof. It follows from (2.3.5.3)–(2.3.5.6) that for $\xi \in \mathbf{K}_f$,

$$\hat{D}_f\xi = \frac{1 - x\lambda}{1 - x}\left\langle\frac{b - c}{1 - x}, \xi\right\rangle + \lambda\left\langle\frac{a}{1 - x\lambda}, \xi\right\rangle.$$

Since

$$\frac{a}{1 - \lambda x} = (a - c) + \frac{c - b}{1 - x} + D_{\tilde{f}}1,$$

we compute

$$\left\langle \frac{a}{1-\lambda x}, 1^* \right\rangle = a - c,$$

$$\left\langle \frac{a}{1-\lambda x}, \left(\frac{1}{1-x}\right)^* \right\rangle = c - b.$$

Thus

$$\hat{D}_f 1^* = (a-c)\lambda,$$

$$\hat{D}_f \left(\frac{1}{1-x}\right)^* = \frac{1-x\lambda}{1-x}(b-c) + \lambda(c-b),$$

from which the assertion follows directly. □

Chapter 3

Transcendental Theory

Let Ω be an algebraically closed field of characteristic p, complete under a fixed non-archimedean valuation with residue class field of characteristic p.

In this chapter, λ will denote an element of Ω, $\lambda \neq 0, 1, \infty$. For $v \in S = \{0, 1, 1/\lambda, \infty\}$ we define the local variable, T_v, and the quantity t_v as in §2.3.5.2. We define

$$\Gamma_v = \mathrm{Min}(1, |t_v|^{-1}) \tag{3.0.1}$$

for each $v \in S$. Thus we have the table,

v	0	1	$1/\lambda$	∞
Γ_v	$\mathrm{Min}\left(1, \frac{1}{\lvert\lambda\rvert}\right)$	$\mathrm{Min}\left(1, \left\lvert\frac{\lambda - 1}{\lambda}\right\rvert\right)$	$\mathrm{Min}(1, \lvert\lambda - 1\rvert)$	$\mathrm{Min}(1, \lvert\lambda\rvert)$

We consider real 4-tuples, $\varepsilon \in \mathbb{R}_+^4$, $\varepsilon = (\varepsilon_0, \varepsilon_1, \varepsilon_{1/\lambda}, \varepsilon_\infty)$ satisfying the condition

$$0 < \varepsilon_v < \Gamma_v, \qquad v \in S. \tag{3.0.2}$$

3.0.2.1 *Remark.* The four disks indexed by $v \in S$

$$\{x \,|\, |T_v| \leq \varepsilon_v\},$$

with ε satisfying (3.0.2) are disjoint.

3.0.3 **Notation.** For each ε satisfying (3.0.2), let

$$(Q_{\varepsilon,\lambda} =) Q_\varepsilon = \{x \in \Omega \,|\, |T_v| > \varepsilon_v \ \forall v \in S\}$$

and let $H(Q_\varepsilon)$ be the space of analytic elements on Q_ε defined over Ω. Thus $H(Q_\varepsilon)$ is the completion under the sup norm on Q_ε of the space of elements of $\Omega(x)$ which have no pole on Q_ε.

3.0.4 If $\tilde{\varepsilon}$ is another 4-tuple also satisfying (3.0.2) we write

$$\tilde{\varepsilon} > \varepsilon$$

if $\tilde{\varepsilon}_v \geq \varepsilon_v$ for all $v \in S$. Under this condition we have the inclusion

$$Q_{\tilde{\varepsilon}} \subseteq Q_{\varepsilon}$$

and hence there is a natural injection

$$H(Q_\varepsilon) \hookrightarrow H(Q_{\tilde{\varepsilon}}).$$

We define L_λ to be "limit" of $H(Q_\varepsilon)$ as ε approaches $\Gamma = (\Gamma_0, \Gamma_1, \Gamma_{1/\lambda}, \Gamma_\infty)$. Thus in particular if $\xi \in L_\lambda$ then there exists ε satisfying (3.0.2) such that ξ may be identified with an element of $H(Q_\varepsilon)$.

With a, b, c satisfying (1.1) we extend the definition of D_f (Chapter 1) to a mapping of L_λ into itself.

3.1 **Theorem.** *Under condition* (1.1),

$$L_\lambda = \Omega \oplus \Omega \frac{1}{1-x} \oplus D_f L_\lambda.$$

Remark. This is a special case of Adolphson [2]. We give an *ad hoc* proof. We need a preliminary result.

3.2 **Lemma.** *Let r, s be strictly positive real numbers, $0 < r < s < \infty$. Let g_+ be an analytic function on the disk, $D(0, s^-)$, having no zero in the annulus*

$$\Delta_{r,s} = \{x \mid r < |x| < s\}$$

but with

$$g'_+/g_+ \in \Omega(x)$$

and that quotient vanishing at infinity.

Let η_- be a function analytic on the complement of $D(0, r^+)$, $\eta_-(\infty) = 0$, and let $\beta \in \mathbb{Q} \cap \mathbb{Z}_p$, $\beta \notin \mathbb{Z}$. Then there exists ξ_- also analytic on the complement of $D(0, r^+)$ such that

$$\left(E + \frac{E(x^\beta g_+)}{x^\beta g_+}\right)\xi_- - \eta_- \in \Omega(x) \tag{3.2.1}$$

and such that the poles of this rational function lie among those of g'_+/g_+ and such that the rational function vanishes at infinity.

Proof. We first find y analytic on $\Delta_{r,s}$ such that

$$\left(E + \frac{E(x^\beta g_+)}{x^\beta g_+}\right) y = \eta_- . \tag{3.2.2}$$

By hypothesis g_+ is analytic on $D(0, s^-)$, η_- is analytic on $\complement D(0, r^+)$. Hence g_+ may be represented by an element of $\Omega[[x]]$ on $D(0, s^-)$ and η_- may be represented by an element of $\Omega[[1/x]]$ on $\complement D(0, r^+)$ and so their product may be represented by a Laurent series

$$g_+ \eta_- = \sum_{j=-\infty}^{+\infty} A_j x^j$$

converging on $\Delta_{r,s}$ and with coefficients in Ω. Since condition (3.2.2) is formally,

$$E(x^\beta g_+ y) = x^\beta g_+ \eta_- ,$$

it is clear that we may obtain a solution by setting

$$y = g_+^{-1} \sum_{j=-\infty}^{\infty} A_j \frac{x^j}{j+\beta} \tag{3.2.3}$$

and this represents a function analytic on $\Delta_{r,s}$ since in the first place g_+ has no zeros in $D(0, s)$ and secondly since

$$\mathrm{ord}_p(j + \beta) = 0(\log j),$$

the factor $1/(j + \beta)$ does not change the radii of convergence of the Laurent series.

Having found y as asserted, we write

$$y = y_+ + y_- , \tag{3.2.4}$$

where y_+ (resp. y_-) lies in $\Omega[[x]]$ (resp. $\Omega[[1/x]]$ and converges for $|x| < s$ (resp. $|x| > r$). Furthermore we may let $y_-(\infty) = 0$. We now choose h_+, h_- each in $\Omega(x)$, having poles only among those of g'/g (and so no poles in $\Delta_{r,s}$) such that

$$\begin{aligned} &y_- \frac{Eg_+}{g_+} - h_+ \text{ is without poles in } \complement D(0, r^+), \\ &y_+ \frac{Eg_+}{g_+} - h_- \text{ is without poles in } D(0, s^-), \end{aligned} \tag{3.2.5}$$

i.e., h_+ (resp. h_-) is the principal part of $y_- Eg_+/g_+$ (resp. $y_+ Eg_+/g_+$) on $\complement D(0, r^+)$ (resp. $D(0, s^-)$). Thus

(3.2.6)
$$h_+ \text{ has no poles in } D(0, s^-),$$
$$h_- \text{ has no poles in } \complement D(0, r^+).$$

Substituting (3.2.4) in (3.2.2),

(3.2.7)
$$\left\{\left(E + \beta + \frac{Eg_+}{g_+}\right)y_- - \eta_- - h_+\right\} + h_-$$
$$= -\left\{\left(E + \beta + \frac{Eg_+}{g_+}\right)y_+ - h_-\right\} - h_+,$$

both sides being analytic on $\Delta_{r,s}$. It follows from (3.2.5) and (3.2.6) that the left side of (3.2.7) is analytic on $\complement D(0, r^+)$ and by the same argument the right side of equation (3.2.7) is analytic on $D(0, s^-)$. Thus each side is represented on $\Delta_{r,s}$ by the same Laurent series which converges for all x and hence must be a constant. Thus there exists a constant, l, such that

(3.2.8)
$$\left(E + \beta + \frac{Eg_+}{g_+}\right)y_- - \eta_- = h_+ - h_- + l.$$

The right side is rational with finite poles only among those of g'/g. However, we may evaluate the left side at ∞ and determine that the rational function vanishes at infinity. This completes the proof of the lemma. □

3.3

The following lemma represents the first step in the proof of Theorem 3.1. We first "specialize" the definition of $\mathbf{L}$ in Chapter 1.

Definition. $\mathbf{L}_\lambda$ is the Ω-space of elements of $\Omega(x)$ with poles only at $x = 0, 1, 1/\lambda, \infty$.

Lemma. $L_\lambda = \mathbf{L}_\lambda + D_f L_\lambda$.

Proof. Inclusion in one direction is trivial. Let $\eta \in L_\lambda$ then we may identify η with an element of $H(Q_\varepsilon)$ for some ε satisfying (3.0.2). Thus we may write

(3.3.1)
$$\eta = \eta_0 + \eta_1 + \eta_{1/\lambda} + \eta_\infty,$$

where for each $v \in S$, η_v is an element of $\Omega[[1/T_v]]$ which converges for $|T_v| > \varepsilon_v$. With no loss in generality we may assume

$$\eta_v = 0 \quad \text{when } T_v = \infty. \tag{3.3.2}$$

Our object is to construct $\xi_v \in \Omega[[1/T_v]]$ and converging for $|T_v| > \varepsilon_v$ such that

$$(D_f \xi_v - \eta_v) \in \Omega(x). \tag{3.3.3}$$

We define (cf. §2.3.5.2)

$$\begin{aligned} g_0 &= (1 - T_0)^{c-b}(1 - t_0 T_0)^{-a}, \\ g_1 &= (1 - T_1)^{b}(1 - t_1 T_1)^{-a}, \\ g_{1/\lambda} &= (1 - T_{1/\lambda})^{b}(1 - t_{1/\lambda} T_{1/\lambda})^{c-b}, \\ g_\infty &= (1 - T_\infty)^{c-b}(1 - t_\infty T_\infty)^{-a}, \end{aligned} \tag{3.3.4}$$

where by means of binomial expansions each of the left sides is to be interpreted as an element of $\Omega[[T_v]]$ with constant term 1. Since $a, b, c \in \mathbb{Z}_p$ it is clear that the series representing g_v converges for

$$|T_v| < \Gamma_v, \qquad v \in S. \tag{3.3.5}$$

We rewrite:

$$\begin{aligned} \frac{T_v}{x} D_f &= T_0 \frac{d}{dT_0} + \frac{T_0(d/dT_0)(T_0^b g_0)}{T_0^b g_0} & v = 0 \\ &= -\left[T_1 \frac{d}{dT_1} + \frac{T_1(d/dT_1)(T_1^{c-b} g_1)}{T_1^{c-b} g_1}\right] & v = 1 \\ &= -\lambda\left[T_{1/\lambda} \frac{d}{dT_{1/\lambda}} + \frac{T_{1/\lambda}(d/T_{1/\lambda})(T_{1/\lambda}^{-a} g_{1/\lambda})}{T_{1/\lambda}^{-a} g_{1/\lambda}}\right], & v = 1/\lambda. \\ D_f &= -\left[T_\infty \frac{d}{dT_\infty} + \frac{T_\infty(d/dT_\infty)T_\infty^{a-c} g_\infty}{T_\infty^{a-c} g_\infty}\right], & v = \infty. \end{aligned} \tag{3.3.6}$$

We observe that $(T_v/x)\eta_v$ is analytic for $|T_v| > \varepsilon_v$ if $v = 0$ but for $v = 1, 1/\lambda$, we must replace this by $(T_v/x)(\eta_v - (\eta_v)_{x=0})$ to obtain a function analytic for

$|T_\nu| > \varepsilon_\nu$. We thus may use Lemma 3.2 to construct ξ_ν analytic for $|T_\nu| > \varepsilon_\nu$ such that

$$D_f\xi_0 - \eta_0 = z_0$$

(3.3.7) $$\frac{T_\nu}{x}(D_f\xi_\nu - (\eta_\nu - (\eta_\nu)_{x=0}) = z_\nu, \qquad \nu = 1, 1/\lambda$$

$$D_f\xi_\infty - \eta_\infty = z_\infty,$$

where z_ν ($\nu \in S$) lies in $\Omega(x)$ and has poles only among those finite values of T_ν which are poles of $(1/g_\nu)(d/dT_\nu)g_\nu$, i.e., $T_\nu = 1, 1/t_\nu$. This gives the following table of values:

ν	0		1		$1/\lambda$		∞	
T_ν	1	t_ν^{-1}	1	t_ν^{-1}	1	t_ν^{-1}	1	t_ν^{-1}
x	1	$1/\lambda$	0	$1/\lambda$	0	1	1	$1/\lambda$

which shows that the poles of the rational functions z_ν are among $\{0, 1, 1/\lambda\}$ and hence the z_ν lie in $\mathbf{L}_\lambda$.

We now put $\xi = \sum_{\nu \in S} \xi_\nu$ ($\in L_\lambda$) and write

(3.3.8)

$$-\eta + D_f\xi = z_0 + z_\infty + \frac{x}{1-x}z_1 + \frac{x}{1-x\lambda}z_{1/\lambda} - (\eta_1)_{x=0} - (\eta_{1/\lambda})_{x=0}.$$

This completes the proof of the lemma since the right side of (3.3.8) clearly lies in $\mathbf{L}_\lambda$. □

3.4 Proof of Theorem 3.1

The reduction formulae used to prove part (i) of Lemma 1.2 may be used to prove the specialized form

(3.4.1) $$\mathbf{L}_\lambda = \Omega + \Omega\frac{1}{1-x} + D_f\mathbf{L}_\lambda$$

and hence with Lemma 3.3 we obtain

(3.4.2) $$L_\lambda = \Omega + \Omega\frac{1}{1-x} + D_f L_\lambda.$$

We must now show this sum to be direct. Thus let $\xi \in L_\lambda$ with

(3.4.3) $$D_f \xi = M \cdot 1 + N \cdot \frac{1}{1-x},$$

where $M, N \in \Omega$. We assert that $M = 0 = N$. We show that $M = 0$, the same proof with minor modifications gives $N = 0$.

In the construction in the proof of Theorem 2.3.5, we obtain 1^* (resp. $(1/(1-x))^*$) by putting $A = 0$, $B = \lambda/(a-c)$ (resp. $A = 1/(b-c)$, $B = \lambda/(c-b)$). These are elements of $(\Omega[[T_\nu]]_{\nu \in S}$ which may be paired with elements of $\mathbf{L}_\lambda$, the specialized form of $\mathbf{L}$. In this sense

$$\left\langle 1^*, \frac{1}{1-x} \right\rangle = 0$$

and so

$$M = \langle 1^*, D_f \xi \rangle,$$

which surely is well defined as $D_f \xi$ lies in $\mathbf{L}_\lambda$ (even though ξ itself need not).

We recall that $\mathbf{L}_\lambda$ is dense in $H(Q_\varepsilon)$. Since 1^* annihilates $D_f \mathbf{L}_\lambda$, it annihilates $D_f H(Q_\varepsilon)$ and hence $D_f L_\lambda$. We conclude that $M = 0$. This completes the proof of the theorem. □

3.5 The Mapping ψ

In this section we impose upon λ the condition

(3.5.0) $$|\lambda - 1| > |\pi|,$$

where $\pi^{p-1} = -p$. Under this hypothesis

(3.5.1) $$|\lambda^p - 1| = |\lambda - 1|^p,$$

(3.5.2) $$\mathrm{Min}(\Gamma_1, \Gamma_{1/\lambda}) > |\pi|.$$

We recall that t_ν and Γ_ν ((2.3.5.2) and (3.0.1)) depend upon λ. We will designate by Γ' the element of $\mathbb{R}_+^4$ obtained by replacing λ by λ^p. It follows from the table of (3.0.1) that

(3.5.3) $$\Gamma'_{\nu^p} = \Gamma_\nu^p.$$

Note that Γ' is a 4-tuple indexed by $S^p = \{\nu^p \mid \nu \in S\}$.

Lemma. *Let* ε *satisfy* (3.0.2) *and the further condition (harmless in view of* (3.5.2))

$$\text{(3.0.2)}' \qquad \operatorname{Min}(\varepsilon_1, \varepsilon_{1/\lambda}) \geq |\pi|.$$

Let ε' *be the* 4*-tuple indexed by* S^p, *defined by*

$$\text{(3.5.4)} \qquad \varepsilon'_{\nu^p} = \varepsilon^p_\nu.$$

We assert that

$$\text{(3.5.5)} \qquad \varepsilon'_{\nu^p} < \Gamma'_{\nu^p}, \qquad \nu \in S$$

and that for $\xi \in H(Q_\varepsilon)$ *the mapping*

$$\text{(3.5.6)} \qquad x \mapsto (\psi\xi)(x) = \sum_{z^p = x} \frac{1}{p}\, \xi(z)$$

is well defined on

$$Q_{\varepsilon', \lambda^p} = \{x \,|\, |T_{\nu^p}| > \varepsilon'_{\nu_p}\ \forall \nu \in S\}$$

and that $\psi\xi \in H(Q_{\varepsilon', \lambda^p})$. *As linear mapping of* $H(Q_{\varepsilon, \lambda})$ *into* $H(Q_{\varepsilon', \lambda^p})$, ψ *is bounded by* $|1/p|$.

Proof. Inequality (3.5.5) follows (3.5.4), (3.5.3), and (3.0.2). The main point of our construction is that the relations

$$\text{(3.5.7)} \qquad z^p = x \in Q_{\varepsilon', \lambda^p}$$

implies that

$$\text{(3.5.8)} \qquad z \in Q_{\varepsilon, \lambda}.$$

This is trivial since (3.5.7) implies that

$$|z^p| > \varepsilon'_0 = \varepsilon^p_0,$$

$$\text{(3.5.9)} \qquad \left.\begin{aligned} |z^p - 1| &> \varepsilon'_1 = \varepsilon^p_1 \geq |\pi|^p, \\ |1 - z^p\lambda^p| &> \varepsilon'_{1/\lambda^p} = \varepsilon^p_{1/\lambda} \geq |\pi|^p, \end{aligned}\right\} \qquad \text{(by (3.0.2)')}$$

$$|1/z^p| > \varepsilon'_\infty = \varepsilon^p_\infty.$$

If $|z - 1| \leq |\pi|$ then $|z^p - 1| \leq |\pi|^p = |p\pi| \leq \varepsilon'_1$, a contradiction. Hence

$$|z - 1| > |\pi|$$

and so by (3.5.1), $|z-1|^p > \varepsilon_1^p$. A similar argument shows that

$$|1 - z\lambda| > \varepsilon_{1/\lambda}.$$

This completes the verification of (3.5.8). It is now clear that $\xi \in Q_{\varepsilon,\lambda}$, the formula (3.5.6) for $\psi\xi$ gives a well-defined function on $Q_{\varepsilon^p,\lambda^p}$. Trivially then

$$\text{(3.5.10)} \qquad \operatorname{Sup}_{x \in Q_{\varepsilon',\lambda^p}} |(\psi\xi)(x)| \le p \cdot \operatorname{Sup}_{x \in Q_{\varepsilon,\lambda}} |\xi(x)|.$$

If $\xi \in \Omega(x)$ then in terms of the trace mapping, $S_{\Omega(z)/\Omega(x)}$, of $\Omega(z)$ onto $\Omega(x)$ where z is the algebraic function, $z^p = x$, we have

$$\text{(3.5.11)} \qquad p \cdot \psi\xi = S_{\Omega(z)/\Omega(x)}\xi(z).$$

This shows that ψ maps $\Omega(x)$ into itself and by (3.5.10), if ξ has no pole in $Q_{\varepsilon,\lambda}$ then $\psi\xi$ has no pole in $Q_{\varepsilon',\lambda^p}$. Thus ψ restricts to a mapping

$$\Omega(x) \cap H(Q_{\varepsilon,\lambda}) \to \Omega(x) \cap H(Q_{\varepsilon',\lambda^p}),$$

which is continuous relative to the sup norms. If ξ lies in $H(Q_{\varepsilon,\lambda})$ then ξ may be uniformly approximated on $Q_{\varepsilon,\lambda}$ by elements of $\Omega(x) \cap H(Q_{\varepsilon,\lambda})$ and so by (3.5.10) $\psi\xi$ may be uniformly approximated on $Q_{\varepsilon',\lambda^p}$ by elements of $\Omega(x) \cap H(Q_{\varepsilon',\lambda^p})$. This completes the proof of the lemma, the bound for the operator norm $\|\psi\|$ of ψ being a consequence of (3.5.10). □

3.6 The Mapping Φ

3.6.1 **Corollary.** *With ε, ε' as in the lemma, the mapping ψ of $H(Q_{\varepsilon,\lambda})$ into $H(Q_{\varepsilon',\lambda^p})$ has a one-sided inverse, Φ mapping $H(Q_{\varepsilon',\lambda^p})$ into $H(Q_{\varepsilon,\lambda})$, defined by*

$$\xi \mapsto (x \mapsto \xi(x^p)).$$

The mapping Φ is continuous in the sup norm *topology and*

$$\psi \circ \Phi = I$$

as mapping of $H(Q_{\varepsilon',\lambda^p})$ into itself.

Proof. The main point here is that $x \mapsto x^p$ maps $Q_{\varepsilon,\lambda}$ into $Q_{\varepsilon',\lambda^p}$, i.e., we assert that

$$\left.\begin{array}{r} |x| > \varepsilon_0 \\ |1-x| > \varepsilon_1 \\ |1-\lambda x| > \varepsilon_{1/\lambda} \\ |1/x| > \varepsilon_\infty \end{array}\right\} \Rightarrow \left\{\begin{array}{r} |x^p| > \varepsilon_0', \\ |1-x^p| > \varepsilon_1', \\ |1-\lambda^p x^p| > \varepsilon_{1/\lambda}', \\ |1/x^p| > \varepsilon_\infty'. \end{array}\right.$$

This follows from (3.0.2)′, and (3.5.4). The remainder of the assertions are trivial. □

3.6.2 *Remark.* The mapping ψ of $H(Q_{\varepsilon,\lambda})$ into $H(Q_{\varepsilon',\lambda^p})$ is surjective. The operator norm of Φ is 1.

3.6.3 **Corollary.** *Subject to* (3.5.0), *the mapping* ψ (resp. Φ) *extends to a mapping of* L_λ *into* L_{λ^p} (resp. L_{λ^p} *into* L_λ). *Again*

$$\psi \circ \Phi = I,$$

with ψ *surjective and* Φ *injective.*

The proof is trivial.

3.7 The Function $G_{f,h}$.

For (a, b, c) satisfying (1.1), let a', b', c' be elements of $\mathbb{Q} \cap \mathbb{Z}_p$ satisfying the conditions

$$\left.\begin{aligned} pa' - a &= \mu_a \\ pb' - b &= \mu_b \\ pc' - c &= \mu_c \end{aligned}\right\} \text{ lie in } \mathbb{Z}. \tag{3.7.0}$$

We put

$$h(x, \lambda) = h = x^{-1+b'}(1-x)^{c'-b'}(1-\lambda x)^{-a'}.$$

We write

$$G_{f,h} = \frac{xf(x,\lambda)}{x^p h(x^p, \lambda^p)} = G_{f,h}^{(0)} G_{f,h}^{(1)} G_{f,h}^{(2)}, \tag{3.7.1}$$

where

$$G_{f,h}^{(0)} = \frac{(1-x)^{\mu_b - \mu_c}(1-\lambda x)^{\mu_a}}{x^{\mu_b}} \in \mathbf{L}_\lambda,$$

$$G_{f,h}^{(1)} = \left(\frac{(1-x)^p}{1-x^p}\right)^{c'-b'},$$

$$G_{f,h}^{(2)} = \left(\frac{(1-\lambda x)^p}{1-\lambda^p x^p}\right)^{-a'}.$$

Lemma. *$G^{(1)}_{f,h}$ (resp. $G^{(2)}_{f,h}$) is analytic and bounded by unity on the set $|T_1| > |\pi|$ (resp. $|T_{1/\lambda}| > |\pi|$).*

Proof.

$$\frac{(1-x)^p}{1-x^p} = \frac{T_1^p}{1-(1-T_1)^p}$$
$$= \left(1 - \binom{p}{1}\frac{1}{T_1} + \cdots + (-1)^{p-1}\binom{p}{p-1}\frac{1}{T_1^{p-1}}\right)^{-1}.$$

Since $c' - b' \in \mathbb{Z}_p$, it is clear that $G^{(1)}_{f,h}$ is analytic on the set

$$\left| p \cdot \frac{1}{T_1^j} \right| < 1, \qquad j = 1, 2, \ldots, p-1.$$

This proves the assertion for $G^{(1)}_{f,h}$. The same argument holds for $G^{(2)}_{f,h}$. □

3.7.2 **Corollary.**

$$G^{(1)}_{f,h} G^{(2)}_{f,h} \in H(Q_{\varepsilon,\lambda}),$$

where

$$\varepsilon = (e\Gamma_0, |\pi|, |\pi|, e\Gamma_\infty), \quad e \textit{ arbitrary in } (0, 1).$$

3.7.3 **Corollary.** $G_{f,h} \in L_\lambda$.

3.7.4 *Remark.* Let $x_0 \in Q_{\varepsilon,\lambda}$ (ε as in Corollary 3.7.2) and let ξ be a branch at $x = x_0^p$ of the algebraic function $xh(x, \lambda^p)$. Then $\xi G_{f,h}$ is a branch at $x = x_0$ of $xf(x, \lambda)$, and the point is that $G_{f,h}$ is uniquely determined. Since there are p choices of x_0 for a fixed choice of x_0^p, this means that branches η of $xf(x, \lambda)$ at $x = x_0, x_0\zeta_p, \ldots, x_0\zeta_p^{p-1}$ are uniquely determined by ξ by setting

$$\eta(x) = \xi(x^p) G_{f,h}.$$

3.8 The Lifting of the Cartier Operator

By setting

$$\alpha_{f,h} = \psi \circ G_{f,h}, \tag{3.8.1}$$

we obtain a mapping of $H(Q_{\varepsilon,\lambda})$ into $H(Q_{\varepsilon',\lambda^p})$ (with ε given by Corollary 3.7.2) which extends to a mapping of L_λ into L_{λ^p}.

Lemma. *As mapping of* $H(Q_{\varepsilon,\lambda})$ *into* $H(Q_{\varepsilon',\lambda^p})$ (resp. *of* L_λ *into* L_{λ^p}),

(3.8.2) $$\alpha_{f,h} \circ D_{f,\lambda} = pD_{h,\lambda^p} \circ \alpha_{f,h}.$$

Remark. Intuitively

$$D_{f,\lambda} = \frac{1}{xf(x,\lambda)} \circ E \circ xf(x,\lambda),$$

$$\alpha_{f,h} = \frac{1}{xh(x,\lambda^p)} \circ \psi \circ xf(x,\lambda),$$

$$D_{h,\lambda^p} = \frac{1}{xh(x,\lambda^p)} \circ E \circ xh(x,\lambda^p).$$

while

$$\psi \circ E = pE \circ \psi.$$

We make this intuitive argument precise.

Proof. Since both sides of (3.8.2) are continuous operators on $H(Q_{\varepsilon,\lambda})$, it is enough to verify it as a mapping of $\mathbf{L}_\lambda$ into $H(Q_{\varepsilon',\lambda^p})$. Since the quantity $e \in (0, 1)$ appearing in (3.7.2) plays no role, we may choose any x_0,

(3.8.2.1) $$0 < x_0 < \Gamma_0^p,$$

choose e so small that $x_0 \in Q_{\varepsilon',\lambda^p}$ and then it is enough to show that for $\xi \in \mathbf{L}_\lambda$, the image of ξ under each side of (3.8.2) coincide on a neighborhood of x_0. To do this let $z_1, \ldots, z_p$ denote the p branches at x_0 of the algebraic function $z = x^{1/p}$. Let h_0 be a branch of $h = h(x, \lambda^p)$ at x_0. We now choose uniquely a branch of $f = f(x, \lambda)$ at $z = z_i(x_0)$ by writing

(3.8.3) $$f_i(z, \lambda) = G_{f,h}(z)z^{p-1}h_0(z^p, \lambda^p).$$

For $\xi \in \mathbf{L}_\lambda$, x near x_0, we have

(3.8.4) $$\begin{aligned} p(\alpha_{f,h}\xi)(x) &= \sum_{i=1}^{p} G_{f,h}(z_i)\xi(z_i) \\ &= \sum_{i=1}^{p} \frac{z_i f_i(z_i,\lambda)}{z_i^p h_0(z_i^p,\lambda^p)} \xi(z_i) \\ &= \frac{1}{xh_0(x,\lambda^p)} \sum_{i=1}^{p} z_i f_i(z_i,\lambda)\xi(z_i). \end{aligned}$$

Now $E(xf)/xf$ lies in $\mathbf{L}_\lambda$ and so its value at any point does not depend on the choice of branch of f. Thus its value at z near $z_i(x_0)$ is given by

$$\frac{z(d/dz)(zf_i(z, \lambda))}{zf_i(z, \lambda)}.$$

Hence the value of $D_f \xi$ at z near $z_i(x_0)$ is given by

$$(3.8.5) \qquad (D_f \xi)(z) = z \frac{d}{dz} \xi(z) + \frac{z(d/dz)(zf_i(z, \lambda))}{zf_i(z, \lambda)} \xi(z)$$

and so for x close to x_0

$$(3.8.6) \quad p(\alpha_{f,h} D_f \xi)(x) = \frac{1}{xh_0(x, \lambda^p)} \sum_{i=1}^{p} z_i f_i(z_i, \lambda) \times \left[z_i \frac{d}{dz_i} \xi(z_i) + \xi(z_i) \frac{z_i(d/dz_i)(z_i f_i(z_i, \lambda))}{z_i f_i(z_i, \lambda)} \right]$$

$$= \frac{1}{xh_0(x, \lambda^p)} \sum_{i=1}^{p} z_i \frac{d}{dz_i} (z_i f_i(z_i, \lambda) \xi(z_i))$$

$$= \frac{1}{xh_0(x, \lambda^p)} px \frac{d}{dx} \sum_{i=1}^{p} z_i f_i(z_i, \lambda) \xi(z_i).$$

Thus

$$(3.8.7) \quad (\alpha_{f,h} D_f \xi)(x) = \frac{1}{xh_0(x, \lambda^p)} x \frac{d}{dx} \left[xh_0(x, \lambda^p) \sum_{i=1}^{p} \frac{z_i f_i(z_i, \lambda)}{xh_0(x, \lambda^p)} \xi(z_i) \right]$$

$$= \frac{1}{xh_0(x, \lambda^p)} x \frac{d}{dx} [xh_0(x, \lambda^p)(p\alpha_{f,h} \xi)(x)]$$

$$= pD_{h,\lambda^p} \alpha_{f,h} \xi.$$

This completes the proof of (3.8.2). $\square$

3.8.8 **Corollary.** *By passage to quotients $\alpha_{f,h}$ induces a mapping $\bar{\alpha}_{f,h}$ of $\mathbf{W}_{f,\lambda}$ into $\mathbf{W}_{h,\lambda^p}$ where*

$$\mathbf{W}_{f,\lambda} = L_\lambda / D_f L_\lambda$$
$$\mathbf{W}_{f,\lambda^p} = L_{\lambda^p} / D_{f,\lambda^p} L_{\lambda^p}$$

3.8.9 **Corollary.** *The mapping $\bar{\alpha}_{f,h}$ is an isomorphism between $\mathbf{W}_{f,\lambda}$ and W_{h,λ^p}.*

Proof. Clearly $1/G_{f,h} \in L_\lambda$ and so

$$\beta_{f,h} = \frac{1}{G_{f,h}} \circ \Phi \tag{3.8.10}$$

is a mapping of L_{λ^p} into L_λ and

$$\alpha_{f,h} \circ \beta_{f,h} = I \tag{3.8.11}$$

as mapping of L_λ into itself. To complete the proof we show that

$$p\beta_{f,h} \circ D_{h,\lambda^p} = D_{f,\lambda} \circ \beta_{f,h}. \tag{3.8.12}$$

For x_0 in the support of $G_{f,h}$ we may choose a branch of $f(x, \lambda)$ (resp. $h(x, \lambda^p)$) in the neighborhood of x_0 (resp. x_0^p) such that

$$\frac{xf(x, \lambda)}{x^p h(x^p, \lambda^p)} = G(x) \qquad (= G_{f,h}) \tag{3.8.13}$$

for all x in the neighborhood of x_0. For $\xi \in L_{\lambda^p}$, x close to x_0, we compute:

(3.8.14)

$$\begin{aligned}
(D_{f,\lambda}\beta_{f,h}\xi)(x) &= \left(E + \frac{E(xf(x, \lambda))}{xf(x, \lambda)}\right)\frac{1}{G(x)}\,\xi(x^p) \\
&= \frac{E(xf(x, \lambda))}{xf(x, \lambda)}\frac{1}{G(x)}\,\xi(x^p) + \frac{1}{G(x)}\,p(E\xi)(x^p) + \xi(x^p)\frac{-EG}{G^2} \\
&= p\beta_{f,h}(E\xi) + \frac{1}{G}\left[\frac{E(xf(x, \lambda))}{xf(x, \lambda)} - \frac{EG}{G}\right]\xi(x^p).
\end{aligned}$$

On the other hand

$$\begin{aligned}
p\beta_{f,h}D_{h,\lambda^p}\xi &= p\beta_{f,h}\left(E\xi + \xi\,\frac{E(xh(x, \lambda^p))}{xh(x, \lambda^p)}\right) \\
&= p\beta_{f,h}(E\xi) + p\,\frac{1}{G}\,\xi(x^p)\Phi\left(\frac{E(xh(x, \lambda^p))}{xh(x, \lambda^p)}\right).
\end{aligned} \tag{3.8.15}$$

The proof of (3.8.12) reduces to the assertion

$$\frac{E(xf(x, \lambda))}{xf(x, \lambda)} - \frac{EG}{G} = p\Phi\,\frac{E(xh(x, \lambda^p))}{xh(x, \lambda^p)}, \tag{3.8.16}$$

which follows from (3.8.13).

Having checked (3.8.12), we obtain by passage to quotients a map $\bar{\beta}_{f,h}$ of $\mathbf{W}_{h,\lambda^p}$ into $\mathbf{W}_{f,\lambda}$. Clearly

$$\bar{\beta}_{f,h} \circ \bar{\alpha}_{f,h} = I$$

as mapping of $\mathbf{W}_{f,\lambda}$ into itself. By Theorem 3.1, $\mathbf{W}_{f,\lambda}$ and $\mathbf{W}_{h,\lambda^p}$ are finite-dimensional spaces. This completes the proof. □

Chapter 4

Analytic Dual Theory

Our object is to construct an analytic form of the dual space suitable for the construction of a dual to $\alpha_{f,h}$. We again assume $\lambda \in \Omega$, $\lambda \neq 0, 1, \infty$. We change notation so as to replace Ω_2 (in Chapter 2) by Ω. For $v \in S$ we define

$$R_v = \begin{cases} \Omega[[T_v]] & \text{if } v \neq \infty, \\ T_\infty \Omega[[T_\infty]] & \text{if } v = \infty. \end{cases}$$

4.1

With Γ_v given by (3.0.1) we define

$\hat{R}_v$ = subring of R_v consisting of elements which converge for $|T_v| < \Gamma_v$.

$\hat{R}'_v$ = ring of Laurent series in T_v with coefficients in Ω which represent function analytic on an annulus

$$0 < \varepsilon_v < |T_v| < \Gamma_v$$

with unspecified ε_v.

We define $\hat{R}$, $\hat{R}'$

$$\hat{R}(\lambda) = \hat{R} = \bigoplus_{v \in S} \hat{R}_v \hookrightarrow \bigoplus \hat{R}'_v = \hat{R}' = \hat{R}'(\lambda).$$

We note that in the notation of Chapter 3, for $v \in S$

$$L_\lambda \subset \hat{R}'_v$$

and we imbed L_λ diagonally in $\hat{R}'$. We define P_v (principal part at v) on $\hat{R}'_v$ by

$$P_v\colon \hat{R}'_v \to L_\lambda,$$

$$\sum_{j=-\infty}^{+\infty} a_j T_v^j \mapsto \begin{cases} \sum_{j=-\infty}^{-1} a_j T_v^j & \text{if } v \neq \infty, \\ \sum_{j=-\infty}^{0} a_j T_\infty^j & \text{if } v = \infty. \end{cases}$$

We again have a natural projection of $\hat{R}'$ into L_λ

$$\gamma_+ : \hat{R}' \to L_\lambda,$$

$$\xi \mapsto \sum_{v \in S} P_v \xi_v.$$

We set

(4.1.1) $$\gamma_- = I - \gamma_+,$$

a mapping of $\hat{R}'$ into $\hat{R}$. Again γ_- annihilates L_λ and so

(4.1.2) $$\hat{R}' = \hat{R} \oplus L_\lambda,$$

with γ_- (resp. γ_+) as the projection onto the first (resp. second) summand.

4.2

We again define

$$\mathrm{Res}_v\colon \hat{R}'_v \mapsto \Omega,$$

the formulae being as in (2.2.0).

4.2.1 For $v \neq \infty$,

$$\mathrm{Res}_v\colon \hat{R}_v \to \{0\},$$

while for $v = \infty$

$$\mathrm{Res}_\infty\colon T_\infty \hat{R}_\infty \to \{0\}.$$

4.2.2 **Lemma.** $(\xi, \eta) \mapsto \mathrm{Res}_v\, \xi\eta$ *gives a pairing of* $\hat{R}'_v$ *with itself which induces a perfect pairing of* $\hat{R}_v$ *with* $\hat{R}'_v/\hat{R}_v$.

Proof. That the above mapping induces a pairing of $\hat{R}_v$ with $\hat{R}'_v/\hat{R}_v$ follows from 4.2.1. If $\xi \in \hat{R}_v$ and $\text{Res}_v\, \xi\eta = 0$ for all $\eta \in \hat{R}'_v$ then $\xi = 0$ as shown by part (a) of (2.2.2) where only η in the form of monomials was needed. Conversely if $\eta \in \hat{R}'_v$ and $\text{Res}_v\, \xi\eta = 0$ for all $\xi \in \hat{R}_v$ then the argument of part (b) of (2.2.2) shows that $\eta \in \hat{R}'_v \cap R_v$, i.e., to $\hat{R}_v$. This completes the proof. □

4.2.3 **Definition.** We pair $\hat{R}'$ with itself by

$$\langle \xi, \eta \rangle = \sum_{v \in S} \text{Res}_v\, \xi_v \eta_v.$$

4.2.4 **Lemma.**

(4.2.4.1) $\qquad$ *For* $\xi \in L_\lambda$, $\qquad \langle 1, \xi \rangle = 0$,

(4.2.4.2) $\qquad$ *For* $\xi, \eta \in L_\lambda$, $\qquad \langle \eta, \xi \rangle = 0$,

(4.2.4.3) $\qquad$ *For* $\xi, \eta \in \hat{R}$, $\qquad \langle \eta, \xi \rangle = 0$.

Proof. Assertion (4.2.4.2) follows from (4.2.4.1). Assertion (4.2.4.3) follows 4.2.1.

By means of the Mittag–Leffler representation of elements of $H(Q_{\varepsilon,\lambda})$ we obtain

(4.2.4.4)
$$H(Q_{\varepsilon,\lambda}) = \mathbf{L}_\lambda + \frac{d}{dx} H(Q_{\varepsilon',\lambda}),$$

where $\varepsilon' > \varepsilon$ and each component $\varepsilon'_v > \varepsilon_v$. The point here is that integration does not change the radius of convergence but does not preserve uniformity of convergence. We deduce

(4.2.4.5)
$$L_\lambda = \mathbf{L}_\lambda + \frac{d}{dx} L_\lambda.$$

Since (4.2.4.1) has been proven for $\mathbf{L}_\lambda$ (cf. (2.2.4.1)), it is enough to show that $\xi \in L_\lambda$ implies $\langle 1, d\xi/dx \rangle = 0$, but in fact we know that

$$\text{Res}_v \frac{d\xi}{dx} = 0.$$

This completes the proof. □

4.2.5 **Lemma.** *The pairing, $\langle\ ,\ \rangle$, of $\hat{R}'$ with itself restricts to a perfect pairing of $\hat{R}$ with L_λ.*

If $\xi \in L_\lambda$ is annihilated by $\hat{R}$, fix $v \in S$ and let $\eta \in \hat{R}$, $\eta_s = 0$ for $s \neq v$, $s \in S$. Thus

$$\mathrm{Res}_v\, \xi z = 0$$

for all $z \in \hat{R}_v$. Hence by Lemma 4.2.2, $\xi \in \hat{R}_v$, i.e.,

$$P_v \xi = 0, \qquad \forall v \in S.$$

Hence by (4.1.2), $\xi = 0$.

Conversely if $\eta \in \hat{R}$ annihilates L_λ, then η may be viewed as an element of R annihilated by $\mathbf{L}_\lambda$. Thus we may use Lemma 2.2.6 to deduce that $\eta = 0$. This completes the proof of the lemma. □

4.2.6 If $\xi \in \hat{R}'$, $\eta \in L_\lambda$ then

$$\langle \xi, \eta \rangle = \langle \gamma_- \xi, \eta \rangle.$$

This follows from (4.2.4.2) and (4.1.2).

4.2.7 **Lemma.** *Let T be an endomorphism of $\hat{R}'$ (as Ω-space) which is stable on L_λ. Let T^* be an endomorphism of $\hat{R}'$ dual to T under $\langle\ ,\ \rangle$. Then $(\gamma_- \circ T^*)|\hat{R}$ is the unique endomorphism of $\hat{R}$ adjoint to $T|L_\lambda$ under the induced pairing of $\hat{R}$ with L_λ.*

Proof. Uniqueness follows from Lemma 4.2.5, that is the perfectness of the pairing of $\hat{R}$ with L_λ. For $\eta \in \hat{R}$, $\xi \in L_\lambda$ we have by §4.2.6

$$\langle \gamma_- T^*\eta, \xi \rangle = \langle T^*\eta, \xi \rangle = \langle \eta, T\xi \rangle,$$

the point being that $\gamma_- \circ T^*$ is stable on $\hat{R}$. □

4.2.8 **Lemma.** $\mathbf{K}_f$ *lies in* $\hat{R}$ *and under* $\langle\ ,\ \rangle$, $\mathbf{K}_f$ *is the dual space of* $L_\lambda/D_f L = \mathbf{W}_f$.

Proof. Corollary 2.3.6 shows that $\mathbf{K}_f$ lies in $\hat{R}$. We know that $\mathbf{K}_f$ and $\mathbf{W}_f$ have the same dimension as Ω-spaces. It is enough to show that $\mathbf{K}_f$ annihilates $D_f L_\lambda$. If $\xi \in L_\lambda$ then $\xi \in H(Q_{\varepsilon,\lambda})$ for some ε satisfying (3.0.2) and then ξ is uniform limit of elements of $\mathbf{L}_\lambda$, in the sup topology of $H(Q_{\varepsilon,\lambda})$. Since D_f is continuous mapping of $H_\lambda(Q_{\varepsilon,\lambda})$ into itself, we conclude that $D_f \xi$ is the uniform limit of elements of $D_f \mathbf{L}_\lambda$ under this topology. If $\eta \in \mathbf{K}_f$ then η annihilates $D_f \mathbf{L}_\lambda$ and so it is enough to check that for η fixed in $\hat{R}$,

$$\xi \to \langle \eta, \xi \rangle \tag{4.2.9}$$

is continuous in the topology of $H(Q_{\varepsilon,\lambda})$. This follows from the estimates for the coefficients of the Mittag–Leffler expansions. An alternate proof of the lemma is based upon showing that $-\gamma_- \circ D_{1/xf}$ is the endomorphism of $\hat{R}$ adjoint to $D_f | L_\lambda$, i.e., by §4.2.7 it is enough to show that for η, $\xi \in \hat{R}'$ we have

(4.2.10) $$\langle -D_{1/xf}\eta, \xi\rangle = \langle \eta, D_f\xi\rangle.$$

We compute

$$\langle \eta, D_f\xi\rangle = \sum_\nu \operatorname{Res}_\nu\left(\eta_\nu E\xi_\nu + \eta_\nu \frac{E(xf)}{xf}\,\xi_\nu\right),$$

$$\eta_\nu E\xi_\nu = \frac{d}{dx}(x\eta_\nu\xi_\nu) - \xi_\nu \frac{d}{dx}(x\eta_\nu),$$

while

$$\operatorname{Res}_\nu \frac{d}{dx}(x\eta_\nu\xi_\nu) = 0.$$

Thus

$$\begin{aligned}\langle \eta, D_f\xi\rangle &= \sum_\nu \operatorname{Res}_\nu \xi_\nu\left[-\frac{d}{dx}(x\eta_\nu) + \eta_\nu\frac{E(xf)}{xf}\right]\\ &= -\langle D_{1/xf}\eta, \xi\rangle\end{aligned}$$

as asserted. This completes the alternate proof of the lemma. □

4.3

We must now consider the dependence of $\hat{R}$ upon λ and for that purpose will use the symbol $\hat{R}(\lambda)$. We now assume that λ satisfies (3.5.0).

Theorem. *The mapping ψ of L_λ into L_{λ^p} may be extended by means of formula (3.5.6) to mapping of $\hat{R}'(\lambda)$ into $\hat{R}'(\lambda^p)$. The mapping Φ of L_{λ^p} into L_λ extends to a mapping of $\hat{R}'(\lambda^p)$ into $\hat{R}'(\lambda)$ by*

(4.3.1) $$\xi = (\xi_{\nu p})_{\nu\in S} \to (\xi_{\nu p}(x^p))_{\nu\in S} = \phi\xi.$$

(*Note. In $\xi_{\nu p}$ we replace x by x^p and* not *$T_{\nu p}$ by T_ν^p.*)

(4.3.2) *Under the pairing $\langle\ ,\ \rangle$, of $\hat{R}'$ with itself, Φ is the dual of $\psi \circ 1/x^{p-1}$.*

Proof. For $\xi \in \hat{R}'(\lambda^p)$, the component ξ_{ν^p} is analytic on the annulus

$$\Delta_{\nu^p}(\varepsilon'_{\nu^p}, \Gamma'_{\nu^p})$$

defined by

$$0 < \varepsilon'_{\nu^p} < |T_{\nu^p}| < \Gamma'_{\nu^p} \qquad (= \Gamma_\nu^p \text{ by (3.5.3)}).$$

By (3.5.2) we may assume that

$$\text{(4.3.3)} \qquad \operatorname{Min}(\varepsilon'_1, \varepsilon'_{1/\lambda^p}) \geq |\pi|^p.$$

We put

$$\varepsilon_\nu = \varepsilon'^{1/p}_{\nu^p} < \Gamma_\nu$$

and we assert that the mapping $x \to x^p$ maps

$$\text{(4.3.4)} \qquad \Delta_\nu(\varepsilon_\nu, \Gamma_\nu) \to \Delta'_{\nu^p}(\varepsilon'_{\nu^p}, \Gamma'_{\nu^p}).$$

For $\nu = 0, \infty$ this is trivial while for $\nu = 1, 1/\lambda$ this follows from equations (3.5.1) and (4.3.3).

Thus composing ξ_{ν^p} with the map $x \to x^p$ gives an analytic function on $\Delta_\nu(\varepsilon_\nu, \Gamma_\nu)$. This gives the extension of Φ to $\hat{R}'(\lambda^p)$.

In the opposite direction, given $\eta \in \hat{R}'(\lambda)$, we have for each $\nu \in S$, a component η_ν analytic on $\Delta_\nu(\varepsilon_\nu, \Gamma_\nu)$ and again by (3.5.2) we may assume

$$\text{(4.3.3')} \qquad \operatorname{Min}(\varepsilon_1, \varepsilon_{1/\lambda}) \geq |\pi|.$$

We assert that $z^p = x \in \Delta_{\nu^p}(\varepsilon_\nu^p, \Gamma_\nu^p)$ implies

$$\text{(4.3.5)} \qquad z \in \Delta_\nu(\varepsilon_\nu, \Gamma_\nu).$$

This is trivial for $\nu = 0, \infty$ while for $\nu = 1$, the assertion follows from (3.5.1) if $|z - 1| > |\pi|$, while $|1 - z| \leq |\pi|$ implies that

$$|\pi|^p \leq \varepsilon_1^p < |1 - z^p| \leq |p\pi|,$$

a contradiction. A similar proof demonstrates (4.3.5) for $\nu = 1/\lambda$.

Thus under (4.3.3'), for $x \in \Delta_{\nu^p}(\varepsilon_\nu^p, \Gamma_\nu^p)$ we define

$$(\psi\eta_\nu)(x) = \frac{1}{p} \sum_{z^p = x} \eta_\nu(z).$$

In this way we obtain $\psi\eta_\nu$, a well-defined function on $\Delta_{\nu p}(\varepsilon_\nu^p, \Gamma_\nu^p)$. This function is analytic since for each r, s such that

$$\varepsilon_\nu < r < s < \Gamma_\nu,$$

η_ν lies in $H(\Delta_\nu(r, s))$, hence can be uniformly approximated by rational functions and so $\psi\eta_\nu$ can be uniformly approximated on $\Delta_{\nu p}(r^p, s^p)$ by rational functions. This completes our extension of ψ.

We now turn to the verification of (4.3.2). We must show that for $\xi \in \hat{R}'(\lambda^p)$, $\eta \in \hat{R}'(\lambda)$,

$$\langle \Phi\xi, \eta \rangle = \left\langle \xi, \psi \frac{1}{x^{p-1}} \eta \right\rangle. \tag{4.3.6}$$

It is enough to show that for $\xi \in \hat{R}'_{\nu p}(\lambda^p)$, $\eta \in \hat{R}'_\nu(\lambda)$, we have

$$\operatorname{Res}_\nu \xi(x^p)\eta(x) = \operatorname{Res}_{\nu p} \xi\psi\left(\frac{1}{x^{p-1}}\eta\right). \tag{4.3.7}$$

We put

$$u = \eta\Phi\xi \in \hat{R}'_\nu(\lambda).$$

Then

$$\psi \frac{1}{x^{p-1}} u = \xi\psi\left(\frac{1}{x^{p-1}}\eta\right)$$

and so (4.3.7) is reduced to

$$\operatorname{Res}_\nu u = \operatorname{Res}_{\nu p} \psi\left(\frac{1}{x^{p-1}} u\right) \tag{4.3.8}$$

for all $u \in \hat{R}'_\nu(\lambda)$.

Now we know that u may be represented by a Laurent series on $\Delta_\nu(\varepsilon_\nu, \Gamma_\nu)$ and hence there exists $v \in R'_\nu(\lambda)$ such that

$$u = \frac{dv}{dx} + \begin{cases} A/T_\nu & \text{if } \nu \neq \infty, \\ AT_\infty & \text{if } \nu = \infty. \end{cases} \tag{4.3.9}$$

Now

$$\begin{aligned} \psi\left(\frac{1}{x^{p-1}}\frac{dv}{dx}\right) &= \frac{1}{p}\sum_{z^p = x} \frac{1}{z^{p-1}} v'(z) = \sum_{z^p = x} \frac{d}{dx} v(z) \\ &= \frac{d}{dx}\sum_{z^p = x} v(z) = p\frac{d}{dx}(\psi v). \end{aligned} \tag{4.3.10}$$

This shows that $\psi \circ 1/x^{p-1}$ is stable on $(d/dx)\hat{R}'_v$. Since $(d/dx)\hat{R}'_v$ lies in the kernel of Res_v, we conclude that (4.3.8) surely holds for $u \in (d/dx)\hat{R}'_v$. It only remains then to check (4.3.8) for

$$u = \begin{cases} 1/T_v & \text{if } v \neq \infty, \\ T_\infty & \text{if } v = \infty. \end{cases}$$

For $v = \infty$

(4.3.11)
$$\psi \frac{1}{x^{p-1}} T_\infty = \psi \frac{1}{x^p} = \frac{1}{x}.$$

For $v = 0$

(4.3.12)
$$\psi \frac{1}{x^{p-1}} \frac{1}{T_0} = \frac{1}{x}.$$

For $v = 1$

(4.3.13)
$$\psi \frac{1}{x^{p-1}} \frac{1}{T_1} = \psi \frac{1}{x^{p-1}(1-x)} = \psi \sum_{j=0}^{\infty} \frac{x^j}{x^{p-1}}$$
$$= \frac{1}{1-x}.$$

For $v = 1/\lambda$

$$\psi \frac{1}{x^{p-1}} \frac{1}{1-\lambda x} = \psi \sum_{j=0}^{\infty} \frac{(\lambda x)^j}{x^{p-1}} = \lambda^{p-1} \sum_{j=0}^{\infty} (\lambda^p x)^j$$
$$= \lambda^{p-1} \frac{1}{1-\lambda^p x}.$$

Thus for $v = \infty, 0$ we are reduced to showing that

$$\mathrm{Res}_v \frac{1}{x} = \mathrm{Res}_{v^p} \frac{1}{x},$$

which is trivial. For $v = 1, 1/\lambda$ we must show

$$\mathrm{Res}_1 \frac{1}{1-x} = \mathrm{Res}_{1^p} \frac{1}{1-x},$$

$$\mathrm{Res}_{1/\lambda} \frac{1}{1-\lambda x} = \mathrm{Res}_{1/\lambda^p} \frac{\lambda^{p-1}}{1-x\lambda^p}.$$

These are trivial. This completes the proof of the theorem. □

4.4

4.4.1 **Corollary.** *If $g \in L_\lambda$ then relative to $\langle\ ,\ \rangle$ the mapping*

$$\xi \mapsto x^{p-1} g \Phi \xi$$

of $\hat{R}'(\lambda^p)$ into $\hat{R}'(\lambda)$ is dual to the map

$$\eta \mapsto \psi(g\eta)$$

of $\hat{R}'(\lambda)$ into $\hat{R}(\lambda^p)$.

Proof. $\psi \circ g$ is the composition of the map ψ of $\hat{R}'(\lambda^p)$ into $\hat{R}'(\lambda)$ with the mapping $\xi \mapsto \xi \cdot g$ of $\hat{R}'(\lambda)$ into itself. The assertion follows from the theorem since multiplication by g is self dual. □

The following corollary follows from §4.2.7.

4.4.2 **Corollary.** *Under the hypothesis of Corollary* 4.4.1,

$$\gamma_- x^{p-1} g \circ \Phi$$

is the unique mapping of $\hat{R}(\lambda^p)$ into $\hat{R}(\lambda)$ dual to the mapping $\psi \circ g$ of L_λ into L_{λ^p}.

4.4.3 **Lemma.**

(i) $\theta = \gamma_- x^{p-1} G_{f,h} \circ \Phi$ *is the unique map of $\hat{R}(\lambda^p)$ into $\hat{R}(\lambda)$ dual to $\alpha_{f,h} (= \psi \circ G_{f,h})$ as mapping of L_λ into L_{λ^p}.*

(ii) *The mapping θ of part* (i) *maps $\mathbf{K}_{h,\lambda^p}$ into $\mathbf{K}_{f,\lambda}$ and putting*

$$\alpha^*_{f,h} = \theta | \mathbf{K}_{h,\lambda^p},$$

we conclude that $\alpha^*_{f,h}$ is the dual of $\bar{\alpha}_{f,h}$ the mapping of $\mathbf{W}_{f,\lambda}$ into $\mathbf{W}_{h,\lambda}$ induced by $\alpha_{f,h}$.

Proof. The first assertion is just a restatement of the previous corollary with g replaced by $G_{f,h}$. For part (ii) let $\xi \in \mathbf{K}_{h,\lambda^p}$, $\eta \in L_\lambda$, then by (3.8.2)

$$\begin{aligned}
\langle \gamma_- x^{p-1} G_{f,h} \Phi \xi, D_{f,\lambda} \eta \rangle &= \langle \xi, \alpha_{f,h} D_{f,\lambda} \eta \rangle \\
&= p \langle \xi, D_{h,\lambda^p} \alpha_{f,h} \eta \rangle \\
&= p \langle D^*_{h,\lambda^p} \xi, \alpha_{f,h} \eta \rangle,
\end{aligned}$$

the last equality following from the alternate proof of Lemma 4.2.8. But $\xi \in \mathbf{K}_{h,\lambda^p}$ and so $\theta(\mathbf{K}_{h,\lambda^p}) \subset \hat{R}(\lambda)$ and annihilates $D_{f,\lambda} L_\lambda$. Since the annihilator of $D_{f,\lambda} L_\lambda$ in $\hat{R}(\lambda)$ is precisely $\mathbf{K}_{f,\lambda}$, we conclude that

$$\theta \mathbf{K}_{h,\lambda^p} \subset \mathbf{K}_{f,\lambda}.$$

The assertion is now clear. □

4.5 The Matrix of $\alpha_{f,h}$

Let us use $\{1^*, (1/(1-x))^*\}$ as bases of $\mathbf{K}_{f,\lambda}$ and of $\mathbf{K}_{h,\lambda^p}$. Of course these symbols refer to completely different objects in the two cases. Let us write

$$\alpha^*_{f,h}\begin{pmatrix}1^*\\(1/(1-x))^*\end{pmatrix}=\begin{pmatrix}B_1 & B_2\\B_3 & B_4\end{pmatrix}\begin{pmatrix}1^*\\(1/(1-x))^*\end{pmatrix} \tag{4.5.1}$$

giving the matrix of the mapping $\alpha^*_{f,h}$ of $\mathbf{K}_{h,\lambda^p}$ into $\mathbf{K}_{f,\lambda}$. We know that this mapping is well defined for λ satisfying (3.5.0).

Theorem. *Each coefficient B_i in* (4.5.1) *is the sum of an element of $Q[\lambda, 1/\lambda]$ with an element of $Q[[1/(\lambda-1)]]$ which converges subject to* (3.5.0).

4.5.1.1 *Remark.* This result is one of the main reasons for the construction of the dual theory.

Let ξ^* be an element of the basis of $\mathbf{K}_{h,\lambda^p}$ and let η be an element of the basis of $\mathbf{W}_{f,\lambda}$. The corresponding matrix coefficient B_i is then given by:

$$\begin{aligned}B_i &= \langle\alpha^*_{f,h}\xi^*, \eta\rangle\\ &= \langle x^{p-1}G_{f,h}\Phi\xi^*, \eta\rangle\\ &= \sum_{v\in S}\operatorname{Res}_v x^{p-1}G_{f,h}\eta\Phi(\xi^*)_{v^p},\end{aligned} \tag{4.5.2}$$

the symbol $\Phi(\xi^*)_{v^p}$ meaning the result of replacing x by x^p in the v^p component of ξ^*. Now η is either 1 or $1/(1-x)$ and hence by (3.7.1),

$$x^{p-1}G_{f,h}\eta = T_0^{k_0}T_1^{k_1}T_{1/\lambda}^{k_2}G_0, \tag{4.5.3}$$

where

$$k_0 = p-1-\mu_b,$$

$$k_1 = \mu_b-\mu_c-\begin{cases}0, & \eta = 1,\\ 1, & \eta = 1/(1-x),\end{cases}$$

$$k_2 = \mu_a,$$

$$G_0 = G^{(1)}_{f,h}G^{(2)}_{f,h}.$$

By Theorem 2.3.5

$$\xi^*_{vP} = \sum_{s=0}^{\infty} T^s_{vP} C_s(t_{vP}) \cdot \begin{cases} 1 & \text{if } v \neq 1/\lambda, \\ \lambda^p & \text{if } v = 1/\lambda, \end{cases} \tag{4.5.4}$$

where C_s is a polynomial (depending upon v) of degree bounded by s (resp. $s + 1$) if $v \neq 0$ (resp. $v = 0$) and with coefficients in $U_{\theta_{vP}, s}$. Note that in applying Table 2.3.5.2 to determine θ_{vP} we must here replace a, b, c by a', b', c'. For our present application it is sufficient to observe that since $\theta_{vP} \in \mathbb{Q}$, there exists a constant κ independent of s such that

$$|z| \leq \kappa s \tag{4.5.5}$$

for each coefficient z of the polynomial C_s. The precise dependence of C_s upon v plays no role in the proof.

We now write

$$\begin{aligned} M_v &= T_0^{k_0} T_1^{k_1} T_{1/\lambda}^{k_2} G_0 \cdot \sum_{s=0}^{\infty} C_s(t_{vP}) \Phi(T^s_{vP}), \\ J_v &= \operatorname{Res}_v M_v. \end{aligned} \tag{4.5.6}$$

We will show that

$$J_0(\lambda) \in \mathbb{Q}[\lambda], \tag{4.5.7}$$

$$J_\infty(\lambda) \in \mathbb{Q}[1/\lambda, \lambda], \tag{4.5.8}$$

(4.5.9) $J_v(\lambda)$ satisfies the conclusions of the theorem for $v = 1, 1/\lambda$.

The theorem obviously follows from these three statements. We consider the four values of v separately.

Case 1. $v = 0$

We have

$$M_0 = T_0^{k_0} T_1^{k_1} T_{1/\lambda}^{k_2} G_0 \cdot \sum_{s=0}^{\infty} C_s(\lambda^p) x^{p^s}, \tag{4.5.10}$$

each factor (except for $T_0^{k_0}$ if $k_0 < 0$) is analytic on the disk $D(0, \Gamma_0)$ and hence $J_0(\lambda)$ is zero unless $k_0 < 0$ and in that case $J_0(\lambda)$ is the coefficient of x^{-k_0-1} in the series expansion of

$$\begin{aligned} &(1 - x)^{k_1}(1 - \lambda x)^{k_2}(1 - x)^{p(c'-b')}(1 - x^p)^{b'-c'}(1 - \lambda x)^{-pa'} \\ &\quad \times (1 - \lambda^p x^p)^{a'} \sum_{s=0}^{\infty} C_s(\lambda^p) x^{ps}. \end{aligned} \tag{4.5.11}$$

This coefficient obviously lies in $\mathbb{Q}[\lambda]$. This completes the verification of (4.5.7).

Case 2. $\nu = \infty$

Here

$$\Phi(\xi^*)_\infty = \sum_{s=1}^{\infty} C_s\left(\frac{1}{\lambda^p}\right) T_\infty^{ps}, \tag{4.5.12}$$

which is analytic on the disk

$$|T_\infty| < \Gamma_\infty = \text{Min}(1, |\lambda|). \tag{4.5.13}$$

In this disk we may represent G_0 by

$$G_0(x) = (1 - T_\infty)^{p(c'-b')}(1 - T_\infty^p)^{b'-c'}\left(1 - \frac{1}{\lambda} T_\infty\right)^{-pa'}\left(1 - \frac{T_\infty^p}{\lambda^p}\right)^{a'}, \tag{4.5.14}$$

and

$$T_0^{k_0} T_1^{k_1} T_{1/\lambda}^{k_2} = (-1)^{k_1+k_2}\lambda^{k_2}(1 - T_\infty)^{k_1}\left(1 - \frac{1}{\lambda} T_\infty\right)^{k_2} \frac{1}{T_\infty^{k_1+k_2+k_0}}.$$

Hence M_∞ is the product of $1/T_\infty^{k_1+k_2+k_0}$ with a power series in T_∞ which converges in the indicated disk. It follows that $J_\infty(\lambda)$ is the coefficient of $T_\infty^{1+k_1+k_2+k_0}$ in the series expansion of

$$(-1)^{k_1+k_2}\lambda^{k_2}(1 - T_\infty)^{k_1}\left(1 - \frac{1}{\lambda} T_\infty\right)^{k_2} G_0\, \phi(\xi^*)_\infty. \tag{4.5.15}$$

It follows from (4.5.12) and (4.5.14) that this coefficient lies in $\lambda^{k_2}\mathbb{Q}[1/\lambda]$. This completes the verification of equation (4.5.8).

Case 3. $\nu = 1$.

Here

$$\Phi(\xi^*)_1 = \sum_{s=0}^{\infty} C_s\left(\frac{\lambda^p}{\lambda^p - 1}\right)(1 - x^p)^s \tag{4.5.16}$$

converges if

$$|1 - x| < \Gamma_1. \tag{4.5.17}$$

(Recall by (3.5.0) that $\Gamma_1 > |\pi|$.)

In this disk

$$|\lambda(1-x)| < |\lambda - 1|$$

and hence

$$(4.5.18) \qquad |1 - x\lambda| = |1 - \lambda + \lambda(1-x)| = |1-\lambda| > |\pi|.$$

Thus $G^{(2)}_{f,h}$ is analytic and bounded by 1 on this disk. We write

$$M_1 = \Theta(x)\rho(x),$$

where

$$(4.5.19) \qquad \begin{aligned} \Theta(x) &= (1-x)^{\mathrm{Min}(0,k_1)} G^{(1)}_{f,h}, \\ \rho(x) &= x^{k_0}(1-x)^{\mathrm{Max}(0,k_1)}(1-\lambda x)^{k_2} G^{(2)}_{f,h} \Phi(\xi^*)_1. \end{aligned}$$

Clearly ρ is analytic on the disk (4.5.17), but we make this more explicit. In particular

$$\frac{(1-\lambda x)^p}{1-\lambda^p x^p} = \frac{(1-\lambda)^p}{1-\lambda^p} \frac{(1-t_1 T_1)^p}{1-(\lambda^p/(\lambda^p-1))(1-(1-T_1)^p)}$$

and so

$$(4.5.20) \qquad G^{(2)}_{f,h}(x) = G^{(2)}_{f,h}(1) \cdot (1 - t_1 T_1)^{-pa'} \left(1 - \frac{\lambda^p}{\lambda^p - 1}(1-(1-T_1)^p)\right)^{a'},$$

$$(4.5.21) \qquad x^{k_0}(1-x)^{\mathrm{Max}(0,k_1)}(1-\lambda x)^{k_2} = (1-\lambda)^{k_2}(1-T_1)^{k_0} T_1^{\mathrm{Max}(0,k_1)}(1-t_1 T_1)^{k_2}.$$

Also

$$(4.5.22) \qquad \Phi(\xi^*)_1 = \sum_{s=0}^{\infty} C_s\left(\frac{\lambda^p}{1-\lambda^p}\right)(1-(1-T_1)^p)^s.$$

It follows from (4.5.20)–(4.5.22) that

$$(4.5.23) \qquad \rho(x) = (1-\lambda)^{k_2} G^{(2)}_{f,h}(1) T_1^{\mathrm{Max}(0,k_1)} \sum_{s=0}^{\infty} A_s T_1^s,$$

with

$$(4.5.24) \qquad A_s \in Q(\lambda), \qquad s \in \mathbb{N}.$$

We assert that with κ as in (4.5.5), λ satisfying (3.5.0),

$$(4.5.25)\qquad |A_s| \le \begin{cases} \dfrac{\kappa}{p}\dfrac{s}{\Gamma_1^s} & \text{if } \dfrac{1}{s} < \log \dfrac{1}{|\pi|}\Gamma_1, \\[2ex] \dfrac{\kappa}{pe|\pi|^s}\dfrac{1}{\log(\Gamma_1/|\pi|)} & \text{if } \dfrac{1}{s} > \log \dfrac{1}{|\pi|}\Gamma_1. \end{cases}$$

Comparing (4.5.23) with (4.5.19) and using (4.5.18) and the fact that $G_{f,h}^{(2)}(x)$ takes unit values for x on the disk (4.5.17), we conclude that for

$$0 < r < \Gamma_1,$$

we have

$$(4.5.26)\qquad |A_s|r^s \le \sup_{|x-1|=r} \left| \sum C_s\left(\frac{\lambda^p}{1-\lambda^p}\right)(1-x^p)^s \right|.$$

Using (4.5.5) we have

$$(4.5.27)\qquad \left| C_s\left(\frac{\lambda^p}{1-\lambda^p}\right) \right| \le \frac{\kappa s}{\Gamma_1^{ps}},$$

while

$$|1-x^p| \le \begin{cases} r^p & \text{if } |1-x| = r \ge |\pi|, \\ |p|r & \text{if } |1-x| = r \le |\pi|. \end{cases}$$

Thus

$$(4.5.28)\qquad |A_s|r^s \le \begin{cases} \sup_{m\in\mathbb{N}} \kappa m\left(\dfrac{r}{\Gamma_1}\right)^{pm}, & \text{if } r \in [|\pi|, \Gamma_1), \\[2ex] \sup_{m\in\mathbb{N}} \kappa m\left(\dfrac{r|p|}{\Gamma_1^p}\right)^{m}, & \text{if } r \in [0, |\pi|]. \end{cases}$$

(*Note.* We only use the first estimate on the right.)

We estimate the right side of (4.5.28):

If $0 < z < 1$ then the real valued function

$$t \to t \cdot z^t, \qquad t \ge 0$$

assumes its maximum value at $t = 1/\log(1/z)$ and the value is $(1/e)1/\log(1/z)$. Thus

(4.5.29)
$$|A_s|r^s \leq \frac{\kappa}{e}\frac{1}{p\log(\Gamma_1/r)}, \qquad r \in [|\pi|, \Gamma_1).$$

We now estimate $|A_s|$ by maximizing $r^s \log(\Gamma_1/r)$ with $r \in (|\pi|, \Gamma_1)$. In the range $(0, \Gamma_1)$ the maximum occurs at $r = \Gamma_1 e^{-1/s}$. If this exceeds $|\pi|$ then substituting this value for r in (4.5.29) we obtain the first case of (4.5.25). For the second case, we simply substitute $|\pi|$ for r in (4.5.29). This completes the verification of (4.5.25).

We now write

(4.5.30)
$$\Theta(x) = T_1^{\mathrm{Min}(0,k_1)} \sum_{s=0}^{\infty} \theta_s \frac{1}{T_1^s},$$

where by Lemma 3.7, $\theta_s \in Q$

(4.5.31)
$$\operatorname{ord} \theta_s \geq \frac{s}{p-1}.$$

We now have

(4.5.32)
$$J_1(\lambda) = G_{f,h}^{(2)}(1)(1-\lambda)^{k_2} \sum_{\substack{s_1, s_2 \geq 0 \\ s_1 = 1 + s_2 + k_1}} \theta_{s_1} A_{s_2}.$$

It follows from (4.5.25) and (4.5.31) that $J_1(\lambda)$ is analytic on the set defined by (3.5.0). This completes the verification of (4.5.9) for $v = 1$.

Case 4. $v = 1/\lambda$

Here

(4.5.33)
$$\begin{aligned}\Phi(\xi^*)_{1/\lambda^p} &= \sum_{s=0}^{\infty} \lambda^p C_s\left(\frac{1}{1-\lambda^p}\right)(1 - \lambda^p x^p)^s \\ &= \sum_{s=0}^{\infty} \lambda^p C_s\left(\frac{1}{1-\lambda^p}\right)(1 - (1 - T_{1/\lambda})^p)^s\end{aligned}$$

converges for

(4.5.34)
$$|T_{1/\lambda}| < \Gamma_{1/\lambda}.$$

We write

(4.5.35)

$$\frac{(1-x)^p}{1-x^p} = \frac{(\lambda-1)^p}{\lambda^p-1}(1-t_{1/\lambda}T_{1/\lambda})^p \bigg/ \left(1 - \frac{1}{\lambda^p-1}[(1-T_{1/\lambda})^p - 1]\right),$$

which shows that $G^{(1)}_{f,h}(x)$ is analytic on the disk (4.5.34) and may be written

$$G^{(1)}_{f,h}(x) = G^{(1)}_{f,h}(\lambda)\cdot \sum_{s=0}^{\infty} \alpha_s T^s_{1/\lambda}, \tag{4.5.36}$$

where $\alpha_s \in \mathbb{Q}(\lambda)$ and the series in $T_{1/\lambda}$ converges and is bounded by 1 on the disk (4.5.34).

We write

$$M_{1/\lambda} = \Theta(x)\rho(x),$$

where

(4.5.37)

$$\Theta(x) = T^{\mathrm{Min}(0,k_2)}_{1/\lambda} G^{(2)}_{f,h}(x),$$

$$\rho(x) = \left(\frac{1}{\lambda}\right)^{k_0}\left(\frac{\lambda-1}{\lambda}\right)^{k_1}(1-T_{1/\lambda})^{k_0}$$
$$\times (1-t_{1/\lambda}T_{1/\lambda})^{k_1} T^{\mathrm{Max}(0,k_2)}_{1/\lambda} G^{(1)}_{f,h}(x)\cdot \Phi(\xi^*)_{1/\lambda^p}.$$

It follows from (4.5.36) and (4.5.33) that

$$\rho(x) = \left(\frac{1}{\lambda}\right)^{k_0}\left(\frac{\lambda-1}{\lambda}\right)^{k_1} G^{(1)}_{f,h}(\lambda) T^{\mathrm{Max}(0,k_2)}_{1/\lambda} \sum A_s T^s_{1/\lambda}, \tag{4.5.38}$$

where

$$A_s \in \mathbb{Q}(\lambda). \tag{4.5.39}$$

Precisely as for (4.5.25), we obtain for λ satisfying (3.5.0)

$$|A_s| \le \frac{\kappa}{p}\frac{s}{\Gamma^s_{1/\lambda}} \quad \text{if}\ \frac{1}{s} < \log\frac{1}{|\pi|}\Gamma_{1/\lambda}. \tag{4.5.40}$$

We write

$$\Theta(x) = T^{\mathrm{Min}(0,k_2)}_{1/\lambda} \sum_{s=0}^{\infty} \theta_s \frac{1}{T^s_{1/\lambda}}, \tag{4.5.41}$$

with

(4.5.42)
$$\begin{cases} \operatorname{ord}\theta_s \geq \dfrac{s}{p-1}, \\ \theta_s \in \mathbb{Q}. \end{cases}$$

We now have

(4.5.43)
$$-\lambda J_{1/\lambda}(\lambda) = G^{(1)}_{f,h}(\lambda)\left(\frac{\lambda-1}{\lambda}\right)^{k_1}\left(\frac{1}{\lambda}\right)^{k_0} \sum_{\substack{s_1 = 1+s_2+k_2 \\ s_1, s_2 \geq 0}} \theta_{s_1} A_{s_2}.$$

The analytic properties of $J_{1/\lambda}$ may be deduced from (4.5.40) and (4.5.42). □

4.6 Deformation

The object is to study the relation between $\mathbf{K}_{f,\lambda}$ and $\mathbf{K}_{f,z}$ for z close to λ.

4.6.1 **Lemma.** *If*

(4.6.1.1)
$$|\lambda - z| < |z| \operatorname{Min}(1, |z-1|),$$

then

$$L_\lambda = L_z \quad \textit{and} \quad \hat{R}'(\lambda) = \hat{R}'(z).$$

Proof. It is enough to show for each of the four values of v that the disk

$$|T_v| < \Gamma_v$$

does not change if we replace λ by z. For $v = 0, 1, \infty$ it is therefore enough to check that

(4.6.1.2)
$$|\lambda| = |z|,$$

(4.6.1.3)
$$|\lambda - 1| = |z - 1|,$$

and for $v = 1/\lambda$ we need

(4.6.1.4)
$$\left|\frac{1}{\lambda} - \frac{1}{z}\right| < \frac{1}{|z|} \operatorname{Min}(1, |z-1|).$$

Assertion (4.6.1.2) follows from $|\lambda - z| < |z|$. Assertion (4.6.1.3) follows from (4.6.1.2) if $|z| > 1$, while for $|z| \leq 1$, the assertion follows from $|\lambda - z| < |z - 1|$ which may be deduced from (4.6.1.1). Finally, assertion (4.6.1.4) follows from (4.6.1.1) by division by $|\lambda z|$ and using (4.6.1.2). □

4.6.2 **Theorem.**

(i) *Let* $\theta(x) = f(x, \lambda)/f(x, z) = ((1 - \lambda x)/(1 - zx))^{-a}$. *Under hypothesis* (4.6.1.1), $\theta \in L_z$ *and*

$$T_{z,\lambda} = \gamma_- \circ \theta$$

maps $\mathbf{K}_{f,z}$ *onto* $\mathbf{K}_{f,\lambda}$. *For* z *fixed,* λ *variable, the diagram*

$$\begin{array}{ccc} \mathbf{K}_z & \xrightarrow{T_{z,\lambda}} & \mathbf{K}_\lambda \\ {\scriptstyle -\frac{\partial}{\partial\lambda}}\big\downarrow & & \big\downarrow{\scriptstyle \sigma_\lambda^*} \\ \mathbf{K}_z & \xrightarrow[T_{z,\lambda}]{} & \mathbf{K}_\lambda \end{array}$$

commutes. (σ_λ^* *is used to denote the symbol* σ_f^* *of* §2.4.)

(ii) *The matrix of* $T_{z,\lambda}$ *relative to the basis* $\langle 1^*, (1/(1 - x))^* \rangle$ *is analytic as function of* λ *in the region* (4.6.1.1).

4.6.2.1 *Remark.* $\mathbf{K}_z$ is independent of z and hence is annihilated by $\partial/\partial\lambda$. The assertion remains valid if $\mathbf{K}_z$ and $\mathbf{K}_\lambda$ are replaced by their tensor products with a field, M, of functions of λ meromorphic in a neighborhood of z.

Proof. Let $T = T_{1/z}$. Clearly

$$\frac{1 - \lambda x}{1 - zx} = \frac{\lambda}{z}\left(1 - \frac{\lambda - z}{\lambda}\frac{1}{T}\right)$$

and hence

(4.6.2.2)
$$\theta = \left(\frac{\lambda}{z}\right)^{-a}\left(1 - \frac{\lambda - z}{\lambda}\frac{1}{T}\right)^{-a},$$

the first factor being analytic function of λ in disk (4.6.1.1) and the second factor is analytic function of x for

(4.6.2.3)
$$|T| > \left|\frac{\lambda - z}{\lambda}\right| \quad \left(= \left|\frac{\lambda - z}{z}\right|\right),$$

the equality being valid by (4.6.1.1). Thus as function of x, θ is analytic for $|T_{1/z}| > \varepsilon$, where

$$\varepsilon = \left|\frac{\lambda - z}{z}\right| < \mathrm{Min}(1, |z - 1|) = \Gamma_{1/z}. \tag{4.6.2.4}$$

This shows that $\theta \in L_z = L_\lambda$ and also gives the Mittag–Leffler decomposition:

$$\theta = \left(\frac{\lambda}{z}\right)^{-a} \sum_{m=0}^{\infty} \binom{-a}{m} \left(\frac{z - \lambda}{\lambda} \frac{1}{T_{1/z}}\right)^{m}. \tag{4.6.2.5}$$

Clearly then multiplication by θ is stable on $\hat{R}'(z)$ which shows that $\gamma_- \circ \theta$ is stable on $\hat{R}(z)$. To show that $\mathbf{K}_{f,z}$ is mapped into $\mathbf{K}_{f,\lambda}$ we show that the diagram

$$\begin{array}{ccc} \hat{R} & \xrightarrow{T_{z,\lambda}} & \hat{R} \\ {\scriptstyle D^*_{f,z}}\downarrow & & \downarrow{\scriptstyle D^*_{f,\lambda}} \\ \hat{R} & \xrightarrow[T_{z,\lambda}]{} & \hat{R} \end{array}$$

commutes, i.e., as operators on $\hat{R}(z) = \hat{R}(\lambda)$,

$$D^*_{f,\lambda} \circ T_{z,\lambda} = T_{z,\lambda} \circ D^*_{f,z}. \tag{4.6.2.6}$$

We write the left side explicitly as

$$\begin{aligned} -D^*_{f,\lambda} \circ T_{z,\lambda} &= \gamma_- D_{1/xf,\lambda} \circ \gamma_- \circ \theta \\ &= \gamma_- D_{1/xf,\lambda} \circ \theta - \gamma_- D_{1/xf,\lambda} \gamma_+ \theta. \end{aligned} \tag{4.6.2.7}$$

Likewise

$$-T_{z,\lambda} \circ D^*_{f,z} = \gamma_- \theta \circ D_{1/xf,z} - \gamma_- \theta \gamma_+ D_{1/xf,z}. \tag{4.6.2.8}$$

Now as operators on $\hat{R}' = \hat{R}'(z)$, both $\gamma_- \circ D_{1/xf,\lambda} \circ \gamma_+$ and $\gamma_- \circ \theta \circ \gamma_+$ are trivial, since γ_+ projects $\hat{R}'$ onto L_z, θ and $D_{1/xf,\lambda}$ are stable on L_z and L_z is annihilated by γ_-.

On the other hand as operators on $\hat{R}'$,

$$\theta \circ D_{1/xf,z} = D_{1/xf,\lambda} \circ \theta, \tag{4.6.2.9}$$

in view of the definition of θ, since

$$D_{1/xf,z} = f(x, z) \circ E \circ \frac{1}{f(x, z)},$$

$$D_{1/xf,\lambda} = f(x, \lambda) \circ E \circ \frac{1}{f(x, \lambda)}.$$

This completes the verification of (4.6.2.6).

Since condition (4.6.1.1) is symmetric in λ and z, and since interchanging λ and z transforms θ into $1/\theta$, we conclude that

$$T_{\lambda,z} = \gamma_- \frac{1}{\theta}$$

maps $\mathbf{K}_{f,\lambda}$ into $\mathbf{K}_{f,z}$. However

(4.6.2.10)
$$T_{\lambda,z} \circ T_{z,\lambda} = \gamma_- \circ \frac{1}{\theta} \circ \gamma_- \circ \theta$$
$$= \gamma_- \circ \frac{1}{\theta} \circ \theta - \gamma_- \circ \frac{1}{\theta} \circ \gamma_+ \circ \theta.$$

As noted before $\gamma_- \circ 1/\theta \circ \gamma_+$ annihilates $\hat{R}'$ while γ_- acts like the identity on $\hat{R}$. Thus $T_{\lambda,z} \circ T_{z,\lambda}$ restricted to $K_{f,z}$ is the identity map. This shows that $T_{z,\lambda}$ maps $\mathbf{K}_{f,z}$ onto $\mathbf{K}_{f,\lambda}$.

We now show as mapping of $M \otimes \hat{R}'(z)$ into itself,

(4.6.2.11)
$$\sigma_\lambda^* \circ T_{z,\lambda} = -T_{z,\lambda} \circ \frac{\partial}{\partial \lambda},$$

i.e.,

$$\gamma_- \circ \sigma_{1/f(x,\lambda)} \circ \gamma_- \circ \theta = \gamma_- \circ \theta \circ \frac{\partial}{\partial \lambda}.$$

The right side is the same as $\gamma_- \circ \sigma_{1/f(x,\lambda)} \circ \theta$ and so we are reduced to checking

(4.6.2.12)
$$\left(\frac{\partial}{\partial \lambda} - \frac{1}{f(x, \lambda)} \frac{\partial}{\partial \lambda} f(x, \lambda)\right) \circ \frac{f(x, \lambda)}{f(x, z)} = \frac{f(x, \lambda)}{f(x, z)} \circ \frac{\partial}{\partial \lambda}.$$

which is trivial. This completes the verification of (i).

(ii) We now give an explicit representation of the matrix representing $T_{z,\lambda}$. A typical matrix coefficient is

$$M_{z,\lambda}(\xi, \eta) = \langle T_{z,\lambda}\xi, \eta\rangle,$$

where ξ is either 1^* or $(1/(1-x))^*$ and η is 1 or $1/(1-x)$. For example,

$$M_{z,\lambda}(\xi, 1) = \sum_{v\in(0,1,1/\lambda,\infty)} \operatorname{Res}_v \xi_v(a, b, c, z, x)\cdot\theta$$

Now in fact $\theta\cdot\xi_v$ lies in R_v for $v = 0, 1$, so the corresponding residues are zero. For $v = \infty$, we observe that

$$\theta \equiv \left(\frac{\lambda}{z}\right)^{-a} \quad \bmod T_\infty$$

and so

$$\operatorname{Res}_\infty \theta\xi_\infty = -\left(\frac{\lambda}{z}\right)^{-a}\cdot\text{leading coefficient of } \xi_\infty = \left(\frac{\lambda}{z}\right)^{-a}\cdot\langle\xi, 1\rangle.$$

Finally to compute $\operatorname{Res}_{1/\lambda}$, we observe that it is the same as $\operatorname{Res}_{1/z}$ and hence by (4.6.2.5) and (2.3.5.10)

$$\operatorname{Res}_{1/\lambda}\theta\cdot\xi_{1/\lambda} = \left(\frac{\lambda}{z}\right)^{-a}\sum_{j=0}^{\infty}\binom{-a}{j+1}\left(\frac{z-\lambda}{\lambda}\right)^{j+1} C_j\left(\frac{1}{1-\lambda}\right).$$

where C_j is a polynomial of degree bounded by j whose coefficients lie in $U_{a,j}$. This shows the analyticity in region (4.6.1.1) of the matrix coefficients, as asserted. □

4.6.3 **Corollary.** *Let $\eta\in\mathbf{K}_{f,z}$ (independent of λ). Then subject to* (4.6.1.1),

(4.6.3.1) $$T_{z,\lambda}\eta = u_1\cdot 1^* + u_2\cdot\left(\frac{1}{1-x}\right)^*,$$

(1^*, $(1/(1-x))^*$ refer to elements of $\mathbf{K}_{f,\lambda}$) where u_1, u_2 are functions of λ satisfying the system of differential equations (2.4.5.1) *with initial conditions given by*

(4.6.3.2) $$\eta = u_1(z)(1^*)_z + u_2(z)\left(\frac{1}{1-x}\right)^*_z,$$

the subscript z being used to remind us that in this relation, 1^* *and* $(1/(1-x))^*$

are elements of $\mathbf{K}_{f,z}$. *In particular,* u_1, u_2 *are power series in* $\lambda - z$ *converging in the region* (4.6.1.1).

Proof. This follows from the theorem and §2.4.5. □

4.6.4 We now give the relation between $\alpha_{f,h}^*$ and $T_{z,\lambda}$.

Theorem. *Under the restrictions* (4.6.1.1) *and*

(3.5.0) $$|z - 1| > |\pi|.$$

The following diagram is commutative:

$$\begin{array}{ccc} \mathbf{K}_{f,z} & \xrightarrow{T_{z,\lambda,f}} & \mathbf{K}_{f,\lambda} \\ \uparrow{\scriptstyle \alpha_z^*} & & \uparrow{\scriptstyle \alpha_\lambda^*} \\ \mathbf{K}_{h,z^p} & \xrightarrow[T_{z^p,\lambda^p,h}]{} & \mathbf{K}_{h,\lambda^p} \end{array}$$

In this diagram α_z^* *is used to indicate the specialization at* $\lambda = z$ *of the mapping* $\alpha_{f,h}^*$ (cf. Lemma 4.4.3).

Proof. Recalling that $G_{f,h}(x) = G_{f,h}(x, \lambda)$ depends on both x and λ, the proof may be reduced to showing that as mappings of $\hat{R}'(z^p)$ into $\hat{R}'(\lambda) = \hat{R}'(z)$,

(4.6.4.1)

$$\gamma_- \frac{f(x, \lambda)}{f(x, z)} \gamma_- x^{p-1} G_{f,h}(x, z) \circ \Phi = \gamma_- x^{p-1} G_{f,h}(x, \lambda) \circ \Phi \circ \gamma_- \frac{h(x, \lambda^p)}{h(x, z^p)}.$$

To show this we observe that $x^{p-1} G_{f,h}(x, z)\Phi$, maps $\hat{R}'(z^p)$ into $\hat{R}'(z)$. To replace the interior "γ_-" on the left side of (4.6.4.1) by $I - \gamma_+$, recall that $f(x, \lambda)/f(x, z)$ is stable on L_z and that hence

$$\gamma_- \circ \frac{f(x, \lambda)}{f(x, z)} \circ \gamma_+ = 0$$

as mapping of $\hat{R}'(z)$ into itself. We conclude that we may drop the interior "γ_-" on the left side of (4.6.4.1). Likewise

$$\gamma_- x^{p-1} G_{f,h}(x, \lambda) \circ \Phi \circ \gamma_+ = 0$$

as mapping of $\hat{R}'(z^p)$ into $\hat{R}'(z)$ and so we may drop the interior "γ_-" on the right side of (4.6.4.1). The verification of (4.6.4.1) is thus reduced to checking that

(4.6.4.2)
$$\frac{f(x, \lambda)}{f(x, z)} G_{f,h}(x, z) = G_{f,h}(x, \lambda) \frac{h(x^p, \lambda^p)}{h(x^p, z^p)},$$

which is an easy consequence of (3.7.1). □

4.7 Symplectic Structure

We recall (§2.5) the map $\hat{D}_f$ of $\mathbf{K}_f$ into $\mathbf{W}_f$. We recall (3.8.10) $\beta_{\tilde{f},\tilde{h}}$ the mapping of $\mathbf{W}_{\tilde{h},\lambda p}$ into $\mathbf{W}_{\tilde{f},\lambda}$. In terms of these operations we explain the duality of our theory. We assume (3.5.0).

Theorem. *The following diagram commutes with a factor p.*

$$\begin{array}{ccc} \mathbf{K}_{f,\lambda} & \xrightarrow{\hat{D}_{f,\lambda}} & \mathbf{W}_{\tilde{f},\lambda} \\ \uparrow {\scriptstyle \alpha^*_{f,h}} & & \uparrow {\scriptstyle \beta_{\tilde{f},\tilde{h}} = \bar{\alpha}^{-1}_{\tilde{f},\tilde{h}}} \\ \mathbf{K}_{h,\lambda^p} & \xrightarrow[\hat{D}_{h,\lambda^p}]{} & \mathbf{W}_{\tilde{h},\lambda^p} \end{array}$$

Explicitly,

(4.7.1)
$$\hat{D}_{f,\lambda} \circ \alpha^*_{f,h} = p\beta_{\tilde{f},\tilde{h}} \circ \hat{D}_{h,\lambda^p}$$

as map of $\mathbf{K}_{h,\lambda^p}$ *into* $\mathbf{W}_{\tilde{f},\lambda}$.

Proof. For $\xi \in \mathbf{K}_{h,\lambda^p}$

$$\begin{aligned} \alpha^*_{f,h}\xi &= \gamma^- x^{p-1} G_{f,h} \xi^{\Phi} \\ &= x^{p-1} G_{f,h} \xi^{\Phi} + \eta, \end{aligned}$$

where

$$\eta = -\gamma_+ x^{p-1} G_{f,h} \xi^{\Phi} \in L_\lambda.$$

Thus assertion (4.7.1) is equivalent to

(4.7.2)

$$(1 - x\lambda) D_{1/xf}(x^{p-1} G_{f,h} \xi^{\Phi} + \eta) \equiv p \frac{1}{G_{\tilde{f},\tilde{h}}} \Phi(1 - x\lambda^p) D_{1/xh,\lambda^p} \xi \mod D_{\tilde{f}} L_\lambda.$$

We assert that

$$(4.7.3) \qquad (1 - x\lambda)D_{1/xf}\eta \in D_{\tilde{f}}L_{\lambda}.$$

This follows trivially from the calculation

$$(4.7.4) \quad (1 - x\lambda)D_{1/xf} = (1 - x\lambda)f \circ E \circ \frac{1}{f} = \frac{1}{x\tilde{f}} \circ E \circ x\tilde{f} \circ (1 - x\lambda)$$
$$= D_{\tilde{f}} \circ (1 - x\lambda).$$

Furthermore, by (3.7.1)

$$G_{f,h} = x^{1-p}F,$$

$$G_{\tilde{f},\tilde{h}} = \frac{1 - x^p\lambda^p}{1 - x\lambda}\frac{1}{F},$$

where $F = f(x, \lambda)/h(x^p, \lambda^p)$. Thus the verification of (4.7.2) reduces to that of

$$(4.7.5) \quad (1 - x\lambda)D_{1/xf,\lambda}F \circ \Phi = p\,\frac{1 - x\lambda}{1 - x^p\lambda^p}F \circ \Phi \circ (1 - x\lambda^p)D_{1/xh,\lambda^p}$$

as mapping of $\hat{R}'(\lambda^p)$ into $\hat{R}'(\lambda)$, and the right side may be simplified to

$$p(1 - x\lambda)F \circ \Phi \circ D_{1/xh,\lambda^p}$$

Now formally the left side of (4.7.5) is

$$(1 - x\lambda)f \circ E \circ \frac{1}{f}F \circ \Phi = (1 - x\lambda)f \circ E \circ \frac{1}{h(x^p, \lambda^p)} \circ \Phi$$
$$= (1 - x\lambda)f \circ E \circ \Phi \circ \frac{1}{h(x, \lambda^p)}$$
$$= p(1 - x\lambda)f \circ \Phi \circ E \circ \frac{1}{h(x, \lambda^p)}$$
$$= p(1 - x\lambda)F \circ h(x^p, \lambda^p) \circ \Phi \circ E \circ \frac{1}{h(x, \lambda^p)}$$
$$= p(1 - x\lambda)F \circ \Phi \circ h(x, \lambda^p) \circ E \circ \frac{1}{h(x, \lambda^p)}$$

and this coincides with the right side of (4.7.5). This formal calculation may be made precise. This concludes the verification of the theorem. □

4.7.6 **Corollary.** *Using the basis* 1, $1/(1-x)$ *for* $W_{\tilde{f},\lambda}$ (resp. $W_{\tilde{h},\lambda^p}$) *and the basis* 1^*, $(1/(1-x))^*$ *for* $\mathbf{K}_{f,\lambda}$ (resp. $\mathbf{K}_{h,\lambda^p}$), *we have*

$$\det \alpha^*_{f,h} \det \alpha_{f,h} = p^2 \frac{\lambda^p(1-\lambda^p)}{\lambda(1-\lambda)} e,$$

$$e = \frac{(b'-c')(a'-c')}{(b-c)(a-c)}.$$

Proof. This follows from (4.7.1), the determinant of the matrices representing $\hat{D}_{f,\lambda}$ and $\hat{D}_{h,\lambda}$ being given by (2.5.2). □

4.7.7 *Remark.* This corollary will be used in Theorem 6.6 below to determine the magnitude of $\det \alpha^*_{f,h}$ under the condition $|\lambda| = 1 = |\lambda - 1|$.

Chapter 5

Basic Properties of ψ Operator

5.1

5.1.1 **Definition.** For $\xi = \sum_{j=0}^{m} a_j x^j \in \Omega[x]$, we define

$$|\xi|_{\text{gauss}} = \sup_j |a_j|.$$

This definition is extended to $\Omega(x)$ by taking quotients.

5.1.2 **Definition.** For ξ analytic on $D(0, R^-)$ we define for $r < R$.

$$|\xi|_0(r) = \sup_{|x|=r} |\xi(x)|.$$

This is extended to functions meromorphic on $D(0, R^-)$ by taking quotients.

We remark that for $\xi \in \Omega(x)$, we have

$$|\xi|_{\text{gauss}} = |\xi|_0(1). \tag{5.1.3}$$

We have noted that the mapping ψ defined by (3.5.6) is stable on $\Omega(x)$. It is clear that $|\psi|_{\text{gauss}}$, the operator norm of ψ is bounded by p.

5.2 **Lemma.** $|\psi|_{\text{gauss}} = 1$.

Proof. Since $\psi(1) = 1$, it is clear that $|\psi|_{\text{gauss}} \geq 1$. We wish to reverse this inequality. Let $\xi \in \Omega(x)$, $|\xi|_{\text{gauss}} \leq 1$. We assert $|\psi\xi|_{\text{gauss}} \leq 1$. We choose $r_0 < 1$ such that ξ has no pole in the annulus

$$\Delta(r_0, 1) = \{x \mid r_0 < |x| < 1\}$$

and thus ξ may be represented by a Laurent series

$$\xi(x) = \sum_{x=-\infty}^{+\infty} \alpha_s x^s \tag{5.2.1}$$

on this annulus. For $r \in (r_0, 1)$,

$$|\xi|_0(r) = \sup_s |\alpha_s| r^s. \tag{5.2.2}$$

Since the mapping $r \mapsto |\xi|_0(r)$ is continuous,

$$|\xi|_0(1) = \sup_s |\alpha_s|.$$

Thus

$$|\psi\xi|_0(1) = \sup_s |\alpha_{sp}| \leq 1.$$

This completes the proof of the lemma. □

5.3 Second Proof of Lemma 5.2

A second proof could be based upon the action of ψ upon the partial function decomposition of ξ. We do not give all the details but we do consider the action of ψ upon positive powers of $1/(1 - x)$. We define rational numbers $A_{s,j}$, $s, j \in \mathbb{N}$ by setting

$$\psi \frac{1}{(x-1)^s} = \sum A_{s,j} \frac{1}{(x-1)^j}. \tag{5.3.1}$$

Lemma.

$$A_{s,j} = \frac{s}{pj} \cdot \textit{coefficient of } z^s \textit{ in } z^j \left(\sum_{i=1}^{p} \binom{p}{i} z^{i-1} \right)^j.$$

Proof. Put $x = 1 + y$, let $t_1, \ldots, t_p$ be the p roots of

$$t_i^p = x. \tag{5.3.2}$$

Let us write

$$t_i = 1 + w_i.$$

Then

$$\psi \frac{1}{(x-1)^s} = \frac{1}{p} \sum_{i=1}^{p} \frac{1}{(t_i - 1)^s} = \frac{1}{p} \sum_{i=1}^{p} \frac{1}{w_i^s}.$$

Thus

$$\frac{1}{p}\sum_{i=1}^{p}\frac{1}{w_i^s} = \sum_j A_{s,j}\frac{1}{y^j}. \tag{5.3.3}$$

Note that

$$(w+1)^p = x = 1 + y,$$

so that

$$w^p + pw^{p-1} + \cdots + pw = y$$

is the equation satisfied by $w_1, \ldots, w_p$. Thus

$$(z - w_1)\cdots(z - w_p) = z^p + pz^{p-1} + \cdots + pz - y$$

and so dividing by $(-1)^p w_1 \cdots w_p = -y$, we obtain

$$\prod_{i=1}^{p}\left(1 - \frac{z}{w_i}\right) = 1 - \frac{1}{y}\left(pz + \binom{p}{2}z + \cdots + \binom{p}{p-1}z^{p-1} + z^p\right). \tag{5.3.4}$$

Taking logarithms and comparing the coefficient of z^s on the two sides

$$\frac{1}{s}\sum_{i=1}^{p}\frac{1}{w_i^s} = \text{coefficient of } z^s \text{ in} \sum_{l=1}^{\infty}\frac{1}{ly^l}[pz + \cdots + pz^{p-1} + z^p]^l.$$

Equation (5.3.3) identifies $(p/s)A_{s,j}$ with the coefficient of $1/y^j$ in this coefficient of z^s. The lemma follows directly. □

5.3.5 **Corollary.** $A_{s,j} = 0$ *unless* $s/p \le j \le s$.

This follows from the fact that in the statement of the lemma we have a polynomial in z of degree pj divisible by z^j.

$$A_{s,s} = p^{s-1},$$

$$A_{s,s/p} = 1 \quad \text{if } p \mid s. \tag{5.3.6}$$

We assert

$$A_{s,j} \equiv \text{coefficient of } z^s \text{ in } \frac{s}{p}\frac{z^{pj}}{j} \mod s\mathbb{Z}_p. \tag{5.3.7}$$

This follows from the lemma since

$$\left(\sum_{i=1}^{p}\binom{p}{i}z^i\right)^j \equiv z^{pj} \mod pj\mathbb{Z}_p[z].$$

Equation (5.3.7) shows that $A_{s,j} \equiv 0 \mod s\mathbb{Z}_p$ unless $s = pj$ and that case is covered by (5.3.6). We conclude

(5.3.8) $$A_{s,j} \in \mathbb{Z}_p \cap \mathbb{Q}, \qquad \forall s, j.$$

This is a special case of Lemma 5.2 from which that lemma could be deduced. We note that Lemma 5.3 implies

(5.3.9) $$A_{s,j} \in p^{j-1}\mathbb{Z}_p \quad \text{if } s < p.$$

5.4 **Lemma.**

$$|A_{s,j}| \leq |\pi|^{pj-s} \cdot p.$$

Proof. By definition

$$\frac{1}{p}\sum_{y^p=x}\frac{1}{(y-1)^s} = \sum_j A_{s,j}\frac{1}{(x-1)^j}.$$

Let $r \in \mathbb{R}$, $r > |\pi|$. If $|x-1| = r^p$ then $|y-1| = r$ and so

$$|A_{s,j}|\frac{1}{r^{pj}} \leq \frac{p}{r^s}.$$

The lemma follows by taking limits as $r \to |\pi|$. □

5.5 **Lemma.**

$$\psi\frac{1}{(x\lambda-1)^s} = \sum_j A_{s,j}\frac{1}{(x\lambda^p-1)^j}.$$

Proof.

$$\psi\frac{1}{(x\lambda-1)^s} = \frac{1}{p}\sum_{z^p=x}\frac{1}{(z\lambda-1)^s}$$
$$= \frac{1}{p}\sum_{v^p=x\lambda^p}\frac{1}{(v-1)^s} = \sum\frac{A_{s,j}}{(x\lambda^p-1)^j}$$

as asserted. □

5.6

5.6.1 **Notation.** Let $\mathbf{Q}$ be the subring $\mathbb{Q}[1/\lambda, 1/(\lambda - 1), \lambda]$ of elements $\mathbb{Q}(\lambda)$ having singularities only at 0, 1, ∞. Let $\mathbf{Q}_0$ be the subring of those $\xi \in \mathbf{Q}$ such that $|\xi|_{\text{gauss}} \leq 1$. Let $\mathbf{Q}'$ (resp. $\mathbf{Q}_0'$) denote the completion of $\mathbf{Q}$ (resp. $\mathbf{Q}_0$) under the gauss norm.

5.6.2 **Lemma.** *Let v be an element of S. Let m_0, m_1, m_2 be arbitrary elements of $\mathbb{Z}$, then the monomial*

$$T_0^{m_0} T_1^{m_1} T_{1/\lambda}^{m_2} \in \mathbf{Q}_0((T_v)).$$

Proof. It is enough to check this for each monomial $T_{v'}^m$ where $v' \in S$, and hence it is enough to check this for $m = \pm 1$. For $v \neq \infty$ it is enough to check that for $v' \neq v, \infty$ we have

$$T_{v'} = A + BT_v,$$

with $A, A^{-1}, B \in \mathbf{Q}_0$. This we may read off from the table following (2.3.5.7). For $v = \infty$, it is enough to check that for $v' \neq \infty$ we have

$$T_{v'} = A \frac{1}{T_\infty} + B,$$

with $A, A^{-1}, B \in \mathbf{Q}_0$. This completes the proof of the lemma. $\square$

5.6.3 **Corollary.** *For each $v \in S$*

$$P_v(T_0^{m_0} T_1^{m_1} T_{1/\lambda}^{m_2}) \in \mathbf{Q}_0\left[\frac{1}{T_v}\right].$$

(P_v denotes principal part.)

5.6.4 *Remark.* We have used $\mathbf{Q}'$ to denote the completion of $\mathbf{Q}$ (cf. 5.6.1) under the gauss norm. The elements of $\mathbf{Q}'$ are thus analytic elements on the set

(5.6.4.1) $$\{\lambda \in \Omega \mid |\lambda| = 1 = |\lambda - 1|\} = \Omega_1.$$

5.6.5 **Lemma.** *Let $M(x)$ be a function of x and λ analytic and bounded by* 1 *on the set*

$$\lambda \in \Omega_1 \quad (\textit{cf.}\ (5.6.4.1)),$$

$$|T_1| > |\pi|,$$

$$|T_{1/\lambda}| > |\pi|,$$

given explicitly by

$$M(x) = \sum_{j=1}^{\infty} \frac{M_{1,j}}{T_1^j} + \sum_{j=1}^{\infty} \frac{M_{1/\lambda,j}}{T_{1/\lambda}^j},$$

where each $M_{1,j}$ (resp. $M_{1/\lambda,j}$) $\in \mathbf{Q}'$. *Let* k_0, k_1, k_2 *be fixed elements of* $\mathbb{Z}$. *Then for each* $v \in S$,

$$P_v(T_0^{k_0} T_1^{k_1} T_{1/\lambda}^{k_2} M) \in \mathbf{Q}'\left[\left[\frac{1}{T_v}\right]\right]$$

and converges for $\lambda \in \Omega_1$, $|T_v| > |\pi|$ (resp. 0) *if* $v = 1, 1/\lambda$ (resp. 0, ∞).

Proof. By hypothesis for $v' = 1, 1/\lambda$,

$$|M_{v',j}|_{\text{gauss}} \leq |\pi|^j,$$

while by 5.6.3 for $v \in S$,

$$P_v(T_0^{k_0} T_1^{k_1} T_{1/\lambda}^{k_2} T_{v'}^{-j}) \in \mathbf{Q}_0\left[\frac{1}{T_v}\right].$$

It is clear that the degree of this polynomial is bounded independently of j if $v \neq v'$ and is bounded by $j + 0(1)$ if $v = v'$. The lemma follows immediately. □

Remark. The radius $|\pi|$ appearing in the above lemma may be replaced by any quantity strictly less than unity.

5.7

We now wish to consider ψ as operator on functions analytic on an "open" annulus of outer radius unity. For this purpose we introduce the following notation. For $z \in \Omega$, $|z| \leq 1$, let W'_z be the space of germs of functions analytic at the boundary of $D(z, 1^-)$, i.e., W'_z is the space of Laurent series in $x - z$ defined over Ω which converge in and are bounded on

$$\Delta_z(r, 1) = \{r < |x - z| < 1\}$$

for some unspecified $r \in (0, 1)$. Clearly W'_z depends only upon the residue class of z and so with no loss in generality we may take $z = 0$ if $|z| < 1$. Likewise we may define W'_∞ as the space of Laurent series in x which converge on an annulus $\Delta_0(1, r)$ for some unspecified $r > 1$. We shall use the norm for $\xi \in W'_z$,

$$\|\xi\|_z = \lim_{r \to 1} |\xi|_z(r).$$

By the methods of Chapter 3, the mapping ψ as given by (3.5.6) defines a mapping of W'_z into W'_{z^p}.

Lemma. *For* $\xi \in W'_z$,

$$\|\psi\xi\|_{z^p} \leq \|\xi\|_z.$$

Proof. The assertion is trivial if $z = 0$ and hence we may suppose $|z| = 1$. By Lemma 5.5 we may with no loss in generality let $z = 1$. Thus

$$\xi = \sum_{s=-\infty}^{+\infty} B_s(1 - x)^s, \tag{5.7.1}$$

where

$|B_s|$ is bounded as $s \to +\infty$ and

$|B_s|r^s$ is bounded as $s \to -\infty$ for some $r \in (0, 1)$.

Here

$$\|\xi\|_1 = \sup_s |B_s|. \tag{5.7.2}$$

Put

$$\begin{aligned} \xi_- &= \sum_{s=1}^{\infty} B_{-s}(1 - x)^{-s}, \\ \xi_+ &= \sum_{s=0}^{\infty} B_s(1 - x)^s. \end{aligned} \tag{5.7.3}$$

It follows from (5.3.5) that

$$\psi\xi_- = \sum_{j=1}^{\infty} (1 - x)^{-j} \sum_{s=j}^{pj} A_{s,j} B_{-s} \tag{5.7.4}$$

and hence by (5.3.8)

$$\|\psi\xi_-\|_1 \leq \|\xi_-\|_1. \tag{5.7.5}$$

For each $r \in (0, 1)$, ξ_+ is approximated on $D(1, r)$ by a partial sum of the form

$$\sum_{s=0}^{N} B_s(1 - x)^s,$$

a polynomial bounded on $D(1, 1^-)$ by $\|\xi\|_1$ and hence $\psi\xi_+$ is approximated on $D(1, r^p)$ by a polynomial bounded on $D(1, 1^-)$ by $\|\xi\|_1$. Letting $r \to 1$ we find $\psi\xi_+$ is bounded on $D(1, 1^-)$ by $\|\xi\|_1$. This completes the proof of the lemma. □

5.8

We define

$$\delta = x\frac{d}{dx} = xD,$$

$$W_z'^{(0)} = \{\xi \in W_z' \mid \|\xi\|_z \le 1\},$$

$$\phi: \xi(x) \to \xi(x^p),$$

so that ϕ is a mapping of W_{z^p}' into W_z' and

$$\psi\phi = I.$$

Proposition.

(5.8.1) $$W_z'^{(0)} = (\operatorname{Ker}\psi \cap W_z'^{(0)}) \oplus \phi W_{z^p}'^{(0)}.$$

(5.8.2) *If* $u \in W_z'^{(0)}$ *and*

$$\psi u \equiv 0 \mod pW_{z^p}'^{(0)},$$

then

$$u \in (\operatorname{Ker}\psi \cap W_z'^{(0)}) + p\phi W_{z^p}'^{(0)}.$$

(5.8.3) *If* $u \in W_z'^{(0)}$ *and*

$$\delta u \equiv 0 \mod pW_z'^{(0)},$$

then

$$u \in (\operatorname{Ker}\psi \cap pW_z'^{(0)}) + \phi W_{z^p}'^{(0)}.$$

Proof. With no loss in generality we may restrict z to be either zero or unity. The sum in (5.8.1) is clearly direct as $\psi(\phi w) = w$. On the other hand for $u \in W_z'^{(0)}$ we have

$$u = (1 - \phi\psi)u + \phi\psi u, \tag{5.8.4}$$

which shows that u has the asserted decomposition.

Assertion (5.8.2) is an immediate consequence of (5.8.4).

It is clear that δ maps $\phi W_{zp}'^{(0)}$ into $pW_z'^{(0)}$. Hence to verify (5.8.3) it is enough to show that

$$\left.\begin{array}{l} u \in W_z'^{(0)} \\ \delta u \equiv 0 \mod pW_z'^{(0)} \\ \psi u \equiv 0 \mod pW_{zp}'^{(0)} \end{array}\right\} \Rightarrow u \in pW_z'^{(0)}. \tag{5.8.5}$$

For $z = 0$, we may write

$$u = \sum_{s=-\infty}^{+\infty} A_s x^s,$$

with

$$|A_s| \le 1,$$
$$|sA_s| \le |p|, \qquad \forall s \in \mathbb{Z}.$$
$$|A_{sp}| \le |p|,$$

Hence $|A_s| \le |p|$ for all s which demonstrates (5.8.5) in that case. If $z = 1$ then

$$u = \sum_{s=-\infty}^{+\infty} A_s(1 - x)^s,$$

with

$$|sA_s| \le |p|, \qquad \forall s \in \mathbb{Z}.$$
$$|A_s| \le 1.$$

Hence

$$u \equiv \sum_{s=-\infty}^{+\infty} A_{sp}(1 - x)^{ps} \mod pW_1'^{(0)}$$
$$\equiv \sum_{s=-\infty}^{+\infty} A_{sp}(1 - x^p)^s \mod pW_1'^{(0)}.$$

Thus

$$\psi u \equiv \sum_{s=-\infty}^{+\infty} A_{sp}(1-x)^s \mod pW_1'^{(0)}.$$

But then by the hypothesis on ψu, we have

$$|A_{sp}| \le |p|, \qquad \forall s \in \mathbb{Z}.$$

This completes the proof of (5.8.5) and hence of the proposition. □

5.9 **Lemma.** *Let* $s \in \mathbb{N}$, $\xi \in W_z'^{(0)}$. *Then*

(5.9.1)
$$\delta\xi \equiv 0 \mod p^s W_z'^{(0)}$$

if and only if

(5.9.2)
$$\xi = \xi_0^{\phi^s} + p\xi_1^{\phi^{s-1}} + \cdots + p^s\xi_s,$$

where for $0 \le j \le s$

$$\xi_j \in W'^{(0)}_{zp^{s-j}},$$

and for $1 \le j \le s$

$$\psi\xi_j = 0.$$

Proof. It is obvious that (5.9.2) implies (5.9.1). For the converse, the case $s = 1$ follows from (5.8.3). We may suppose $s > 1$ and use induction on s. By (5.8.3) condition (5.9.1) implies that

(5.9.3)
$$\xi = u^\phi + pw,$$

with $u \in W_{zp}'^{(0)}$, $w \in \operatorname{Ker} \psi \cap W_z'^{(0)}$. Hence by (5.9.1)

(5.9.4)
$$p(\delta u)^\phi + p\delta w \equiv 0 \mod p^s W_z'^{(0)}.$$

Applying ψ we obtain

(5.9.5)
$$\delta u \equiv 0 \mod p^{s-1} W_{zp}'^{(0)},$$

since

$$\psi\delta w = p\delta\psi w = 0.$$

We now apply the induction hypothesis and obtain

(5.9.6) $$u = \xi_0^{\phi^{s-1}} + p\xi_1^{\phi^{s-2}} + \cdots + p^{s-1}\xi_{s-1},$$

where for $0 \leq j \leq s-1$,

$$\xi_j \in W'^{(0)}_{zp^{s-j}},$$

and for $1 \leq j \leq s-1$

$$\Psi\xi_j = 0$$

Substituting in (5.9.3) we obtain

(5.9.7) $$\xi - (\xi_0^{\phi^s} + p\xi_1^{\phi^{s-1}} + \cdots + p^{s-1}\xi_{s-1}^{\phi}) = pw$$

and hence by (5.9.1)

$$\delta w \equiv 0 \mod p^{s-1}.$$

Hence again by induction,

(5.9.8) $$w = w_0^{\phi^{s-1}} + pw_1^{\phi^{s-2}} + \cdots + p^{s-1}w_{s-1},$$

where for $0 \leq j \leq s-1$

$$w_j \in W'^{(0)}_{zp^{s-1-j}}.$$

Among all such representations of w, let us choose one such that

(5.9.9) $$w = p^a w_a^{\phi^{s-1-a}} + p^{a+1}w_{a+1}^{\phi^{s-a-2}} + \cdots + p^{s-1}w_{s-1},$$

with a maximal, $a \leq s-1$. Applying ψ to this relation and recalling that $w \in \operatorname{Ker}\psi$, we obtain if $a < s-1$

(5.9.10) $$0 \equiv p^a w_a^{\phi^{s-a-1}} \mod p^{a+1}W'^{(0)}_z$$

and so applying ψ^{s-a-1},

(5.9.11) $$w_a \equiv 0 \mod pW'^{(0)}_{zp^{s-1-a}}.$$

Thus $w_a = pv$ with $v \in W'^{(0)}_{zp^{s-1-a}}$ and

(5.9.12) $$w = p^{a+1}(v^{\phi} + w_{a+1})^{\phi^{s-a-2}} + p^{a+2}w_{a+2}^{\phi^{s-a-3}} + \cdots + p^{s-1}w_{s-1}.$$

This contradicts the hypothesis of maximality of a in the representation (5.9.9) and so we conclude that $a = s - 1$. This means that equation (5.9.9) has the form

$$w = p^{s-1}\xi_s, \tag{5.9.13}$$

where $\xi_s \in W_z'^{(0)}$ and by (5.9.3)

$$\psi\xi_s = 0.$$

The verification of (5.9.2) is now completed by substituting (5.9.13) in (5.9.7). This completes the proof of the lemma. □

5.10 **Lemma.** *Let* $\xi \in W_z'^{(0)}$, $s \in \mathbb{N}$, *then*

$$\psi^i\xi \equiv 0 \mod p^i W_{zp^i}'^{(0)}, \qquad i = 0, 1, \ldots, s \tag{5.10.1}$$

if and only if there exists $\eta \in W_z'^{(0)}$ *such that*

$$\delta\eta \equiv \xi \mod p^s W_z'^{(0)}. \tag{5.10.2}$$

Proof. It is enough to show that (5.10.1) implies (5.10.2).

With no loss in generality we may restrict z to be 0, 1. For $z = 0$, let us write

$$\xi = \sum_{m=-\infty}^{+\infty} A_m x^m \tag{5.10.3}$$

and conclude that for $m \in \mathbb{Z}$,

$$\operatorname{ord} A_m \geq \operatorname{Min}(s, \operatorname{ord} m), \tag{5.10.4}$$

so in particular $\operatorname{ord} A_0 \geq s$.

Thus modulo p^s,

$$\xi \equiv \sum_{p^s \nmid m} A_m x^m = \delta\eta, \tag{5.10.5}$$

where

$$\eta = \sum_{p^s \nmid m} A_m \frac{x^m}{m}.$$

It is clear that if ξ converges in $\Delta_0(r, 1)$, then the same holds for η which by (5.10.4) lies in $W_z'^{(0)}$. This completes the proof. □

We now consider the case $z = 1$. The assertion is proven by induction on s. It is trivial for $s = 0$. We may assume $s > 0$ and the lemma valid for $s - 1$. Thus certainly there exists η, Λ in $W_1'^{(0)}$ such that

$$\xi = \delta\eta + p^{s-1}\Lambda. \tag{5.10.6}$$

We know that

$$\psi^s\Lambda \equiv 0 \mod pW_1'^{(0)}. \tag{5.10.7}$$

Let us write

$$\Lambda = \Lambda_+ + \Lambda_-, \tag{5.10.8}$$

where

$$\Lambda_+ \in \Omega[[x - 1]],$$

$$\Lambda_- \in \frac{1}{x-1}\Omega\left[\left[\frac{1}{x-1}\right]\right].$$

Clearly (cf. definition 9.5.2.1 below)

$$\psi^s\Lambda_+ \equiv 0 \mod pW_1^{(0)}. \tag{5.10.9}$$

Let us write

$$\Lambda_+ = \sum_{m=0}^{\infty} A_m(x-1)^m \tag{5.10.10}$$

all A_m bounded by unity. We rewrite this in the form

$$\begin{aligned}\Lambda_+ &= \sum_{m=0}^{\infty}(x-1)^{mp^s}\sum_{j=0} A_{mp^s+j}(x-1)^j \\ &\equiv \sum_{m=0}^{\infty}(x^{p^s}-1)^m\sum_{j=0}^{p^s-1} A_{mp^s+j}(x-1)^j \mod p.\end{aligned} \tag{5.10.11}$$

Thus

$$\begin{aligned}\psi^s\Lambda_+ &\equiv \sum_{m=0}^{\infty}(x-1)^m\sum_{j=0}^{p^s-1} A_{mp^s+j}\psi^s(x-1)^j \mod p \\ &= \sum_{m=0}^{\infty}(x-1)^m\sum_{j=0}^{p^s-1} A_{mp^s+j}(-1)^j\end{aligned}$$

and so by (5.10.9), for all $m \in \mathbb{N}$,

$$\sum_{j=0}^{p^s-1} A_{mp^s+j}(-1)^j = pB_m, \tag{5.10.12}$$

with B_m bounded by 1.

Thus

$$\begin{aligned}
\Lambda_+ - \sum_{m=0}^{\infty} pB_m(x-1)^{p^s m} &= \sum_{m=0}^{\infty} (x-1)^{p^s m} \sum_{j=0}^{p^s-1} A_{mp^s+j}[(x-1)^j - (-1)^j] \\
&= x \sum_{m=0}^{\infty} (x-1)^{p^s m} \sum_{j=1}^{p^s-1} A_{mp^s+j} \\
&\quad \times [(x-1)^{j-1} + (x-1)^{j-2}(-1) + \cdots + (-1)^{j-1}] \\
&\equiv \delta \sum_{m=0}^{\infty} (x-1)^{p^s m} \sum_{j=1}^{p^s-1} A_{mp^s+j} \sum_{i+k=j-1} \frac{(x-1)^{i+1}}{i+1}(-1)^k \mod pW_1^{(0)}.
\end{aligned}$$

This shows that

$$\Lambda_+ \in pW_1^{(0)} + \delta W_1^{(0)} \cdot p^{1-s} \tag{5.10.13}$$

and so $p^{s-1}\Lambda_+$ is of the desired form.

5.10.14 *Remark.* If Λ_+ is a polynomial then $\Lambda_+ = pu + \delta v p^{1-s}$ with u, v polynomials in $W_1^{(0)}$.

We now consider Λ_-. We first recall that Λ_- is analytic on the complement of $D(1, r^+)$ for some $r < 1$ and hence writing

$$\Lambda_- = \sum_{m=1}^{\infty} B_m \frac{1}{(x-1)^m}$$

there exists m_0 such that

$$|B_m| < |p|, \qquad \forall m > m_0.$$

Thus putting

$$\Lambda_2 = \sum_{m=m_0+1}^{\infty} B_m \frac{1}{(x-1)^m},$$

$$\Lambda_1 = \sum_{m=1}^{m_0} B_m \frac{1}{(x-1)^m},$$

we have

$$\Lambda_2 \in pW_1'^{(0)},$$

$$\Lambda_- = \Lambda_1 + \Lambda_2,$$

and we may conclude from Lemma 5.7 that

$$\psi^s \Lambda_1 \equiv 0 \mod pW_1'^{(0)}.$$

We now choose $N > s$ such that $p^N > m_0$. Of course

$$(x - 1)^{p^N} \equiv H(x^{p^s}) \mod pW_1^{(0)},$$

with $H = (x^{p^{N-s}} - 1)$. Clearly

$$\psi^s(x - 1)^{p^N}\Lambda_1 \equiv H(x)\psi^s\Lambda_1 \equiv 0 \mod p.$$

But $(x - 1)^{p^n}\Lambda_1$ is a polynomial in $x - 1$ and hence by 5.10.14,

$$(x - 1)^{p^N}\Lambda_1 = \delta\eta_1 p^{1-s} + p\xi_1,$$

where η_1 and ξ_1 are polynomials in $x - 1$ (with coefficients bounded by 1) and so

$$\Lambda_1 \in \delta\left(\frac{\eta_1 p^{1-s}}{(x - 1)^{p^N}}\right) + pW_1'^{(0)}.$$

This completes the proof of the lemma since $p^{s-1}\Lambda_-$ has the desired form.□

5.11 Congruences for Binomial Coefficients

Theorem (F. Baldassarri). *Let* $a, b \in \mathbb{N}, i = -1, 0, 1, \ldots, p - 2$ *then* mod $pb\mathbb{Z}_p$

(5.11.1)
$$\binom{ap + i}{bp - 1} \equiv \begin{cases} \binom{a - 1}{b - 1} & \textit{if } i = -1, \\ 0 & \textit{if } i = 0, 1, \ldots, p - 2. \end{cases}$$

Proof.

$$(1 + x)^p \equiv 1 + x^p \mod p\mathbb{Z}_p[x]$$

and hence for $\gamma \in \mathbb{N}$

(5.11.2)
$$(1 + x)^{p\gamma} \equiv (1 + x^p)^\gamma \mod p\gamma\mathbb{Z}_p[x],$$

which shows that for $\beta \in \mathbb{N}$,

$$\binom{p\gamma}{p\beta} \equiv \binom{\gamma}{\beta} \mod p\gamma\mathbb{Z}_p. \tag{5.11.3}$$

We now consider

$$\binom{p\gamma + i}{p\beta - 1} \qquad \text{for } i = -1, 0, 1, 2, \ldots, p - 2.$$

$i = -1$ Using (5.11.3)

$$\binom{p\gamma - 1}{p\beta - 1} = \binom{p\gamma}{p\beta}\frac{\beta}{\gamma} \equiv \binom{\gamma}{\beta}\frac{\beta}{\gamma} \mod p\beta\mathbb{Z}_p \tag{5.11.4}$$
$$= \binom{\gamma - 1}{\beta - 1}.$$

$i = 0$

$$\binom{p\gamma}{p\beta - 1} = \binom{p\gamma}{p\beta}\frac{p\beta}{p\gamma - p\beta + 1} \equiv 0 \mod p\beta\mathbb{Z}_p.$$

$i = 1, 2, \ldots, p - 2$

$$\binom{p\gamma + i}{p\beta - 1} = \binom{p\gamma}{p\beta - 1}\frac{(p\gamma + i)(p\gamma + i - 1)\cdots(p\gamma + 1)}{(p\gamma - p\beta + i + 1)(p\gamma - p\beta + i)\cdots(p\gamma - p\beta + 2)}$$
$$\equiv 0 \mod p\beta\mathbb{Z}_p.$$

This completes the proof of the theorem. □

5.12 **Corollary.** *If $a, b, t \in \mathbb{N}$, $-1 \le i \le p^t - 2$, then*

$$\binom{ap^t + i}{bp^t - 1} \equiv \begin{cases} \binom{a - 1}{b - 1} & \mod pb\mathbb{Z}_p, \quad i = -1, \\ 0 & \quad 0 \le i \le p^t - 2. \end{cases}$$

Proof. The corollary has been verified for $t = 1$. We use induction on t. Let $t \ge 2$ and suppose the assertion is valid for $t - 1$. Thus let us write

$$i = p^{t-1}i_0 + j, \qquad -1 \le j \le p^{t-1} - 2, \quad 0 \le i_0 \le p - 1$$

and so by induction,

$$\binom{ap^t + i}{bp^t - 1} = \binom{(ap + i_0)p^{t-1} + j}{(bp)p^{t-1} - 1} \equiv 0 \mod pb\mathbb{Z}_p, \tag{5.12.1}$$

unless $j = -1$, in which case we obtain

$$\binom{ap + i_0 - 1}{bp - 1} \mod pb\mathbb{Z}_p. \tag{5.12.2}$$

But by the theorem this is congruent to zero mod $pb\mathbb{Z}_p$ unless $i_0 = 0$, (i.e., unless $i = -1$) and in that case we obtain $\binom{a-1}{b-1}$ mod $pb\mathbb{Z}_p$. This completes the proof of the lemma.

5.13 **Corollary.** *For $h \geq \mu \geq 0$, $h, \mu, i, s, a \in \mathbb{N}$*

$$\binom{s}{ip^h - 1} \equiv 0 \mod p^{\mu+1} \quad \text{if } s \not\equiv -1 \mod p^{h-\mu}\mathbb{Z}_p,$$

$$\binom{ap^{h-\mu} - 1}{ip^h - 1} \equiv \binom{a - 1}{ip^\mu - 1} \mod p^{\mu+1}\mathbb{Z}_p.$$

Proof. Put $b = ip^\mu$, $t = h - \mu$ and apply Corollary 5.12. □

5.14 **Lemma.** *Let i, h, μ be as in Corollary 5.13. As operators on $W_z'^{(0)}$*

$$\frac{D^{ip^h - 1}}{(ip^h - 1)!} \equiv \phi^{h-\mu} \circ \frac{D^{ip^\mu - 1}}{(ip^\mu - 1)!} \frac{1}{x} \circ \psi^{h-\mu} \circ x \mod p^{\mu+1}. \tag{5.14.1}$$

Proof. Each element of $W_z'^{(0)}$ reduces modulo $p^{\mu+1}W_z'^{(0)}$ to an element of $\Omega((x - z))$, thus it is enough to verify the assertion for elements of $\Omega((x - z) \cap W_z'^{(0)}$. We show that there exists N such that as operators on $W_z'^{(0)}$,

$$(x - z)^{p^N} \circ \frac{D^{ip^h - 1}}{(ip^h - 1)!} \equiv \frac{D^{ip^h - 1}}{(ip^h - 1)!} \circ (x - z)^{p^N} \mod p^{1+\mu}, \tag{5.14.2}$$

$$(x - z)^{p^N} \circ \phi^{h-\mu} \circ \frac{D^{ip^\mu - 1}}{(ip^\mu - 1)!} \circ \frac{1}{x} \psi^{h-\mu} \circ x \tag{5.14.3}$$

$$\equiv \phi^{h-\mu} \circ \frac{D^{ip^\mu - 1}}{(ip^\mu - 1)!} \circ \frac{1}{x} \psi^{h-\mu} \circ x \circ (x - z)^{p^N} \mod p^{1+\mu}.$$

Indeed

$$\frac{D^s}{s!} \circ u = \sum_{k+j=s} \frac{D^k u}{k!} \frac{D^j}{j!} \equiv u \circ \frac{D^s}{s!} \mod p^{1+\mu} \tag{5.14.4}$$

provided for $k \in [1, s]$ we have

$$\frac{D^k u}{k!} \equiv 0 \mod p^{1+\mu}.$$

Since

$$\frac{D^k u}{k!} = \frac{D^{k-1}}{(k-1)!} \frac{u'}{k},$$

it is enough if for $1 \leq k \leq s$ we have

$$\frac{u'}{k} \equiv 0 \mod p^{1+\mu}. \tag{5.14.5}$$

To apply this to (5.14.2), we take $s = ip^h - 1$, $u = (x - z)^{p^N}$ and so it is enough if $p^N/k \in p^{1+\mu}\mathbb{Z}_p$ for all $k \leq ip^h$ which is clearly true if $N > c$ where

$$c = 1 + \mu + h + \left[\frac{\log i}{\log p}\right]. \tag{5.14.6}$$

This completes the proof of (5.14.2). For (5.14.3), we choose N so large that

$$(x - z)^{p^N} \equiv H(x^{p^c}) \mod p^{1+\mu}, \tag{5.14.7}$$

where $H \in \mathbb{Z}[z, x]$. Then

$$\frac{1}{x} \psi^{h-\mu} \circ x \circ (x - z)^{p^N} \equiv H(x^{p^{c-h+\mu}}) \circ \frac{1}{x} \circ \psi^{h-\mu} \circ x \mod p^{1+\mu}, \tag{5.14.8}$$

while

$$\frac{D^{ip^\mu - 1}}{(ip^\mu - 1)!} \circ H(x^{p^{c-h+\mu}}) \equiv H(x^{p^{c-h+\mu}}) \circ \frac{D^{ip^\mu - 1}}{(ip^\mu - 1)!} \mod p^{1+\mu} \tag{5.14.9}$$

by (5.14.5) (with $s = ip^\mu - 1$, $u = H^{\phi^{c-h+\mu}}$).

Finally

$$\phi^{h-\mu} \circ H^{\phi^{c-h+\mu}} = H^{\phi^c} \circ \phi^{h-\mu} \equiv (x - z)^{p^N} \circ \phi^{h-\mu} \mod p^{1+\mu}. \tag{5.14.10}$$

This completes the verification of (5.14.2) and (5.14.3).

With N so chosen we may write each $\xi \in W_z'^{(0)} \cap \Omega((x - z))$ in the form

$$\xi = \sum_{m=-r}^{\infty} (x - z)^{mp^N} h_m(x), \tag{5.14.11}$$

where for each m, $h_m \in W_z'^{(0)} \cap \Omega[x]$. By means of (5.14.2) and (5.14.3), we reduce the question to the validity of (5.14.1) as operator on the h_m, i.e., as operators on polynomials. For polynomials it is enough to check the validity on monomials and this is assured by Corollary 5.13. This completes the proof of the lemma. □

Chapter 6

Calculation Modulo p of the Matrix of $\alpha_{f,h}$

The object of this section is to determine conditions under which we may replace $G^{(1)}_{f,h}$, $G^{(2)}_{f,h}$ by unity in the calculation of the coefficients of the matrix $\alpha_{f,h}$ (relative to the basis $1, 1/(1-x)$) modulo p. Since these coefficients vary with λ, and indeed may become large as $\lambda \to 0, 1, \infty$, we specify here that $\lambda \in \Omega_1$ (cf. (5.6.4.1)).

6.1 **Lemma.** *Let M be a function analytic and bounded by* 1 *on the set*

(6.1.1)
$$|T_1| > |\pi|,$$
$$|T_{1/\lambda}| > |\pi|.$$

Assume further that

(6.1.2)
$$|M(x)| \leq |p|$$

on the set

(6.1.3)
$$|T_0| = |T_1| = |T_{1/\lambda}| = 1.$$

Let $k_0, k_1, k_2 \in \mathbb{Z}$. We write

(6.1.4)
$$\psi M T_0^{k_0} T_1^{k_1} T_{1/\lambda}^{k_2} \equiv A_k + B_k \frac{1}{1-x} \quad \bmod D_{h,\lambda^p} L_{\lambda^p}.$$

Let

(6.1.5)
$$\begin{cases} -l_0 = \mathrm{Min}(0, k_0), \\ -l_1 = \mathrm{Min}(0, k_1, k_2), \\ l_\infty = \mathrm{Sup}(0, k_0 + k_1 + k_2), \\ \gamma = \mathrm{Sup}(|b' - c'|_\infty, |a'|_\infty), \\ d = \textit{least common denominator of } a, b, c. \end{cases}$$

We define $j_0, j_1, j_\infty \in \mathbb{N}$ *by*

$$pj_0 > l_0, \quad j_0 \textit{ minimal,}$$

(6.1.6)
$$pj_\infty > l_\infty, \quad j_\infty \textit{ minimal,}$$

$$\frac{pj - l_1}{p - 1} \geq 2 + \frac{\log d(j - 1 + \gamma)}{\log p} \quad \textit{for all } j \geq j_1.$$

We assert that

(6.1.7)
$$|p| \geq \mathrm{Sup}(|A_k|, |B_k|)$$

subject to (1.1), (3.7.0) *and the further conditions*

$(6.1.8)_0$ $\quad m - b'$ *is unit in* $\mathbb{Z}_p$ *for* $m = 1, 2, \ldots, j_0 - 1$

vacuous if $j_0 < 2$

$(6.1.8)_1$ $\quad m + b' - c'$ *is unit in* $\mathbb{Z}_p$ $\quad m = 0, 1, \ldots, j_1 - 2$

$(6.1.8)_{1/\lambda}$ $\quad m + a'$ *is unit in* $\mathbb{Z}_p$ $\quad m = 0, 1, \ldots, j_1 - 2$

$(6.1.8)_\infty$ $\quad m + c' - a'$ *is unit in* $\mathbb{Z}_p$ $\quad m = 0, 1, \ldots, j_\infty - 1.$

Proof. Let $\varepsilon \in \mathbb{R}_+^4$, $\varepsilon = (\varepsilon_0, \varepsilon_1, \varepsilon_{1/\lambda}, \varepsilon_\infty)$ and

(6.1.9)
$$\varepsilon_\nu \leq \Gamma_\nu, \qquad \forall \nu \in S.$$

(In the present context $\Gamma_\nu = 1$.)

We define Q_ε ($= Q_{\varepsilon,\lambda}$) as in (3.0.3) and define $\hat{H}(Q_\varepsilon)$ to be the space of functions analytic and bounded on Q_ε and defined over Ω. Precisely as in §3.5, under (3.0.2)′ and (3.5.4), formula (3.5.6) defines a mapping of $\hat{H}(Q_{\varepsilon,\lambda})$ into $\hat{H}(Q_{\varepsilon',\lambda^p})$ with operator norm bounded by p, the operator norm being in terms of the sup norms on $Q_{\varepsilon,\lambda}$ (resp. $Q_{\varepsilon',\lambda^p}$). We put

(6.1.10)
$$\varepsilon_1 = |\pi| = |\varepsilon_{1/\lambda}|,$$

so M is bounded by 1 on $Q_{\varepsilon,\lambda}$. On the same set

(6.1.11)
$$|T_0^{k_0} T_1^{k_1} T_{1/\lambda}^{k_2}| \leq \frac{1}{|\pi|^{l_1} \varepsilon_0^{l_0} \varepsilon_\infty^{l_\infty}}$$

since the four disks

$$|T_\nu| < 1$$

are mutually disjoint.

We write as element of $\hat{H}(Q_{\varepsilon', \lambda^p})$

$$\psi(MT_0^{k_0}T_1^{k_1}T_{1/\lambda}^{k_2}) = K_\infty + K_1 + K_{1/\lambda^p} + K_0, \tag{6.1.12}$$

with

$$\begin{cases} K_0 = \sum_{s=1}^{\infty} \mu_{0,s} \dfrac{1}{T_0^s}, \\ K_1 = \sum_{s=1}^{\infty} \mu_{1,s} \dfrac{1}{T_1^s}, \\ K_{1/\lambda^p} = \sum_{s=1}^{\infty} \mu_{1/\lambda^p, s} \dfrac{1}{T_{1/\lambda^p}^s}, \\ K_\infty = \sum_{s=1}^{\infty} \mu_{\infty,s} x^s. \end{cases} \tag{6.1.13}$$

By our estimates on $Q_{\varepsilon', \lambda^p}$, we have

$$\left\{ \begin{array}{c} |\mu_{0,m}|/\varepsilon_0^{pm} \\ |\mu_{\infty,m}|/\varepsilon_\infty^{pm} \\ |\mu_{1,m}|/|\pi|^{pm} \\ |\mu_{1/\lambda^p,m}|/|\pi|^{pm} \end{array} \right\} \leq \frac{p}{|\pi|^{l_1}\varepsilon_0^{l_0}\varepsilon_\infty^{l_\infty}}, \qquad \forall m \in \mathbb{N}. \tag{6.1.14}$$

Letting $\varepsilon_0 \to 0$ and $\varepsilon_\infty \to 0$ we obtain

$$\begin{cases} \mu_{0,s} = 0 \quad \text{for } ps > l_0, \\ \mu_{\infty,s} = 0 \quad \text{for } ps > l_\infty. \end{cases} \tag{6.1.15}$$

We now let $\varepsilon_0 \to 1$, $\varepsilon_\infty \to 1$ and deduce

$$|\mu_{\nu^p,m}| \leq p|\pi|^{pm-l_1}, \qquad \nu = 1, 1/\lambda. \tag{6.1.16}$$

We now restrict our attention to the set (6.1.3). Here $M(x)T_0^{k_0}T_1^{k_1}T_{1/\lambda}^{k_2}$ is a uniform limit of rational functions bounded by $|p|$. Thus for λ fixed in Ω_1 and using the gauss norm relative to x (cf. Lemma 5.2), we conclude that

$$|\mu_{\nu^p,m}| \leq 1/p, \qquad \forall \nu \in S, m \in \mathbb{N}. \tag{6.1.17}$$

We now follow the proof of Theorem 4.5 and compute

$$A_k = \sum_{v \in S} \mathrm{Res}_{v^p} K_{v^p} \cdot (1^*)_{v^p, \lambda^p}, \tag{6.1.18.1}$$

where $(1^*)_{v^p, \lambda^p}$ refers to the v^p component of the element $1^* \in \mathbf{K}_{h, \lambda^p}$. Likewise

$$B_k = \sum_{v \in S} \mathrm{Res}_{v^p} K_{v^p} \left(\frac{1}{1-x}\right)^*_{v^p, \lambda^p}. \tag{6.1.18.2}$$

Note that we do not use the Φ mapping in this treatment. We use Theorem 2.3.5 to write

$$(1^*)_{v^p, \lambda^p} = \sum_{m=0}^{\infty} T^m_{v^p} C_{v^p, m}, \qquad \forall v \in S, \tag{6.1.19}$$

but now by (5.6.4.1),

$$C_{v^p, m} \in U_{\theta_{v^p}, m} (\text{resp. } U_{\theta_\infty, m-1}) \quad \text{if } v \neq \infty. \tag{6.1.20}$$

v	0	1	$1/\lambda$	∞
θ_{v^p}	$1 - b'$	$b' - c'$	a'	$c' - a'$

A formula similar to (6.1.19) holds for $(1/(1-x))^*$. We compute the four residues separately.

Case 1. $v = 0$

$$\mathrm{Res}_0\, K_0 (1^*)_{0, \lambda^p} = \sum_{m=1} \mu_{0,m} C_{0,m-1}$$

and by (6.1.15) the sum is from $m = 1$ to $m = j_0 - 1$ and so by (6.1.17) the sum lies in $pU_{0, j_0 - 2}$. Thus the residue is zero if $j_0 < 2$ and if $j_0 \geq 2$ then the residue is trivial mod p if $1 - b' + s$ is a unit for $0 \leq s \leq j_0 - 2$. This is condition $(6.1.8)_0$.

Case 2. $v = \infty$

$$-\mathrm{Res}_\infty K_\infty (1^*)_{\infty, \lambda^p} = \sum_{m=0} \mu_{\infty, m} C_{\infty, m+1} \tag{6.1.21}$$

and by (6.1.15) the sum is over $0 \leq m \leq j_\infty - 1$ and so by (6.1.17) the sum lies in $pU_{\infty, j_\infty - 1}$, and hence is zero if $j_\infty \leq 0$. If $j_\infty \geq 1$ then the residue is trivial mod p if $c' - a' + m$ is a unit for $0 \le m \le j_\infty - 1$. This is condition $(6.1.8)_\infty$.

Case 3. $v = 1$

$$\text{Res}_1 K_1(1^*)_{1,\lambda^p} = \sum_{m=1} \mu_{1,m} C_{1,m-1}, \tag{6.1.22}$$

$$C_{1,m-1} \in \sum_{j=0}^{m-1} \mathbb{Z}_p \frac{1}{j + b' - c'}.$$

Hence

$$|C_{1,m-1}|_p \le ((m-1) + |b' - c'|_\infty)d. \tag{6.1.23}$$

We deduce from (6.1.16) that

$$|\mu_{1,m} C_{1,m-1}| \le |p| \tag{6.1.24}$$

if

$$p|\pi|^{pm - l_1} d((m-1) + |b' - c'|_\infty) \le 1/p.$$

By definition of j_1, this holds for all $m \ge j_1$. Hence mod p the residue at 1 is congruent to

$$\sum_{m=1}^{j_1 - 1} \mu_{1,m} C_{1,m-1}. \tag{6.1.25}$$

By (6.1.17) this lies in $p \sum_{j=0}^{j_1-2} \mathbb{Z}_p[1/(j + b' - c')]$. This is trivial mod p subject to condition $(6.1.8)_1$.

Case 4. $v = 1/\lambda$

This case proceeds precisely as Case 3 above. □

6.2

If we introduce a hypothesis involving nonramification then the integer j_1 (cf. (6.1.6)) may in the above lemma be replaced by the smaller integer j_2 defined by the condition that for all $j \ge j_2, j \in \mathbb{N}$

$$\left\{\frac{pj - l_1}{p - 1}\right\} > 1 + \frac{\log d(j - 1 + \gamma)}{\log p}, \tag{6.2.1}$$

the symbol $\{x\}$ denoting for $x \in \mathbb{R}_+$

$$\{x\} = \begin{cases} x & \text{if } x \in \mathbb{N}, \\ [x] + 1 & \text{if } x \notin \mathbb{N}. \end{cases}$$

We make this explicit in the following corollary.

Corollary. *Suppose that the function M in Lemma* 6.1 *satisfies the conditions of Lemma* 5.6.5. *Then in the statement of Lemma* 6.1 *we may replace j_1 by the integer j_2 given by equation* (6.2.1).

Proof. It follows from Lemmas 5.6.5 and 5.5, and Corollary 5.3.5 that the coefficients $\mu_{\nu p, m}$ in (6.1.13) lie in $\mathbf{Q}'$. Now the estimate (6.1.16) is valid for the norm of this coefficient in the sense of supremum over Ω_1. This norm is unramified, and hence we may replace (6.1.16) by

$$\frac{\log|\mu_{\nu p, m}|}{\log p} \leq 1 - \left\{\frac{pm - l_1}{p - 1}\right\}. \tag{6.2.2}$$

A similar reasoning holds for the estimate for $|\mu_{1,m} C_{1,m}|$ in equation (6.1.24) and hence we have

$$|\mu_{1,m} C_{1,m-1}| \leq |p|,$$

if

$$|\mu_{1,m} C_{1,m-1}| < 1,$$

i.e., if

$$\frac{\log(m - 1 + |b' - c'|_\infty)d}{\log p} + 1 - \left\{\frac{pm - l_1}{p - 1}\right\} < 0.$$

This gives condition (6.2.1). $\square$

6.3 **Theorem.** *In the notation of* (3.7.1) *let μ_a, μ_b, μ_c all lie in the interval* $[1, p - 1]$. *Let a', b' lie in the interval* $(0, 1)$ *and* $c' \in [0, 1]$. *Let*

$$\left.\begin{aligned} \psi T_0^{-\mu_b} T_1^{\mu_b - \mu_c} T_{1/\lambda}^{\mu_a} &\equiv \chi_1 + \chi_3 \frac{1}{1 - x} \\ \psi T_0^{-\mu_b} T_1^{\mu_b - \mu_c - 1} T_{1/\lambda}^{\mu_a} &\equiv \chi_2 + \chi_4 \frac{1}{1 - x} \end{aligned}\right\} \bmod D_{h,\lambda^p} \mathbf{L}_{\lambda^p}. \tag{6.3.1}$$

Let

$$\text{(6.3.2)}\quad \alpha_{f,h}\begin{pmatrix}1\\ 1/(1-x)\end{pmatrix} \equiv \begin{pmatrix}\mathbf{A}_1 & \mathbf{A}_3\\ \mathbf{A}_2 & \mathbf{A}_4\end{pmatrix}\begin{pmatrix}1\\ 1/(1-x)\end{pmatrix} \quad \operatorname{mod} D_{h,\lambda^p}L_{\lambda^p}.$$

Then on Ω_1

$$\text{(6.3.3)}\qquad \mathbf{A}_i \equiv \chi_i \mod p, \qquad i = 1, 2, 3, 4$$

provided

$$\text{(6.3.4)}\qquad p \geq 2d \quad (\text{resp. } p \geq \sqrt{3d}),$$

(6.3.5)

$a', b' - c', c' - a'$ *are units (in* $\mathbb{Z}_p$*)* (resp. $1 + b' - c', 1 + a'$ *are also units).*

Proof. In the present application of Lemma 6.1, we take

$$\text{(6.3.6)}\qquad M = G^{(1)}_{f,h}G^{(2)}_{f,h} - 1.$$

By Lemma 3.7, $G^{(1)}_{f,h}$ (resp. $G^{(2)}_{f,h}$) has a Mittag–Leffler representation of the form

$$G^{(1)}_{f,h} = \sum_{m\geq 0} \frac{\alpha_m}{T_1^m},$$

$$\left(\text{resp. } G^{(2)}_{f,h} = \sum_{m\geq 0} \frac{\beta_m}{T_{1/\lambda}^m}\right),$$

with α_m, β_m in $\mathbb{Q} \cap \mathbb{Z}_p$,

$$\alpha_0 = \beta_0 = 1,$$

$$\operatorname{Min}(\operatorname{ord} \alpha_m, \operatorname{ord} \beta_m) \geq \frac{m}{p-1}.$$

It follows that M satisfies the condition of Corollary 6.2 as well as those of Lemma 6.1. In the present application

$$\text{(6.3.7)}\qquad \begin{aligned} k_0 &= -\mu_b,\\ k_1 &= \mu_b - \mu_c - \begin{cases} 0,\\ 1,\end{cases}\\ k_2 &= \mu_a.\end{aligned}$$

Hence

$$l_0 \leq p - 1,$$
$$(6.3.8) \qquad l_1 \leq p - 1,$$
$$l_\infty \leq p - 1.$$

Thus $j_0, j_\infty = 1$ and so condition $(6.1.8)_0$ is vacuous while $(6.1.8)_\infty$ reduces to part of (6.3.5). We assert that (6.2.1) is satisfied for $j \geq 2$ (resp. $j \geq 3$) if $p \geq 2d$ (resp. $p > \sqrt{3d}$). We know that

$$\left\{\frac{pj - l_1}{p - 1}\right\} \geq \left\{\frac{pj - (p - 1)}{p - 1}\right\} \geq \left\{j - 1 + \frac{j}{p - 1}\right\} \geq j,$$

while since $\gamma \in (0, 1)$

$$1 + \frac{\log d(j - 1 + \gamma)}{\log p} < 1 + \frac{\log jd}{\log p}.$$

Thus it is enough to check that

$$(6.3.9) \qquad \frac{1}{j} p^{j-1} \geq d$$

for all $j \geq 2$ (resp. $j \geq 3$) if $p \geq 2d$ (resp. $p \geq \sqrt{3d}$). This is clear since for $p \geq 2$ the left side of (6.3.9) has positive derivative with respect to j if $j \geq 2$. □

6.4 Calculation of χ

Lemma. *Let μ_a, μ_b, μ_c all lie in the interval $[1, p - 1]$. We assume*

$$(6.4.1) \qquad \frac{p^m}{m + b' - c'} \equiv 0 \mod p \quad \textit{for } 1 \leq m \leq 1 + \mu_c - \mu_b.$$

$$(6.4.2) \qquad \mu_a \neq \mu_c \neq \mu_b.$$

(*Note.* Condition (6.4.1) is vacuous if, for example, $\mu_b > \mu_c$.)
We tabulate 4 cases

Case	Condition
1	$\text{Min}(\mu_a, \mu_b) > \mu_c$
2	$\mu_c > \text{Max}(\mu_a, \mu_b)$
3	$\mu_b < \mu_c < \mu_a$
4	$\mu_a < \mu_c < \mu_b$

We define the matrix

$$\chi = \begin{pmatrix} \chi_1 & \chi_2 \\ \chi_3 & \chi_4 \end{pmatrix}$$

as in (6.3.1). We assert that the matrix coefficients χ_i lie in $\mathbf{Q}_0$, and

(6.4.3) $$|\det \chi|_{\text{gauss}} = 1 \quad \text{in Case 3,}$$

(6.4.4) $$\chi = 0 \quad \text{in Case 4,}$$

(6.4.5) $$\det \chi \equiv 0 \mod p\mathbf{Q}_0 \quad \text{in Cases 1 and 2.}$$

Furthermore in Cases 1 and 2 the matrix χ satisfies the following:

(6.4.6)

	Case 1	Case 2
χ_1, χ_2	nontrivial polynomials mod $p\mathbf{Q}_0$ of degree $<p$	trivial mod $p\mathbf{Q}_0$
χ_3, χ_4	trivial mod $p\mathbf{Q}_0$	nontrivial polynomials mod $p\mathbf{Q}_0$ of degree $<p$

Proof. The calculation involves certain error terms which will be evaluated separately. In the following the symbol e denotes 0 (resp. 1) if we are calculating χ_1, χ_3 (resp. χ_2, χ_4).

Case 1. $0 < \mu_c < \mathrm{Min}(\mu_a, \mu_b)$

(6.4.7) $$\psi \frac{(1-x)^{\mu_b-\mu_c-e}(1-\lambda x)^{\mu_a}}{x^{\mu_b}} = \text{coefficient of } x^{\mu_b} \text{ in the numerator,}$$

$(1-x)^{\mu_b-\mu_c-e}(1-\lambda x)^{\mu_a}$, since $0 < \mu_b < p$ and the degree of the numerator is strictly less than $\mu_b + p$. We thus obtain $\chi_3 = 0 = \chi_4$

(6.4.8)

$$\chi_1 \text{ (resp. } \chi_2) = (-1)^{\mu_b} \sum_{i+j=\mu_b} \binom{\mu_b - \mu_c - e}{i}\binom{\mu_a}{j}\lambda^j \quad \text{with } e = 0 \text{ (resp. } e = 1).$$

This completes the verification (6.4.5) and (6.4.6) for Case 1.

Case 2. $0 < \mathrm{Max}(\mu_a, \mu_b) < \mu_c$

$$\psi \frac{T_1^{\mu_b - \mu_c - e} T_{1/\lambda}^{\mu_a}}{T_0^{\mu_b}} = \frac{1}{T_1} \psi(\Lambda_2 + \Lambda_{2,0}), \tag{6.4.9}$$

where

$$\Lambda_2 = \frac{T_1^{p + \mu_b - \mu_c - e} T_{1/\lambda}^{\mu_a}}{T_0^{\mu_b}},$$

$$\Lambda_{2,0} = \frac{[(1 - x^p) - (1 - x)^p] T_{1/\lambda}^{\mu_a}}{T_0^{\mu_b} T_1^{e + \mu_c - \mu_b}}.$$

We first calculate

$$\begin{aligned} \frac{1}{T_1} \psi \Lambda_2 &= \frac{1}{T_1} \psi \frac{(1 - x)^{p + \mu_b - \mu_c - e}(1 - \lambda x)^{\mu_a}}{x^{\mu_b}} \\ &= \frac{1}{T_1} \cdot \text{coefficient of } X^{\mu_b} \text{ in the numerator of } \Lambda_2, \end{aligned} \tag{6.4.10}$$

since $0 < \mu_b < p$ and since the degree of the numerator is strictly less than $p + \mu_b$. We will show (§6.5 below)

$$\frac{1}{T_1} \psi \Lambda_{2,0} \in p\mathbf{Q}_0 + p\mathbf{Q}_0 \frac{1}{T_1} + D_{h, \lambda^p} \mathbf{L}_{\lambda^p}. \tag{6.4.11}$$

Subject to the verification of 6.4.11, we have

$$\chi_1 \equiv \chi_2 \equiv 0 \mod p\mathbf{Q}_0,$$

$$\chi_3 \text{ (resp. } \chi_4) \equiv \sum_{i + j = \mu_b} (-1)^{\mu_b} \binom{p + \mu_b - \mu_c - e}{i} \binom{\mu_a}{j} \lambda^j \mod p\mathbf{Q}_0 \tag{6.4.12}$$

with $e = 0$ (resp. $e = 1$).

Aside from the verification of (6.4.11) this completes the verification of (6.4.6) and of (6.4.5) for Case 2.

Case 3. $0 < \mu_b < \mu_c < \mu_a$

$$\psi \frac{T_1^{\mu_b - \mu_c - e} T_{1/\lambda}^{\mu_a}}{T_0^{\mu_b}} = \frac{1}{T_1} \psi \Lambda_3 + \frac{1}{T_1} \psi \Lambda_{3,0}, \tag{6.4.13}$$

where

$$\Lambda_3 = \frac{T_1^{p+\mu_b-\mu_c-e}T_{1/\lambda}^{\mu_a}}{T_0^{\mu_b}},$$

$$\Lambda_{3,0} = \frac{T_1^{\mu_b-\mu_c-e}[(1-x^p)-(1-x)^p]T_{1/\lambda}^{\mu_a}}{T_0^{\mu_b}}$$

The degree of the numerator of Λ_3 is at least equal to $p+\mu_b$ and hence letting $S_{1,e}$ (resp. $R_{1,e}$) be the coefficient of x^{μ_b} (resp. $x^{p+\mu_b}$) in the numerator of Λ_3, we obtain

$$\text{(6.4.14)} \qquad \begin{aligned}\frac{1}{T_1}\psi\Lambda_3 &= \frac{1}{T_1}(S_{1,e}+xR_{1,e}), \qquad e=0,1\\ &= -R_{1,e}+(S_{1,e}+R_{1,e})\frac{1}{T_1}.\end{aligned}$$

We will show (§6.5 below) that

$$\text{(6.4.15)} \qquad \frac{1}{T_1}\psi\Lambda_{3,0}\in p\mathbf{Q}_0 + p\mathbf{Q}_0\frac{1}{T_1} + D_{h,\lambda^p}\mathbf{L}_{\lambda^p}.$$

Subject to the verification of (6.4.15), we have

$$\text{(6.4.16)} \qquad \chi \equiv \begin{pmatrix} -R_{1,0} & R_{1,0}+S_{1,0}\\ -R_{1,1} & R_{1,1}+S_{1,1}\end{pmatrix} \quad \text{mod } p\mathbf{Q}_0.$$

It is easy to calculate

$$\text{(6.4.17)} \qquad \begin{aligned}(-1)^{\mu_b}S_{1,e} &= \sum_{i+j=\mu_b}\binom{p+\mu_b-\mu_c-e}{i}\binom{\mu_a}{j}\lambda^j,\\ (-1)^{p+\mu_b}R_{1,e} &= \sum_{i+j=\mu_b+p}\binom{p+\mu_b-\mu_c-e}{i}\binom{\mu_a}{j}\lambda^j.\end{aligned}$$

Now

$$(-1)^{\mu_b}S_{1,e} \equiv \binom{p+\mu_b-\mu_c-e}{\mu_b} \quad \text{mod } \lambda,$$

while the term of minimal degree in $(-1)^{p+\mu_b}R_{1,e}$ is given by

$$i = p+\mu_b-\mu_c-e, \qquad j=\mu_c+e,$$

so that

$$(-1)^{p+\mu_b}R_{1,e} \equiv \binom{\mu_a}{\mu_c + e}\lambda^{\mu_c+e} \quad \mathrm{mod}\ \lambda^{1+\mu_c+e}.$$

Thus

$$\text{(6.4.18)} \qquad \det \chi \equiv -\binom{p + \mu_b - \mu_c - 1}{\mu_b}\binom{\mu_a}{\mu_c}\lambda^{\mu_c} \quad \mathrm{mod}\ \lambda^{\mu_c+1}.$$

The binomial coefficients in this last formula are not divisible by p and thus aside from the verification of (6.4.15), the verification of (6.4.3) for Case 3 is complete.

Case 4. $\mu_a < \mu_c < \mu_b$

$$\text{(6.4.19)} \qquad \psi \frac{T_1^{\mu_b-\mu_c-e}T_{1/\lambda}^{\mu_a}}{T_0^{\mu_b}} = 0$$

as the numerator $T_1^{\mu_b-\mu_c-e}T_{1/\lambda}^{\mu_a}$ is a polynomial of degree $\mu_b - (e + (\mu_c - \mu_a))$ $< \mu_b$. This completes the treatment of this case and hence of (6.4.4) in Case 4.

Indeed this completes the demonstration of the lemma except for the verification of relations (6.4.11) and (6.4.15).

6.5 Estimates for Errors

The object of this section is to verify relations (6.4.11) and (6.4.15). For this purpose we define $\xi_1 \in x\mathbf{Q}_0[x]$, polynomials of degree $p - 1$,

$$\text{(6.5.1)} \qquad \Phi T_1 - T_1^p = (1 - x^p) - (1 - x)^p = p\xi_1,$$

We write (using the subscript to distinguish between the conditions)

		Conditions
(6.5.2)	$\Theta_2 = \dfrac{T_{1/\lambda}^{\mu_a}\xi_1}{T_1^{\mu_c-\mu_b+e}T_0^{\mu_b}}$	$\mu_a, \mu_b < \mu_c$
	$\Theta_3 = \Theta_2$	$\mu_b < \mu_c < \mu_a$.

We must apply ψ to Θ_2 or Θ_3 then divide by T_1 and then reduce mod D_{h,λ^p}. For $\Theta = \Theta_2$ or Θ_3 we write

$$\psi\Theta = \sum_{v\in S} \psi P_v(\Theta). \tag{6.5.3}$$

By (5.6.3)

$$P_v(\Theta) \in \mathbf{Q}_0\left[\frac{1}{T_v}\right], \qquad v \in S \tag{6.5.4}$$

and is divisible by $1/T_v$ for $v \neq \infty$. Furthermore the degree of this polynomial is bounded as follows

$$\begin{aligned} \deg_{1/T_v} P_v(\Theta) &\le \mu_b, & v &= 0,\\ &\le e + \mu_c - \mu_b, & v &= 1,\\ &< 0, & v &= 1/\lambda,\\ &\le p - 1 + \mu_a - \mu_c - e, & v &= \infty. \end{aligned} \tag{6.5.5}$$

Thus

$$\psi P_v(\Theta) = 0 \quad \text{for } v = 0,\ 1/\lambda, \tag{6.5.6}$$

while for $v = 1$, we have by (5.3.9)

$$\psi P_1(\Theta) \in \frac{1}{p}\,\mathbf{Q}_0\left[\frac{p}{T_1}\right], \tag{6.5.7}$$

and the degree is bounded (cf. Corollary 5.3.5) by

$$\deg_{1/T_1} \psi P_v(\Theta) \le e + \mu_c - \mu_b. \tag{6.5.8}$$

Finally

$$\psi P_\infty(\Theta) \in \mathbf{Q}_0[x], \tag{6.5.9}$$

$$\deg_x \Psi P_\infty(\Theta_j) \le \begin{cases} 0 & \text{if } j = 2,\\ 1 & \text{if } j = 3. \end{cases}$$

Since $\Lambda_{j,0} = p\Theta_j$ ($j = 2, 3$), it follows from (6.5.9), that $(1/T_1)\Psi\Lambda_{j,0}$ has contributions from $P_\infty\Theta_j$ which lie in the $p\mathbf{Q}_0$ span of $1/T_1$ and x/T_1, i.e., in $p\mathbf{Q}_0 + p\mathbf{Q}_0(1/T_1)$.

It only remains to show that ($j = 2, 3$)

$$\frac{p}{T_1}\psi P_1(\Theta_j) \in p\mathbf{Q}_0 + p\mathbf{Q}_0\frac{1}{T_1} + D_{h,\lambda^p}\mathbf{L}_{\lambda^p}. \tag{6.5.10}$$

Thus by (6.5.7) and (6.5.8) is enough to show that for $1 \le m \le e + \mu_c - \mu_b$

$$\frac{1}{T_1}\left(\frac{p}{T_1}\right)^m \in p\mathbf{Q}_0 + p\mathbf{Q}_0\frac{1}{T_1} + D_{h,\lambda^p}\mathbf{L}_{\lambda^p}. \tag{6.5.11}$$

The reduction formulae of Lemma 1.2 show that $1/T_1^{1+m}$ may be reduced to $1/T_1^m$ and terms of lower degree by division by $m + b' - c'$. It follows that (6.5.11) certainly holds if

$$\frac{p^m}{(1 + b' - c')_m} \equiv 0 \mod p \tag{6.5.12}$$

for $1 \le m \le e + \mu_c - \mu_b$. Now in fact there exists at most one integer m in the range $[1, e + \mu_c - \mu_b]$ such that

$$b' - c' + m \equiv 0.$$

This shows that condition (6.5.12) is implied by (6.4.1). This completes the proof of Lemma 6.4. □

6.6

The object of this section is to summarize our calculations and put them in a form suitable for application.

Theorem. *Let μ_a, μ_b, μ_c all be in the interval $[1, p - 1]$. Suppose*

$$\frac{p^m}{m + b' - c'} \equiv 0 \mod p \quad \text{for } 1 \le m \le 1 + \mu_c - \mu_b, \tag{6.6.1}$$

$$\frac{p^m}{m + c' - b'} \equiv 0 \mod p \quad \text{for } 1 \le m \le 1 + \mu_b - \mu_c.$$

$$\mu_a \ne \mu_c \ne \mu_b, \tag{6.6.2}$$

$$c' - a', b' - c', a', 1 - a' \qquad \text{are all units,} \tag{6.6.3}$$

$$p \ge 2d, \tag{6.6.4}$$

(*or* (6.6.4′)
$$p \geq \sqrt{3d},$$

and $1 + a'$, $b' - c' \pm 1$ *are also units).*

Then in terms of the completion $\mathbf{Q}_0'$ *of* $\mathbf{Q}_0$, *we have the following results concerning the matrix* $\mathbf{A}$ *of* $\alpha^*_{\tilde{f},h}$ *relative to the basis* 1^*, $(1/(1-x))^*$

Case 1. $\mathrm{Min}(\mu_a, \mu_b) > \mu_c$

$$\mathbf{A} = \begin{pmatrix} 1 & 0 \\ 0 & p \end{pmatrix} A.$$

Case 2. $\mu_c > \mathrm{Max}(\mu_a, \mu_b)$

$$\mathbf{A} = \begin{pmatrix} p & 0 \\ 0 & 1 \end{pmatrix} A.$$

Case 3. $\mu_b < \mu_c < \mu_a$

$$\mathbf{A} = A.$$

Case 4. $\mu_a < \mu_c < \mu_b$

$$\mathbf{A} = pA,$$

where in each case,

$$A = \begin{pmatrix} A_1 & A_2 \\ A_3 & A_4 \end{pmatrix}.$$

(i) *each* A_i *is holomorphic in the region* (3.5.0) *except for a possible pole at* 0 *and* ∞;
(ii) *each* $A_i \in \mathbf{Q}_0'$;
(iii) $1/\det A \in \mathbf{Q}_0'$.

Furthermore in Case 1 (*resp.* Case 2) A_1 *and* A_2 (*resp.* A_3 *and* A_4) mod $p\mathbf{Q}_0'$ *are polynomials in* λ *of degree bounded by* p.

Proof. The theorem follows by applying Theorem 6.3 and Lemma 6.4 to both f and $\tilde{f}$ and then applying Corollary 4.7.6. For example in Case 1, we know from Theorem 6.3 and Lemma 6.4 that $\det \alpha_{f,h}$ and $\det \alpha_{\tilde{f},\tilde{h}}$ are congruent zero mod $p\mathbf{Q}_0'$. This together with Corollary 4.7.6 shows that for $\lambda \in \Omega_1$,

$$\mathrm{ord} \det \alpha_{f,h} + \mathrm{ord} \det \alpha_{\tilde{f},\tilde{h}} = 2.$$

This shows that for $\lambda \in \Omega_1$,

$$\text{ord det } \alpha_{f,h} = 1.$$

In Case 3, we observe that by Lemma 6.4, the matrix of $\alpha_{\bar{f},\bar{h}}$ is congruent to zero mod $p\mathbf{Q}'_0$ and hence by Corollary 4.7.6,

$$\text{ord det } \alpha_{f,h} = 0$$

for all $\lambda \in \Omega_1$. Cases 2 and 4 are treated likewise. □

6.6.5 *Remark.* We have made no use of (6.4.3) whose verification depended on (6.4.18). In the treatment of the present theorem, (6.4.3) is replaced by Corollary 4.7.6 together with (6.4.4).

Chapter 7

Hasse Invariants

We have now completed our description of the action of Cartier upon $\mathbf{W}_f$. It is our intention to translate this into an action upon the solutions of equation (2.4.5.1). We know from Corollary 4.6.3 that the solutions of (2.4.5.1) at $z \neq 0, 1, \infty$ converge in the region (4.6.1.1).

7.1 **Lemma.** *Let* λ, z *satisfy* (4.6.1.1) *and* (3.5.0). *Let* $C_f(z, \lambda)$ *denote the solution matrix of* (2.4.5.1) *at* z *which is normalized by*

$$C_f(z, z) = I_2. \tag{7.1.1}$$

Likewise let $C_h(z, \lambda)$ *be the solution matrix of* $(2.4.5.1)_{a',b',c'}$ *normalized in a similar way. Let* $\mathbf{A}_\lambda$ *denote the matrix of* α_λ^* $(= \alpha_{f,h;\lambda}^*)$ *relative to the bases* 1^*, $(1/(1-x))^*$ *of* K_{h,λ^p} *and* $K_{f,\lambda}$. *Then*

$$C_h(z^p, \lambda^p)\mathbf{A}_\lambda = \mathbf{A}_z C_f(z, \lambda). \tag{7.1.2}$$

Proof. This follows from Theorems 4.6.4 and 4.6.2, and equation (2.4.5). □

We now introduce the notation

$$M_f = M_{(a,b,c)} = \begin{pmatrix} \dfrac{-c}{\lambda} & \dfrac{c-a}{1-\lambda} \\ \dfrac{c-b}{\lambda} & \dfrac{a+b-c}{1-\lambda} \end{pmatrix}. \tag{7.1.3}$$

Thus (2.4.5.1) may be written

$$\frac{\partial C_f}{\partial \lambda} = C_f M_f. \tag{7.1.4}$$

In the following lemma, $\mathbf{Q}_0'$ is as defined in §5.6.1.

7.2 **Lemma.** *Under the hypothesis of Theorem* 6.6,

$$\frac{\partial \mathbf{A}_\lambda}{\partial \lambda} \equiv \mathbf{A}_\lambda M_f \quad \operatorname{mod} p\mathbf{Q}'_0.$$

Proof. Applying $\delta = \lambda(\partial/\partial\lambda)$ to both sides of (7.1.2) gives

$$p(\delta C_h)(z^p, \lambda^p)\mathbf{A}_\lambda + C_h(z^p, \lambda^p)\delta\mathbf{A}_\lambda = \mathbf{A}_z \delta C_f(z, \lambda), \tag{7.2.1}$$

while by (7.1.4)

$$(\delta C_h)(z^p, \lambda^p) = \lambda^p C_h(z^p, \lambda^p) M_h(\lambda^p),$$

$$\delta C_f(z, \lambda) = \lambda C_f(z, \lambda) M_f(\lambda).$$

Substituting these in (7.2.1) and multiplying on the left by $(C_h(z^p, \lambda^p))^{-1}$ and applying (7.1.2) gives

$$p\lambda^{p-1} M_h(\lambda^p)\mathbf{A}_\lambda + \frac{\partial \mathbf{A}_\lambda}{\partial \lambda} = \mathbf{A}_\lambda M_f(\lambda). \tag{7.2.2}$$

Now by Theorem 6.6, the coefficients of $\mathbf{A}_\lambda$ lie in $\mathbf{Q}'_0$ and so the lemma follows from (7.2.2). □

7.2.3 *Remark.* In Cases 3 and 4 of Theorem 6.6 we obtain

$$\frac{\partial}{\partial \lambda} A_\lambda \equiv A M_f \quad \operatorname{mod} p\mathbf{Q}'_0.$$

This last lemma permits the interpretation of (A_1, A_2) (resp. (A_3, A_4)) mod $p\mathbf{Q}'_0$ in Case 1 (resp. Case 2) as the unique polynomial solutions mod p of degree less than p of (2.4.5.1).

On the other hand det A is determined (by (7.1.2)) up to a constant factor as the ratio of the wronskians, i.e.,

$$\det A = \lambda^{\mu_c}(1 - \lambda)^{\mu_a + \mu_b - \mu_c}\left(\frac{1 - \lambda^p}{(1 - \lambda)^p}\right)^{a' + b' + c'} \cdot \kappa,$$

where κ is a unit independent of λ, but depending presumably upon a, b, c, as will be made explicit in Chapter 25.

Chapter 8

The $a \to a'$ Map

Let $a \in \mathbb{Q} \cap \mathbb{Z}_p$, $a \in (0, 1)$. We define $a' \in \mathbb{Q} \cap \mathbb{Z}_p$ by condition (3.7.0) with the restriction that

$$\mu_a = pa' - a \in [0, p - 1]. \tag{8.1}$$

This fixes a' uniquely as μ_a is the unique representative mod p of $-a$ in $[0, p - 1]$. By hypothesis a is a ratio of integers, the denominator being prime to p and hence a divisor of $p^l - 1$ for some $l \in \mathbb{N}$. Thus

$$a = \frac{a_0 + pa_1 + \cdots + p^{l-1}a_{l-1}}{p^l - 1}, \tag{8.2}$$

with $0 \leq a_i < p$, $a_i \in \mathbb{N}$. Clearly $\mu_a = a_0$ and so

$$a' = \frac{a_1 + pa_2 + \cdots + p^{l-2}a_{l-1} + p^{l-1}a_0}{p^l - 1}. \tag{8.3}$$

This shows that a' lies in the interval (0, 1). Furthermore, we may define successively

$$a^{(i)} = (a^{(i-1)})' \tag{8.4}$$

and here we have $a^{(l)} = a$.

We refer to the denominator of a as the minimal $d \in \mathbb{N}$ such that $da \in \mathbb{Z}$. It follows from our definitions that

$$(p^l - 1)a = \mu_a + p\mu_{a'} + \cdots + p^{l-1}\mu_{a^{(l-1)}}. \tag{8.5}$$

We observe that if $d = \text{denom}\, a$ then $a = m/d$ with $m \in \mathbb{N}$. Thus

$$p\, da' = m + d\mu_a.$$

This shows that denom $a' \mid pd$, but $a' \in \mathbb{Z}_p$ and so

$$\text{denom } a' \mid d.$$

On the other hand if $d' = \text{denom } a'$ then $d'a = d'pa' - d'\mu_a \in \mathbb{N}$ and so $d \mid d'$. This shows that

$$\text{(8.6)} \qquad \text{denom } a' = \text{denom } a.$$

8.7 **Lemma.** *If* $p \geq d = \text{denom } a$ *then* $a_0, a_1, \ldots, a_{l-1}$ (in (8.1)) *are all distinct from zero.*

Proof. We first show that $a_{l-1} \neq 0$. Suppose otherwise, then

$$\text{(8.7.1)} \qquad (p^l - 1)a \leq (p - 1)(1 + p + \cdots + p^{l-2}) = p^{l-1} - 1$$

and so

$$\text{(8.7.2)} \qquad \frac{1}{d} \leq a \leq \frac{p^{l-1} - 1}{p^l - 1} < \frac{1}{p},$$

which contradicts the hypothesis. This shows that $a_{l-1} \neq 0$, as asserted, but for $0 \leq i < l - 1$

$$\text{(8.7.3)} \qquad a^{(i+1)} = \frac{a_{i+1} + pa_{i+2} + \cdots + p^{l-1}a_i}{p^l - 1}$$

and since denom $a^{(i+1)} = d \leq p$, the previous argument shows that $a_i \neq 0$. This completes the proof. □

8.8 **Corollary.** *Let* $a, b \in \mathbb{Q} \cap \mathbb{Z}_p$, $a, b \in (0, 1)$, $a \neq b$. *Let*

$$d = [\text{denom}, a, \text{denom } b] < p,$$

(*i.e.*, $p >$ *least common multiple of* denom a, denom b) *then for each* $i \in \mathbb{N}$,

$$\text{(8.8.1)} \qquad \mu_{a^{(i)}} \neq \mu_{b^{(i)}}.$$

Proof. Let $b > a$, $d \mid (p^l - 1)$, then a and b may both be represented as in (8.1) and

$$(p^l - 1)(b - a) = \sum_{j=0}^{l-1} p^j(b_j - a_j).$$

If $b_{l-1} = a_{l-1}$ then

$$(p^l - 1)(b - a) \leq p^{l-1} - 1$$

and so,

$$\frac{1}{d} \leq b - a < \frac{1}{p},$$

a contradiction which shows that $b_{l-1} \neq a_{l-1}$. By cyclically permitting the digits we show that for $i \in \mathbb{N}$,

$$b_i \neq a_i,$$

the point being that $b \neq a$ implies $b^{(i)} \neq a^{(i)}$. □

Chapter 9

Normalized Solution Matrix

In this chapter we assume

(9.0.1) $$a, b, c \in \mathbb{Q} \cap \mathbb{Z}_p,$$

(9.0.2) $$a, b \in (0, 1), \qquad c \in (0, 1]$$

(9.0.3) $$a \neq c \neq b,$$

(9.0.4) $$p > d = \text{lcm}(\text{denom } a, \text{denom } b, \text{denom } c).$$

(9.0.5) The hypothesis of Theorem 6.6 holds for $(a^{(i)}, b^{(i)}, c^{(i)})$ for all $i \in \mathbb{N}$, in the terminology of Chapter 8.

(9.0.6) $$(p^l - 1)(a, b, c) \in \mathbb{Z}^3.$$

We write

$$f_i = x^{b(i)-1}(1 - x)^{c(i)-b(i)}(1 - \lambda x)^{-a(i)}$$

and we write

(9.0.7) $$\alpha_i^* = \alpha_{f_i, f_{i+1}}^*$$

for the mapping of $\mathbf{K}_{f_{i+1}, \lambda^p}$ into $\mathbf{K}_{f_i, \lambda}$ studied in Chapters 4 and 6. The matrices of this map will always be understood to be relative to the basis $\{1^*, (1/(1 - x))^*\}$.

We will say that α_i^* and its matrix is of *type* 1 (resp. 2, resp. 3, resp. 4) if the mapping falls under Case 1 (resp. 2, resp. 3, resp. 4) of Theorem 6.6. The first two types are said to be *split*. Types 3 and 4 are said to be *unsplit*.

The matrix of α_i^* will be designated by $\mathbf{A}_i$ but our calculations will usually involve $\mathbf{B}_i$.

(9.1.1.1) $$\mathbf{B}_i = \begin{cases} (1/p)\mathbf{A}_i & \text{if } \alpha_i^* \text{ is of type 4,} \\ \mathbf{A}_i & \text{if } \alpha_i^* \text{ is not of type 4.} \end{cases}$$

Let

$$B = \mathbf{B}_{l-1}^{\phi^{l-1}}\mathbf{B}_{l-2}^{\phi^{l-2}} \cdots \mathbf{B}_1^{\phi}\mathbf{B}_0, \tag{9.1.1.2}$$

the matrix of the mapping

$$\frac{1}{p^m}\alpha_0^* \circ \cdots \circ \alpha_{l-1}^*$$

of $\mathbf{K}_{f_0, \lambda^{p^l}}$ into $\mathbf{K}_{f_0, \lambda}$. ($\phi$ is used in the sense of §5.8; m = number of α_i^* of type 4.)

For $i \in \mathbb{N}$

$$B^{(i)} = \mathbf{B}_{l-1}^{\phi^{i-1}}\mathbf{B}_{l-2}^{\phi^{i-2}} \cdots \mathbf{B}_{l-i}, \tag{9.1.1.3}$$

the symbol making sense for $i > l$ if we interpret the subscripts modulo l.

9.1 **Lemma.**

9.1.2 *If each α_i^* is nonsplit then B is invertible in $\mathbf{Q}_0'$.*

9.1.3 *If α_{l-1}^* is split of type* 1 (resp. *type* 2) *then the coefficients of $B^{(i)}$ lie in $\mathbf{Q}_0'$ and more precisely*

$$B^{(i)} = \begin{pmatrix} B_1^{(i)} & B_2^{(i)} \\ B_3^{(i)} & B_4^{(i)} \end{pmatrix},$$

where $B_3^{(i)}$ and $B_4^{(i)}$ (resp. *$B_1^{(i)}$ and $B_2^{(i)}$*) *lie in $p\mathbf{Q}_0'$ and neither $B_1^{(i)}$ nor $B_2^{(i)}$* (resp. *neither $B_3^{(i)}$ nor $B_4^{(i)}$*) *are trivial* mod $p\mathbf{Q}_0'$.

9.1.4 *In particular under the hypothesis of* 9.1.3 *we have*

$$B = \begin{pmatrix} B_1 & B_2 \\ B_3 & B_4 \end{pmatrix},$$

with B_3 and B_4 (resp. *B_1 and B_2*) *lying in $p\mathbf{Q}_0'$ and neither B_1 nor B_2* (resp. *B_3 nor B_4*) *trivial* mod $p\mathbf{Q}_0'$.

9.1.5 *Remark.* In 9.1.4 we make no assertion if α_{l-1}^* is nonsplit.

Proof. Assertion 9.1.2 follows trivially from Theorem 6.6. We prove 9.1.3 by induction on i, the case of $i = 1$ being a consequence of Theorem 6.6. By definition and the induction hypothesis,

$$(9.1.6)\quad B^{(i+1)} = B^{(i)\phi}\mathbf{B}_{l-i-1}$$
$$= \begin{pmatrix} B_1^{(i)} & B_2^{(i)} \\ B_3^{(i)} & B_4^{(i)} \end{pmatrix}^{\phi} \mathbf{B}_{l-i-1}$$
$$\equiv \begin{cases} \begin{pmatrix} B_1^{(i)} & B_2^{(i)} \\ 0 & 0 \end{pmatrix}^{\phi} \mathbf{B}_{l-i-1} & \text{if } \alpha_{l-1}^* \text{ is of type 1,} \\ & \bmod p\mathbf{Q}_0' \\ \begin{pmatrix} 0 & 0 \\ B_3^{(i)} & B_4^{(i)} \end{pmatrix}^{\phi} \mathbf{B}_{l-i-1} & \text{if } \alpha_{l-1}^* \text{ is of type 2.} \end{cases}$$

If $\mathbf{B}_{l-i-1}$ is of type 1 (resp. type 2) then

$$(9.1.7)\qquad \mathbf{B}_{l-i-1} \equiv \begin{pmatrix} \theta_1 & \theta_2 \\ 0 & 0 \end{pmatrix} \left(\text{resp.} \begin{pmatrix} 0 & 0 \\ \theta_1 & \theta_2 \end{pmatrix}\right) \bmod p\mathbf{Q}_0',$$

where θ_1, θ_2 are elements of $\mathbf{Q}_0'$ not trivial mod $p\mathbf{Q}_0'$. Also neither $B_1^{(i)}$ nor $B_2^{(i)}$ (resp. $B_3^{(i)}$ nor $B_4^{(i)}$) is trivial mod $p\mathbf{Q}_0'$ if α_{l-1}^* is of type 1 (resp. type 2). It is now easy to verify assertion 9.1.3 for $B^{(i+1)}$ if $\mathbf{B}_{l-i-1}$ is split, i.e., of type 1 or type 2. We must now consider the case in which α_{l-i-1}^* is not split, and so $\mathbf{B}_{l-i-1}$ is invertible in $\mathbf{Q}_0'$. We complete the verification by showing that if (ξ_1, ξ_2) represents the unique nontrivial mod $p\mathbf{Q}_0'$ row of $B^{(i)}$ then

$$(9.1.8)\qquad (\xi_1, \xi_2)^{\phi}\mathbf{B}_{l-i-1} = (\zeta_1, \zeta_2),$$

where ζ_1 and ζ_2 are each nontrivial mod $p\mathbf{Q}_0'$. By Lemma 7.2, $\mathbf{B}_{l-i-1}$ is a solution matrix mod $p\mathbf{Q}_0'$ of (2.4.5.1) with (a, b, c) replaced by

$$(a^{(l-i-1)}, b^{(l-i-1)}, c^{(l-i-1)})$$

and hence ζ_1, ζ_2 also satisfies this equation mod $p\mathbf{Q}_0'$. Writing this equation in the form

$$(9.1.9)\qquad \frac{d}{d\lambda}(\zeta_1, \zeta_2) \equiv (\zeta_1, \zeta_2)\begin{pmatrix} g_1 & g_2 \\ g_3 & g_4 \end{pmatrix},$$

where

$$g_2 = \frac{c^{(l-i-1)} - a^{(l-i-1)}}{1-\lambda},$$

$$g_3 = \frac{c^{(l-i-1)} - b^{(l-i-1)}}{\lambda},$$

we know that ζ_1 and ζ_2 are not both trivial mod $p\mathbf{Q}'_0$ and if $\zeta_1 \equiv 0$ (resp. $\zeta_2 \equiv 0$) then $g_3\zeta_2$ (resp. $g_2\zeta_1$) $\equiv 0$ and hence either g_3 or g_2 are zero mod $p\mathbf{Q}'_0$. This is impossible by Corollary 8.8. This completes the proof of 9.1.3 from which we obtain 9.1.4. □

9.2 **Theorem.** *If each* α_i^* *in Lemma* 9.1 *is nonsplit then the solutions of* $(2.4.5.1)_{a,b,c}$ *are bounded on each nonsingular residue class. (In fact,* $C_{f_0}(z, \lambda)$ *is bounded by unity.)*

Note. Analyticity on each class is known by Corollary 4.6.3.

Proof. We write equation (7.1.2) for each α_i^* in the form

(9.2.1) $$C_{f_i}(z^p, \lambda^p)\mathbf{B}_{i-1}(\lambda) = \mathbf{B}_{i-1}(z)C_{f_{i-1}}(z, \lambda), \qquad 1 \le i \le l,$$

where $z \in \Omega_1$, $\lambda \in D(z, 1^-)$. From these equations and (9.1.1.2) we deduce

(9.2.2) $$C_{f_0}(z^{p^l}, \lambda^{p^l})B(\lambda) = B(z)C_{f_0}(z, \lambda).$$

In the present situation each B_i is invertible in $\mathbf{Q}'_0$ and hence the same holds for B. Thus $B(\lambda)$ is bounded by 1 on $D(z, 1^-)$ and $B(z)^{-1}$ is bounded by 1. By iteration, for all $s, s' \in \mathbb{N}$,

(9.2.2′) $$C_{f_0}(z^{p^{lss'}}, \lambda^{p^{lss'}})B(\lambda^{p^{l(ss'-1)}}) \cdots B(\lambda) = B(z^{p^{(ss'-1)}}) \cdots B(z)C_{f_0}(z, \lambda).$$

We now specialize to $z \in \Omega_1$, $z^{p^{ls}} = z$. This shows that $C_{f_0}(z, \lambda)$ is bounded by $C_{f_0}(z, \lambda^{p^{lss'}})$. Let $s' \to \infty$ so for fixed $\lambda \in D(z, 1^-)$, $\lambda^{p^{lss'}} \to z$ showing that $C_{f_0}(z, \lambda)$ is bounded by unity. This proves the assertion for each residue class algebraic over $\mathbb{F}_p$. In fact boundedness of all solutions on one residue class implies boundedness on all nonsingular residue classes. But one can construct a direct proof for transcendental residue classes based upon (9.2.2) and the fact that we may construct a Teichmüller representative for such classes. □

9.3

To explain our construction of a normalized solution matrix in the non-totally nonsplit case, we introduce the field Ω_0, the maximal nonramified extension of $\mathbb{Q}_p$ in Ω. The Frobenius automorphism in the residue class field of Ω is lifted to an automorphism σ of Ω_0. Given any power series

(9.3.1) $$\xi(\lambda) = \sum_{j=0}^{\infty} e_j(\lambda - z)^j \in \Omega_0[[\lambda - z]],$$

with $z \in \Omega_0$, we define

$$\xi^{\sigma\phi}: \lambda \to \xi^{\sigma}(\lambda^p) = \sum_{j=0}^{\infty} e_j^{\sigma}(\lambda^p - z^{\sigma}). \tag{9.3.2}$$

In particular if z is a Teichmüller representative,

$$z^{\sigma} = z^p \tag{9.3.3}$$

and if ξ represents the germ of an analytic function at z then so does $\xi^{\sigma\phi}$.

9.3.4 **Definition.** We shall say that a property holds for a *general residue class* if the property holds for all but a finite set of residue classes.

9.4 **Theorem.** *Let $l_0 \geq 1$ be the number of split steps among the maps $\alpha_0^*, \ldots, \alpha_{l-1}^*$.*

9.4.1 *In particular let α_{l-1}^* be split of type* 1 (resp. type 2), *and let z be the Teichmüller representative of a general residue class. We assert the existence of a solution matrix $\mathbf{Y}_0$ of* $(2.4.5.1)_{a,b,c}$ *at z defined over Ω_0 such that*

$$\mathbf{Y}_0^{\sigma^l}(\lambda^{p^l})B = \begin{pmatrix} 1 & 0 \\ 0 & p^{l_0} \end{pmatrix} \mathbf{Y}_0(\lambda), \tag{9.4.1.1}$$

$$\det \mathbf{Y}_0(z) = \text{unit}, \tag{9.4.1.2}$$

(9.4.1.3) *the first row (u_1, u_2) of $\mathbf{Y}_0$ has $|u_1(z)| = 1$*

(resp. $|u_2(z)| = 1$) *and* $1 \geq \mathrm{Max}(|u_1(\lambda)|, |u_2|(\lambda)|)$ *for all $\lambda \in D(z, 1^-)$.*

9.4.2 *These conditions fix $\mathbf{Y}_0$ up to multiplication by a diagonal matrix with coefficients in the group of units of $\mathbb{Z}_p[\zeta]$ where ζ is a primitive $p^l - 1$ root of unity.*

9.4.3 *For $i = 1, 2, \ldots, l - 1$, we set*

$$\mathbf{Y}_{l-i} = \begin{pmatrix} 1 & 0 \\ 0 & p \end{pmatrix}^{-k(i)} \mathbf{Y}_0^{(\sigma\phi)^i} B^{(i)}, \tag{9.4.3.0}$$

where $B^{(i)}$ is defined by (9.1.1.3) *and $k(i)$ is the number of split maps among the maps $\alpha_{l-1}^*, \ldots, \alpha_{l-i}^*$. Then $\mathbf{Y}_{l-i}$ is normalized solution matrix of*

(2.4.5.1)$_{(a^{(l-i)}, b^{(l-i)}, c^{(l-i)})}$ *in the sense that conditions* (9.4.1.1)–(9.4.1.3) *are satisfied with B replaced by*

(9.4.4) $$\mathbf{B}^{(l-i)} = \mathbf{B}_{l-i-1}^{\phi^{l-1}} \mathbf{B}_{l-i-2}^{\phi^{l-2}} \cdots \mathbf{B}_0^{\phi^i} \mathbf{B}_{l-1}^{\phi^{i-1}} \cdots \mathbf{B}_{l-i}.$$

Note. 9.1.4 applies to B but not necessarily to $\mathbf{B}^{(l-i)}$.

Proof. Equation (9.2.2) is valid here and by the choice of z, we deduce

(9.4.5) $$C_{f_0}^{(\sigma\phi)^l} B(\lambda) = B(z) C_{f_0}(z, \lambda).$$

We know that

(9.4.6) $$|\det B(z)| = |p^{l_0}|.$$

Hence by 9.1.4 and Lemma 12.1, there exists $X \in \mathrm{GL}(2, \Omega_0)$ such that

(9.4.7) $$X^{\sigma^l} B(z) = \begin{pmatrix} 1 & 0 \\ 0 & p^{l_0} \end{pmatrix} X.$$

Putting $\chi = X C_{f_0}(z, \lambda)$ we deduce from (9.4.5)

(9.4.8) $$\chi^{\sigma^l \phi^l} B = \begin{pmatrix} 1 & 0 \\ 0 & p^{l_0} \end{pmatrix} \chi.$$

The first row $u = (u_1, u_2)$ of χ thus satisfies

(9.4.9) $$u^{\sigma^l \phi^l} B = u$$

and hence by iteration for $s \geq 1$

(9.4.10) $$u = u^{\sigma^{sl} \phi^{sl}} B^{\phi^{l(s-1)}} B^{\phi^{l(s-2)}} \cdots B.$$

Since the coefficients of B are bounded by 1 on Ω_1, we see that for λ fixed in $D(z, 1^-)$, $u(\lambda)$ is dominated by $u^{\sigma^{sl}}(\lambda^{p^{ls}})$ and for each $r \in (0, 1)$ we may choose s such that $\lambda^{p^{ls}} \in D(z^{\sigma^{sl}}, r^-)$. However $u^{\sigma^{sl}}$ on $D(z^{\sigma^{sl}}, r^-)$ is bounded by the maximum of u on $D(z, r^-)$. Thus letting $r \to 0$, we see that u on $D(z, 1^-)$ is dominated by $\mathrm{Max}(|u_1(z)|, |u_2(z)|)$. Specializing (9.4.9) at $\lambda = z$ gives

(9.4.11) $$u(z) = (u(z))^{\sigma^l} \begin{pmatrix} B_1 & B_2 \\ B_3 & B_4 \end{pmatrix},$$

i.e.,

$$u_2(z) - u_2(z)^{\sigma^l} B_4(z) = u_1(z)^{\sigma^l} B_2(z).$$

If α_{l-1}^* is of type 1, then by 9.1.4, $|B_4(z)| \leq |p|$ and hence

$$|u_2(z)| = |u_1(z)^{\sigma^l} B_2(z)| \leq |u_1(z)|. \tag{9.4.12}$$

If α_{l-1}^* is of type 2 then $|B_1(z)| \leq |p|$ and hence

$$|u_1(z)| = |u_1(z) - u_1(z)^{\sigma^l} B_1(z)| = |u_2(z)^{\sigma^l} B_3(z)| \leq |u_2(z)|.$$

Since $u_1(z)$ (resp. $u_2(z)$) is a nonzero element of Ω_0 we may normalize u by multiplication by an element of $\mathbb{Q}_p$ so that $u_1(z)$ (resp. $u_2(z)$) is a unit. Similarly we may multiply the second row by an element of $\mathbb{Q}_p$ so as to assure validity of (9.4.1.2). This completes the verification of (9.4.1.3) and completes the construction of $\mathbf{Y}_0$.

To verify the uniqueness of $\mathbf{Y}_0$ we observe that any other solution matrix of (2.4.5.1) defined over Ω_0 is of the form $J\mathbf{Y}_0$ with $J \in \mathrm{GL}(2, \Omega_0)$. Condition (9.4.1.1) implies

$$(J^{\sigma^l})^{-1} \begin{pmatrix} 1 & 0 \\ 0 & p^{l_0} \end{pmatrix} J = \begin{pmatrix} 1 & 0 \\ 0 & p^{l_0} \end{pmatrix}, \tag{9.4.13}$$

which shows that J is diagonal with coefficients in $\mathbb{Q}_p(\zeta)$. Then condition (9.4.1.3) shows that J_{11} is a unit and (9.4.1.2) shows that det J is a unit. This completes the verification of uniqueness.

Now

(9.4.14)

$$\begin{aligned}
\mathbf{Y}_{l-i}^{\sigma^l \phi^l} \mathbf{B}^{(l-1)} &= \begin{pmatrix} 1 & 0 \\ 0 & p \end{pmatrix}^{-k(i)} (\mathbf{Y}_0^{\sigma^i \phi^i} B^{(i)})^{\sigma^l \phi^l} \mathbf{B}^{(l-i)} \\
&= \begin{pmatrix} 1 & 0 \\ 0 & p \end{pmatrix}^{-k(i)} \mathbf{Y}_0^{\sigma^{i+l} \phi^{i+l}} (\mathbf{B}_{l-1}^{\phi^{i-1}} \cdots \mathbf{B}_{l-i})^{\phi^l} \mathbf{B}_{l-i-1}^{\phi^{l-1}} \cdots \mathbf{B}_0^{\phi^i} \cdots \mathbf{B}_{l-i} \\
&= \begin{pmatrix} 1 & 0 \\ 0 & p \end{pmatrix}^{-k(i)} (\mathbf{Y}_0^{\sigma^l \phi^l} B)^{\sigma^i \phi^i} \mathbf{B}_{l-1}^{\phi^{i-1}} \cdots \mathbf{B}_{l-i} \\
&= \begin{pmatrix} 1 & 0 \\ 0 & p \end{pmatrix}^{-k(i)} \begin{pmatrix} 1 & 0 \\ 0 & p^{l_0} \end{pmatrix} \mathbf{Y}_0^{\sigma^i \phi^i} B^{(i)} \\
&= \begin{pmatrix} 1 & 0 \\ 0 & p^{l_0} \end{pmatrix} \mathbf{Y}_{l-i}.
\end{aligned}$$

This completes the verification of the analogue of (9.4.1.1) for $\mathbf{Y}_{l-i}$. Letting (u_1, u_2) be the first row of $\mathbf{Y}_0$ then the first row of $\mathbf{Y}_{l-i}$ is by 9.4.3

$$(y_1, y_2) = (u_1, u_2)^{\sigma^i \phi^i} \begin{pmatrix} B_1^{(i)} & B_2^{(i)} \\ B_3^{(i)} & B_4^{(i)} \end{pmatrix}. \tag{9.4.15}$$

If α^*_{l-1} is of type 1 (resp. type 2) then

$$(9.4.16) \qquad (B_3^{(i)}(z), B_4^{(i)}(z)) \equiv 0 \mod p$$

$$(\text{resp. } (B_1^{(i)}(z), B_2^{(i)}(z)) \equiv 0 \mod p)$$

and so

$$(9.4.17) \qquad y_1(z) \equiv u_1(z)^{\sigma^i} B_1^{(i)}(z) \mod p$$

$$(\text{resp. } y_2(z) \equiv u_2(z)^{\sigma^i} B_4^{(i)}(z) \mod p).$$

Thus $y_1(z)$ (resp. $y_2(z)$) is a unit. Furthermore (9.4.15) and (9.4.1.3) shows that y_1 and y_2 are dominated by 1 on $D(z, 1^-)$. This shows that (9.4.1.3) holds for $\mathbf{Y}_{l-i}$. Again it follows from the definition and (9.4.1.2) that det $\mathbf{Y}_{l-i}(z)$ is a unit. This completes the proof of the theorem. □

Remark. If we define $\mathbf{Y}_i$ for i outside the range $[0, l-1]$ by means of $\mathbf{Y}_{i+l} = \mathbf{Y}_i$, then (9.4.3.0) remains valid for $i \in \mathbb{N}$.

9.5

We retain the notation of Theorem 9.4. For each $i \in \mathbb{N}$ we write

$$(9.5.1) \qquad \mathbf{Y}_i = \begin{pmatrix} 1 & 0 \\ \tau_i & 1 \end{pmatrix} \begin{pmatrix} u_i & 0 \\ 0 & \hat{u}_i \end{pmatrix} \begin{pmatrix} 1 & \eta_i \\ 0 & 1 \end{pmatrix},$$

$$\begin{pmatrix} 0 & 1 \\ 1 & 0 \end{pmatrix} \mathbf{Y}_i = \overline{\mathbf{Y}}_i = \begin{pmatrix} 1 & \bar{\tau}_i \\ 0 & 1 \end{pmatrix} \begin{pmatrix} \hat{\bar{u}}_i & \\ & \bar{u}_i \end{pmatrix} \begin{pmatrix} 1 & 0 \\ \bar{\eta}_i & 1 \end{pmatrix}.$$

Then $(u_i, u_i\eta_i)$ is the first row of $\mathbf{Y}_i$, the second row is $(\tau_i u_i, *)$ and

$$\det Y_i = u_i \hat{u}_i.$$

This shows how $u_i, \eta_i, \tau_i, \hat{u}_i$ are uniquely determined by $\mathbf{Y}_i$. Similarly considerations hold for $\bar{u}_i, \hat{\bar{u}}_i, \bar{\tau}_i, \bar{\eta}_i$.

Lemma. *If each of the maps in the sequence*

$$\mathbf{K}_{f_i} \xleftarrow{\alpha_i^*} \mathbf{K}_{f_{i+1}} \leftarrow \cdots \leftarrow \mathbf{K}_{f_{j-1}} \xleftarrow{\alpha_{j-1}^*} \mathbf{K}_{f_j}$$

is unsplit except for α_{j-1}^* *which is of type* 1 (resp. *type* 2) *and if the class of* z *is general then*

(9.5.2)
$$p\tau_i \equiv \tau_j^{(\sigma\phi)^{j-i}} \mod pW_z^{(0)}$$
$$(\text{resp. } p\bar{\tau}_i \equiv \bar{\tau}_j^{(\sigma\phi)^{j-i}} \mod pW_z^{(0)}).$$

9.5.2.1 **Notation.** W_z is the ring of bounded analytic functions on $D(z, 1^-)$. $W_z^{(0)}$ is the subring of elements of W_z which are bounded by unity.

Note. The result here depends upon the type of α_{j-1}^* and not upon the type of α_{i-1}^*.

Proof. It follows from (9.4.3.0) and the definitions that

(9.5.3)
$$\begin{pmatrix} 1 & 0 \\ 0 & p \end{pmatrix} \mathbf{Y}_i = \mathbf{Y}_j^{(\sigma\phi)^{j-i}} H,$$

where

(9.5.4)
$$H = \mathbf{B}_{j-1}^{\phi^{j-i-1}} \mathbf{B}_{j-2}^{\phi^{j-i-2}} \cdots \mathbf{B}_i$$

and so by 9.1.3

(9.5.5)
$$H = \begin{pmatrix} 1 & 0 \\ 0 & p \end{pmatrix} \begin{pmatrix} K_1 & K_2 \\ K_3 & K_3 \end{pmatrix}$$
$$\left(\text{resp. } \begin{pmatrix} p & 0 \\ 0 & 1 \end{pmatrix} \begin{pmatrix} K_1 & K_2 \\ K_3 & K_4 \end{pmatrix}\right),$$

where neither K_1 nor K_2 is trivial mod $p\mathbf{Q}_0'$ (resp. neither K_3 nor K_4 is trivial mod $p\mathbf{Q}_0'$), and all K_i lie in $\mathbf{Q}_0'$.

We deduce from (9.5.1), (9.5.3), and (9.5.5) that

(9.5.6)
$$(p\tau_i - \tau_j^{(\sigma\phi)^{j-i}}) \frac{u_i u_j^{(\sigma\phi)^{j-i}}}{(\det Y_j)^{(\sigma\phi)^{j-i}}} = pK_3 \qquad (\alpha_{j-1}^* \text{ is type 1}),$$
$$(p\bar{\tau}_i - \bar{\tau}_i^{(\sigma\phi)^{j-i}}) \frac{\bar{u}_i \bar{u}_j^{(\sigma\phi)^{j-i}}}{(\det \mathbf{Y}_j)^{(\sigma\phi)^{j-i}}} = pK_2 \qquad (\alpha_{j-1}^* \text{ is type 2}).$$

Now $(u_i, \bar{u}_i)$ is the first row of $\mathbf{Y}_i$ and by 9.4.3 if $\bar{z}$ is a general residue class then this row is dominated by 1 and $u_i(z)$ (resp. $\bar{u}_i(z)$) is a unit if α_{l-1}^* is of type 1 (resp. type 2). However we assert that the same conclusions hold in terms of the type of the split map α_{j-1}^*. This follows from the uniqueness of the normalized solution matrices. These solution matrices could have been constructed starting with $\mathbf{Y}_j$ (rather than $\mathbf{Y}_0$) and so by 9.4.2 $u_i(z)$, $u_j(z)^{\sigma^{j-i}}$ (resp. $\bar{u}_i(z)$, $\bar{u}_j^{\sigma^{j-i}}$) are units. Of course det $\mathbf{Y}_j(z)^{\sigma^{j-i}}$ is a unit. This completes the proof of the lemma. □

9.6

We now return to a detailed consideration of $\mathbf{Y}_0$. The fact that α_{l-1}^* is split will play a role here. We use the notation of 9.1.4.

Theorem. *Let $\bar{z} \neq 0, 1, \infty$ be a residue class at which the* mod p *reduction of B_1* (resp. *B_4*) *does not vanish (if α_{l-1}^* is of type* 1 (resp. *type* 2)). *Then there exists a normalized solution matrix $\mathbf{Y}_0$ on $D(z, 1^-)$ (so here we make more explicit the meaning of "general" residue class in the statement of parts* 9.4.1, 9.4.2 *of Theorem* 9.4). *Furthermore*

(9.6.1) $$\eta_0 \text{ (resp. } \bar{\eta}_0) \quad \textit{and} \quad u_0^{1-\sigma^l\phi^l} \text{ (resp. } \bar{u}_0^{1-\sigma^l\phi^l})$$

are analytic elements on

$$S_H = \{\lambda \,|\, |B_1(\lambda)| = 1 \text{ (resp. } |B_4(\lambda)| = 1)\} \cap \Omega_1.$$

Furthermore

(9.6.2) $$\tau_0^{\sigma^l}(\lambda^{p^l}) = p^{l_0}\tau_0 + p\Lambda$$

$$\text{(resp. } \bar{\tau}_0^{\sigma^l}(\lambda^{p^l}) = p^{l_0}\bar{\tau}_0 + p\bar{\Lambda}),$$

where Λ (resp. *$\bar{\Lambda}$*) *lies in $W_z^{(0)}$, the space of elements of $\Omega_0[[x - z]]$ which converge and are bounded by* 1 *on $D(z, 1^-)$.*

Proof. Let $H_0(S_H)$ denote the ring of analytic elements on S_H which are bounded by unity. The mapping

(9.6.3) $$\rho\colon w \mapsto \frac{B_2 + B_4 w^{\sigma^l\phi^l}}{B_1 + B_3 w^{\sigma^l\phi^l}}$$

$$\left(\text{resp. } \bar{\rho}\colon w \mapsto \frac{B_3 + B_1 w^{\sigma^l\phi^l}}{B_4 + B_2 w^{\sigma^l\phi^l}}\right)$$

is stable on $H_0(S_H)$ since B_1 (resp. B_4) is a unit on S_H while B_3 and B_4 (resp. B_1, B_2) are bounded by $|p|$. We may calculate

$$\rho(w_1) - \rho(w_2) = \frac{(B_1 B_4 - B_3 B_2)(w_1 - w_2)^{\sigma^l \phi^l}}{(B_1 + B_3 w_1^{\sigma^l \phi^l})(B_1 + B_3 w_2^{\sigma^l \phi^l})} \tag{9.6.4}$$

to conclude that ρ (resp. $\bar{\rho}$) is contractive and hence has a unique fixed point, η, in $H_0(S_H)$. On the other hand we may extend ρ (resp. $\bar{\rho}$) to $W_z^{(0)}$, and again obtain a contractive map. Since $H_0(S_H) \hookrightarrow W_z^{(0)}$, we conclude that the unique fixed point in $W_z^{(0)}$ is an element of $H_0(S_H)$. It follows from (9.4.1.3) that the first row (y_1, y_2) of $\mathbf{Y}_0$ has the property that y_2/y_1 (resp. y_1/y_2) is bounded by unity. This ratio is the same as η_0 (resp. $\bar{\eta}_0$) in the notation of (9.5.1). Furthermore by (9.4.1.1), η_0 (resp. $\bar{\eta}_0$) is a fixed point of ρ (resp. $\bar{\rho}$). This proves assertion (9.6.1) for η_0 (resp. $\bar{\eta}_0$). We now substitute equation (9.5.1) with $i = 0$ in (9.4.1.1) to obtain

$$\begin{pmatrix} 1 & \eta_0 \\ 0 & 1 \end{pmatrix}^{\phi^l} B \begin{pmatrix} 1 & -\eta_0 \\ 0 & 1 \end{pmatrix} \tag{9.6.5}$$

$$= \begin{pmatrix} 1/u_0 & 0 \\ 0 & 1/\hat{u}_0 \end{pmatrix}^{\sigma^l \phi^l} \begin{pmatrix} 1 & 0 \\ -\tau_0 & 1 \end{pmatrix}^{\sigma^l \phi^l} \begin{pmatrix} 1 & 0 \\ 0 & p^{l_0} \end{pmatrix} \begin{pmatrix} 1 & 0 \\ \tau_0 & 1 \end{pmatrix} \begin{pmatrix} u_0 & 0 \\ 0 & \hat{u}_0 \end{pmatrix}$$

to conclude that

$$u_0^{1-\sigma^l \phi^l} = B_1 + B_3 \eta_0^{\phi^l} \tag{9.6.6}$$

$$(\text{resp. } \bar{u}_0^{1-\sigma^l \phi^l} = B_4 + B_2 \bar{\eta}_0^{\phi^l}),$$

$$\frac{u_0}{\hat{u}^{\sigma^l \phi^l}} (p^{l_0} \tau_0 - \tau_0^{\sigma^l \phi^l}) = B_3 \tag{9.6.7}$$

$$\left(\text{resp. } \frac{\bar{u}_0}{\hat{\bar{u}}_0^{\sigma^l \phi^l}} (p^{l_0} \bar{\tau}_0 - \bar{\tau}_0^{\sigma^l \phi^l}) = B_2\right).$$

The analytic continuation of $u_0^{1-\sigma^l \phi^l}$ (resp. $\bar{u}_0^{1-\sigma^l \phi^l}$) follows from equation (9.6.6). Finally we deduce (9.6.2) from (9.6.7) with

$$p\Lambda = -B_3 \frac{\det \mathbf{Y}_0^{\sigma^l \phi^l}}{u_0^{1+\sigma^l \phi^l}}$$

$$\left(\text{resp. } p\bar{\Lambda} = +B_2 \frac{\det \mathbf{Y}_0^{\sigma^l \phi^l}}{\bar{u}_0^{1+\sigma^l \phi^l}}\right).$$

This completes the proof since by (9.4.1.3) u_0 (resp. $\bar{u}_0$) is a unit everywhere in $D(z, 1^-)$. ☐

9.7 **Corollary.** *Under the hypothesis of Theorems 9.4 and 9.6 the second row* $v = (v_1, v_2)$ *of* $\mathbf{Y}_0$ *has growth*

$$k_2 \frac{1}{(\log(1/r))^{l_0/l}} < |v|_z(r) < k_1 \frac{1}{(\log(1/r))^{l_0/l}}$$

as $r \to 1$ *where* k_1, k_2 *are strictly positive and*

$$|v|_z(r) = \text{Max}(|v_1|_z(r), |v_2|_z(r)).$$

Proof. We may with no loss in generality restrict our attention to Case 1. We deduce from (9.6.2) that for $s \in \mathbb{N}$,

(9.7.1) $$\tau_0^{(\sigma^l\phi^l)^s} = p^{l_0 s}\tau_0 + \theta^{(\sigma\phi)^{l(s-1)}} + p^{l_0}\theta^{(\sigma\phi)^{l(s-2)}} + \cdots + p^{l_0(s-1)}\theta,$$

where $\theta = p\Lambda$. There are two possibilities

(a) There exists $r_0 \in (|\pi|, 1)$ such that

$$|\tau_0|_z(r_0) > |p|,$$

(b) $$|\tau_0|_z(1) \leq |p|.$$

In case (a), for $r^{p^{ls}} > r_0$, we know that

$$|\lambda - z| = r \quad \text{implies } |\lambda^{\phi^{ls}} - z^{p^{ls}}| = r^{p^{ls}}$$

and hence

(9.7.2) $$|p^{l_0 s}|\,|\tau_0|_z(r) = |\tau_0|_z(r^{p^{ls}}).$$

From this relation we deduce the strong form of the growth of τ_0 (and hence of v).

In case (b), we conclude that

(9.7.3) $$|\tau_0'|_z(1) \leq |p|.$$

Now recall that if we write (2.4.5.1) in the form

$$(u_1, u_2)' = (u_1, u_2)\begin{pmatrix} g_1 & g_2 \\ g_3 & g_4 \end{pmatrix}$$

and if

$$\begin{pmatrix} y_1 & y_2 \\ v_1 & v_2 \end{pmatrix}$$

is the solution matrix $\mathbf{Y}_0$ then

$$y_1' = y_1 g_1 + y_2 g_3,$$

$$v_1' = v_1 g_1 + v_2 g_3,$$

and so

$$\det\begin{pmatrix} y_1 & y_1' \\ v_1 & v_1' \end{pmatrix} = g_3 \det Y_0. \tag{9.7.4}$$

On the other hand, $v_1 = y_1\tau_0$, and so the left side of (9.7.4) reduces to $y_1^2\tau_0'$, which shows that

$$y_1^2\tau_0' = g_3 \det \mathbf{Y}_0. \tag{9.7.5}$$

Specializing this at $\lambda = z$ where y_1 and $\det \mathbf{Y}_0$ take on unit values (by construction), we find that $\tau_0'(z)$ is also a unit since by (2.4.5.1), (9.0.3) and Corollary 8.8, $g_3(z)$ is a unit. This contradicts (9.7.3). This completes the proof of the corollary. □

9.8

We now consider $\mathbf{Y}_i$ in greater detail in the situation in which α_i^* is not split. For this purpose we may consider the situation of Lemma 9.5.

By the same method as that used to obtain (9.5.6), we find if $\bar{z}$ is a general residue class

$$\frac{u_i}{u_j^{(\sigma\phi)^{j-i}}} = K_1 + pK_3\eta_j^{\phi^{j-i}}. \tag{9.8.1}$$

$$\left(\text{resp. } \frac{\bar{u}_i}{\bar{u}_j^{(\sigma\phi)^{j-i}}} = K_4 + pK_2\bar{\eta}_j^{\phi^{j-i}}\right)$$

if α_{j-1}^* is of type 1 (resp. type 2). Let equation $(2.4.5.1)_{a(i),b(i),c(i)}$ be rewritten in the form

$$\lambda\frac{d}{d\lambda}(u_i, u_i\eta_i) = (u_i, u_i\eta_i)\begin{pmatrix} g_{i1} & g_{i2} \\ g_{i3} & g_{i4} \end{pmatrix}. \tag{9.8.2}$$

Then

$$\lambda\frac{u_i'}{u_i} = g_{i1} + \eta_i g_{i3},$$

$$\lambda\frac{\bar{u}_i'}{\bar{u}_i} = g_{i4} + \bar{\eta}_i g_{i2}, \tag{9.8.3}$$

and taking logarithmic derivatives of the two sides of equation (9.8.1),

$$g_{i1} + \eta_i g_{i3} - p^{j-i}(g_{j1} + \eta_j g_{j3})^{\phi^{j-i}} = \frac{\lambda K_1' + p\lambda(K_3\eta_j^{\phi^{j-i}})'}{K_1 + pK_3\eta_j^{\phi^{j-i}}} \tag{9.8.4}$$

$$\left(\text{resp. } g_{i4} + \bar{\eta}_i g_{i2} - p^{j-i}(g_{j4} + \bar{\eta}_j g_{j2})^{\phi^{j-i}} = \frac{(\lambda K_4' + p\lambda(K_2\bar{\eta}_j^{\phi^{j-i}})')}{(K_4 + pK_2\bar{\eta}_j^{\phi^{j-i}})}\right).$$

Since α_{j-1}^* is split, we know that η_j (resp. $\bar{\eta}_j$) has an analytic continuation. This equation demonstrates the analytic continuation of η_i (resp. $\bar{\eta}_i$).

Lemma. *On a general residue class η_i and $\bar{\eta}_i$ assume unit values, and hence likewise for u_i, $u_i\eta_i$ (valid also for $i = j$).*

Proof. We know from (9.5.1) that $(u_i, \bar{u}_i)$ is the first row of $\mathbf{Y}_i$ and hence by (9.4.3) either u_i or $\bar{u}_i$ assumes only unit values and dominates the other. Suppose u_i dominates $\bar{u}_i$, so η_i is bounded by 1 on $D(z, 1^-)$. However from (9.8.2) multiplying on the right by $\binom{-\eta_i}{1}$ we deduce the Riccati equations

$$\lambda\eta_i' = (1, \eta_i)\begin{pmatrix} g_{i1} & g_{i2} \\ g_{i3} & g_{i4} \end{pmatrix}\begin{pmatrix} -\eta_i \\ 1 \end{pmatrix}. \tag{9.8.5}$$

For later use we note that dividing by η_i^2 we obtain

$$\lambda\bar{\eta}_i' = (\bar{\eta}_i, 1)\begin{pmatrix} g_{i1} & g_{i2} \\ g_{i3} & g_{i4} \end{pmatrix}\begin{pmatrix} 1 \\ -\bar{\eta}_i \end{pmatrix}.$$

Since g_{i2} assumes unit values on general disks, we conclude that η_i cannot be bounded by $1 - \varepsilon$ on a general disk $D(z, 1^-)$. This means that the gauss norm of η_i is unity and hence η_i assumes only unit values on general disks. Now by equation (9.5.1)

$$\eta_i\bar{\eta}_i = 1$$

and hence $\bar{\eta}_i$ also assumes unit values on general disks. Thus there is no need to repeat the argument starting with $\bar{\eta}_i$. □

9.8.6 **Corollary.** *On a general residue class* $\tau_i - \bar{\tau}_i$ *is bounded by unity (valid also for* $i = j$*).*

Proof. Equation (9.5.1) shows that

$$\bar{\tau}_i - \tau_i = \frac{\det \mathbf{Y}_i}{u_i \bar{u}_i} \tag{9.8.6.1}$$

and by the lemma, u_i and $\bar{u}_i$ assume unit values on a general class. □

9.9

We now consider the situation of §9.4, but mark the split steps in sequence by subscripts, $s_0 < s_1 < s_2$, etc. Thus $\alpha^*_{s_i - 1}$ is split and we define s_{i+1} to be the minimal value of $j > s_i$ such that α^*_{j-1} is split. To distinguish between τ_j and $\bar{\tau}_j$ we use the symbol e to designate the variable operation which runs over two possibilities: either

$$(\tau, \bar{\tau}) \to (\tau, \bar{\tau})$$

or

$$(\tau, \bar{\tau}) \to (\bar{\tau}, \tau).$$

We *define* e_{s_i} to be the first operation (resp. the second operation) if $\alpha_{s_i - 1}$ is of type 1 (resp. type 2).

Throughout this section we assume that z lies in a general residue class.

Theorem. *For each* $r \in \mathbb{N}$, $r \geq 1$ *there exist* $\Lambda_0, \Lambda_1, \ldots, \Lambda_{r-1}$ *in* $W_z^{(0)}$ *such that*

$$e_{s_0} \tau_{s_0} = \sum_{j=0}^{r-1} \frac{1}{p^j} \Lambda_j^{(\sigma\phi)^{s_j - s_0}} + \frac{1}{p^r} (e_{s_r} \tau_{s_r})^{(\sigma\phi)^{s_r - s_0}}. \tag{9.9.1}$$

Proof. By (9.5.2)

$$e_{s_1} \tau_{s_0} - \frac{1}{p} (e_{s_1} \tau_{s_1})^{(\sigma\phi)^{s_1 - s_0}} \in W_z^{(0)}. \tag{9.9.2}$$

By Corollary 9.8.6

$$e_{s_0} \tau_{s_0} - e_{s_1} \tau_{s_0} \in W_z^{(0)}.$$

By adding these we obtain the assertion for $r = 1$. Hence for $r \in \mathbb{N}$, there exists $\Lambda_r \in W_z^{(0)}$ such that

$$e_{s_r}\tau_{s_r} = \Lambda_r + \frac{1}{p}(e_{s_{r+1}}\tau_{s_{r+1}})^{(\sigma\phi)^{s_{r+1}-s_r}}. \tag{9.9.3}$$

Assuming (9.9.1) for some fixed $r \geq 1$, the statement for $r + 1$ is obtained by substituting equation (9.9.3) in (9.9.1). This completes the proof of the theorem. □

9.9.4 **Notation** (cf. (11.2)).

$$\frac{1}{m!}\frac{d^m}{d\lambda^m} \equiv X_m^{(i)}\frac{d}{d\lambda} + Y_m^{(i)} \quad \mathrm{mod}(2.4.5.2)_{(a^{(i)},b^{(i)},c^{(i)})}$$

$$\equiv \bar{X}_m^{(i)}\frac{d}{d\lambda} + \bar{Y}_m^{(i)} \quad \mathrm{mod}(2.4.5.3)_{a^{(i)},b^{(i)},c^{(i)}},$$

where $X_m^{(i)}$, $Y_m^{(i)}$, $\bar{X}_m^{(i)}$, $\bar{Y}_m^{(i)}$ are elements of $\mathbb{Q}(\lambda)$ with poles only at $\lambda = 0, 1$.

9.10 **Corollary.**

$$\frac{(\varepsilon_{s_r}\tau'_{s_r})^{(\sigma\phi)^{s_r-s_0}}}{\varepsilon_{s_0}\tau'_{s_0}} = p^r(e_{s_0}X^{(s_0)}_{p^{s_r}-s_0}) \quad \mathrm{mod}\, pW_z^{(0)}.$$

Proof. Let $s = s_r - s_0$. By Lemma 11.3 below

$$p^r X_{p^s}^{(s_0)}\tau'_{s_0} = \sum_{i+j=p^s,\, i>0} \frac{p^r}{i}\frac{\tau'^{(i-1)}_{s_0}}{(i-1)!}\frac{u_{s_0}^{(j)}}{u_{s_0}j!} \tag{9.10.1}$$

and this remains valid if each X, τ, u is replaced by the corresponding barred letter. In general by (9.7.5):

$$\tau'_i = \frac{1}{u_i^2}\det \mathbf{Y}_i \frac{c^{(i)} - b^{(i)}}{\lambda}$$

and hence in view of Lemma 9.8, the ith term in the sum on the right side of (9.10.1) lies in $p^r i^{-1} W_z^{(0)}$ and hence lies in $pW_z^{(0)}$ if $p^r \nmid i$. Thus mod $pW_z^{(0)}$ the sum may be restricted to $i = p^r k$, $1 \leq k \leq p^{s-r}$. By (9.9.1)

$$e_{s_0}\frac{1}{k}\frac{\tau_{s_0}^{(kp^r)}}{(p^r k - 1)!} = \sum_{j=0}^{r-1}\frac{D^{kp^r}\Lambda_j^{(\sigma\phi)^{s_j-s_0}}}{(p^r k)!}p^{r-j} + \frac{D^{kp^r}(e_{s_r}\tau_{s_r})^{(\sigma\phi)^s}}{p^r k(kp^r - 1)!}. \tag{9.10.2}$$

Trivially, each of the terms indexed by j in the sum on the right lies in $pW_z^{(0)}$. Now to analyze the last term on the right side of (9.10.2) write

$$\text{(9.10.3)} \qquad \frac{1}{(kp^r)!} D^{kp^r} \circ (\sigma\phi)^s = \frac{1}{kp^r} \frac{D^{kp^r-1}}{(kp^r-1)!} \frac{1}{\lambda} p^s(\sigma\phi)^s \lambda D.$$

Since $\lambda D\tau_{s_r} \in W_z^{(0)}$, this shows that the last term on the right side of (9.10.2) lies in $(p^s/kp^r)W_z^{(0)}$ and hence in $pW_z^{(0)}$ if $\operatorname{ord}(kp^r) < s$. Since $k \le p^{s-r}$, we conclude that the left side of (9.10.2) lies in $pW_z^{(0)}$ unless $kp^r = p^s$. This means that in the sum on the right side of (9.10.1), if we apply e_{s_0} then each term lies in $pW_z^{(0)}$ except for the term indexed by $j = 0$, $i = p^s$. We then obtain

$$\text{(9.10.4)} \qquad \begin{aligned} p^r(e_{s_0} X_{p^s}^{(s_0)}) e_{s_0} \tau'_{s_0} &\equiv \frac{p^r}{p^s!} D^{p^s} e_{s_0} \tau_{s_0} \\ &\equiv \frac{1}{p^s!} D^{p^s} (e_{s_r} \tau_{s_r})^{(\sigma\phi)^s} \quad \operatorname{mod} pW_z^{(0)} \\ &\equiv \frac{D^{p^s-1}}{(p^s-1)!} \frac{1}{\lambda} (\lambda D e_{s_r} \tau_{s_r})^{(\sigma\phi)^s}. \end{aligned}$$

We now apply Lemma 5.14 with $\mu = 0$, to this last expression and obtain

$$p^r(e_{s_0} X_{p^s}^{(s_0)})(e_{s_0} \tau'_{s_0}) \equiv \phi^s \frac{1}{\lambda} \psi^s \lambda \frac{1}{\lambda} (\lambda D e_{s_r} \tau_{s_r})^{(\sigma\phi)^s} \quad \operatorname{mod} pW_z^{(0)},$$

which is equivalent to Corollary 9.10. □

9.10.5 **Theorem.**

$$p^r e_{s_0} X_{p^{s_r}-s_0}^{(s_0)} \equiv \prod_{j=0}^{r-1} (p e_{s_j} X_{p^{s_{j+1}}-s_j}^{(s_j)})^{\phi^{s_j-s_0}} \quad \operatorname{mod} p.$$

Proof. By Corollary 9.10,

$$\begin{aligned} p^r e_{s_0} X_{p^{s_r}-s_0}^{(s_0)} &\equiv \frac{(e_{s_r} \tau'_{s_r})^{(\sigma\phi)^{s_r-s_0}}}{e_{s_0} \tau'_{s_0}} \\ &= \prod_{j=0}^{r-1} \left(\frac{e_{s_{j+1}} \tau'^{(\sigma\phi)^{s_{j+1}-s_j}}_{s_{j+1}}}{e_{s_j} \tau'_{s_j}} \right)^{(\sigma\phi)^{s_j-s_0}} \\ &= \prod_{j=0}^{r-1} (p e_{s_j} X_{p^{s_{j+1}}-s_j}^{(s_j)})^{\phi^{s_j-s_0}} \end{aligned}$$

as asserted. □

9.10.6 The point of the previous theorem is that equation $(2.4.5.1)_{a,b,c}$ has a bounded solution on a residue class, $D(z, 1^-)$ if and only if that class is not a zero mod p of the factors on the right side of (9.10.5). (This follows by the method of Robba [31].)

9.11 Solutions mod powers of p

We say that Y is an *invertible solution matrix* mod p^m of $(2.4.5.1)_{a,b,c}$ if $Y \in \mathrm{GL}(2, W_z^{(0)})$ and Y satisfies the equation mod $p^m W_z^{(0)}$. It follows from Lemma 7.2 that $\mathbf{B}_0$ is an invertible solution matrix mod p of $(2.4.5.1)_{a,b,c}$ if α_0^* is nonsplit, i.e., if μ_c lies between μ_a and μ_b.

9.11.1 **Proposition.** *There is no invertible solution matrix* mod p *if μ_c fails to be between μ_a and μ_b.*

Proof. For solutions modulo p we may restrict our attention to polynomial solutions mod p. These are given by the classical type recursion formulas starting from a mod p solution of the indicial polynomial at zero. Thus there can be two independent solutions mod p only if the following conditions hold

(i) The minimal positive integral solution for s of

$$(a)_s(b)_s \equiv 0 \mod p$$

is less than the corresponding solution of

$$(c)_s \equiv 0 \mod p.$$

(ii) The minimal positive integral solution of

$$(a+1-c)_s(b+1-c)_s \equiv 0 \mod p$$

is less than that of

$$(2-c)_s \equiv 0 \mod p.$$

These conditions imply that μ_c lies between μ_a and μ_b. □

9.11.2 **Lemma.** *Under the hypothesis of Lemma 9.5, the equation* $(2.4.5.1)_{(a^{(i)},b^{(i)},c^{(i)})}$ *has an invertible solution matrix* mod p^{j-i-1} *but no invertible solution matrix* mod p^{j-i}.

Proof. Let $m = j - i$. We follow the argument of Lemma 7.2. Let

$$K = \mathbf{B}_{j-2}^{\phi^{m-2}} \mathbf{B}_{j-3}^{\phi^{m-3}} \cdots \mathbf{B}_i,$$

then in the notation of (9.4.3.0)

$$\mathbf{Y}_i = \mathbf{Y}_{j-1}^{(\sigma\phi)^{m-1}} K \tag{9.11.2.1}$$

and so in the notation of (7.1.3)

$$\begin{aligned}
\lambda \mathbf{Y}_{j-1}^{(\sigma\phi)^{m-1}} K M_{f_i} &= \mathbf{Y}_i \lambda M_{f_i} = \lambda \frac{d}{d\lambda} \mathbf{Y}_i \\
&= \lambda \frac{d}{d\lambda} (\mathbf{Y}_{j-1}^{(\sigma\phi)^{m-1}} K) \\
&= \mathbf{Y}_{j-1}^{(\sigma\phi)^{m-1}} \lambda \frac{dK}{d\lambda} + p^{m-1} \left(\lambda \frac{d}{d\lambda} Y_{j-1} \right)^{(\sigma\phi)^{m-1}} K \\
&= \mathbf{Y}_{j-1}^{(\sigma\phi)^{m-1}} \left[\lambda \frac{dK}{d\lambda} + p^{m-1} (M_{f_{j-1}} \lambda)^{\phi^{m-1}} K \right]
\end{aligned}$$

and so

$$K M_{f_i} = \frac{dK}{d\lambda} + p^{m-1} \frac{1}{\lambda} (\lambda M_{f_{j-1}})^{\phi^{m-1}} K. \tag{9.11.2.2}$$

But K is invertible in $\mathbf{Q}_0'$ since $\mathbf{B}_i, \ldots, \mathbf{B}_{j-2}$ have this property. This shows that K is the required invertible solution matrix mod p^{m-1}.

We now show that $(2.4.5.1)_{(a^{(i)}, b^{(i)}, c^{(i)})}$ has no invertible solution matrix mod p^{j-i}. Suppose otherwise. Let NK be such a solution matrix then

$$\frac{d(NK)}{d\lambda} \equiv NKM_{f_i} \mod p^m W_z^{(0)}$$

and so

$$N'K \equiv N(KM_{f_i} - K') \mod p^m W_z^{(0)}. \tag{9.11.2.3}$$

But then by (9.11.2.2)

$$N' \equiv 0 \mod p^{m-1} W_z^{(0)}. \tag{9.11.2.4}$$

Thus by Lemma 5.9

$$N = N_0^{\phi^{m-1}} + p N_1^{\phi^{m-2}} + \cdots + p^{m-1} N_{m-1},$$

where each N_s is a two by two matrix with coefficients in $W_z^{(0)}$. Since N is invertible in $W_z^{(0)}$, the same must be true of N_0, and so

$$N = V_0^{\phi^{m-1}} (I + p V_1^{\phi^{m-2}} + \cdots + p^{m-1} V_{m-1}), \tag{9.11.2.5}$$

where $V_0 \in \mathrm{GL}(2, W_z^{(0)})$ and the V_s have coefficients in $W_z^{(0)}$. Thus with $\delta = \lambda(d/d\lambda)$,

(9.11.2.6)

$$\frac{1}{p^{m-1}}\,\delta N \equiv (\delta V_0)^{\phi^{m-1}} + V_0^{\phi^{m-1}}[(\delta V_1)^{\phi^{m-2}} + \cdots + \delta V_{m-1}] \mod pW_z^{(0)}.$$

But comparing (9.11.2.3) with (9.11.2.2) gives

$$\delta N \equiv Np^{m-1}(\lambda M_{f_{j-1}})^{\phi^{m-1}} \mod p^m W_z^{(0)},$$

$$\frac{1}{p^{m-1}}\,\delta N \equiv \lambda^{\phi^{m-1}} N(M_{f_{j-1}})^{\phi^{m-1}} \mod pW_z^{(0)},$$

and so by (9.11.2.5)

(9.11.2.7) $$\frac{1}{p^{m-1}}\,\delta N \equiv \lambda^{\phi^{m-1}}(V_0 M_{f_{j-1}})^{\phi^{m-1}} \mod pW_z^{(0)}.$$

Comparison with (9.11.2.6) gives

(9.11.2.8)

$$(\lambda V_0 M_{f_{j-1}})^{\phi^{m-1}} \equiv (\delta V_0)^{\phi^{m-1}} + V_0^{\phi^{m-1}} \sum_{s=1}^{m-1} (\delta V_s)^{\phi^{m-1-s}} \mod pW_z^{(0)}.$$

We now apply ψ^{m-1} and obtain

(9.11.2.9) $$\lambda V_0 M_{f_{j-1}} \equiv \delta V_0 \mod pW_z^{(0)},$$

since $V_1, \ldots, V_s$ have coefficients in $W_z^{(0)}$. This shows that $(2.4.5.1)_{a(j-1), b(j-1), c(j-1)}$ has an invertible solution matrix mod p, contradicting Proposition 9.11.1 and the hypothesis that α_{j-1}^* is split. This completes the proof of the lemma. □

9.12

In this paragraph we find a new relation between τ_i and $\bar{\tau}_i$.

Lemma. *Under the hypothesis of Lemma* 9.5 (*i.e., in*

$$\mathbf{K}_{f_i} \xleftarrow{\alpha_i^*} \mathbf{K}_{f_{i+1}} \leftarrow \cdots \leftarrow \mathbf{K}_{f_{j-1}} \xleftarrow{\alpha_{j-1}^*} \mathbf{K}_{f_j}$$

the indicated sequence each map is unsplit except for α_{j-1}^{*}), *let* $m = j - i$, *then there exists a constant,* γ_i (*independent of* λ *and of* z) *such that*

$$\gamma_i \cdot \frac{\bar{\tau}_i'}{\tau_i'} \equiv \frac{pX_{p^m}^{(i)}}{p\bar{X}_{p^m}^{(i)}} \mod pW_z^{(0)}. \tag{9.12.1}$$

Proof. It follows from Lemma 9.11.2 that the equation

$$\frac{d}{d\lambda}(u_1, u_2) = (u_1, u_2)M_{f_i} \tag{9.12.2}$$

has an invertible solution matrix mod p^{m-1} but not mod p^m. Hence by Lemma 11.6 there exist $C, \bar{C} \in W_z^{(0)}$, neither in $pW_z^{(0)}$ such that

$$\begin{aligned} pX_{p^m}^{(i)} &\equiv \frac{C^{\phi^m}}{\tau_i'} \\ & \qquad\qquad \mod pW_z^{(0)}, \\ p\bar{X}_{p^m}^{(i)} &\equiv \frac{\bar{C}^{\phi^m}}{\bar{\tau}_i'} \end{aligned} \tag{9.12.3}$$

and so letting $C_1 = C/\bar{C}(\in W_z'^{(0)})$ we have

$$\frac{\bar{\tau}_i'}{\tau_i'} \equiv \frac{pX_{p^m}^{(i)}}{p\bar{X}_{p^m}^{(i)}} C_1^{\phi^m} \mod pW_z'^{(0)}. \tag{9.12.4}$$

From (9.5.1) we have

$$\mathbf{Y}_i = \begin{pmatrix} u_i & \bar{u}_i \\ u_i\tau_i & \bar{u}_i\bar{\tau}_i \end{pmatrix}. \tag{9.12.5}$$

Writing

$$\begin{pmatrix} g_1 & g_2 \\ g_3 & g_4 \end{pmatrix}$$

for λM_{f_i}, equation $(2.4.5.1)_{(a^{(i)}, b^{(i)}, c^{(i)})}$ gives

$$\lambda \begin{pmatrix} u_i' & \bar{u}_i' \\ (u_i\tau_i)' & (\bar{u}_i\bar{\tau}_i)' \end{pmatrix} = \begin{pmatrix} u_i & \bar{u}_i \\ u_i\tau_i & \bar{u}_i\bar{\tau}_i \end{pmatrix} M_{f_i}. \tag{9.12.6}$$

Multiplying on the left by $(-\tau_i, 1)$ gives

$$\lambda(u_i\tau_i', *) = (0, \bar{u}_i(\bar{\tau}_i - \tau_i))M_{f_i} \tag{9.12.7}$$

and multiplying on the left by $(-\bar{\tau}_i, 1)$ gives

$$\lambda(*, \bar{u}_i \bar{\tau}_i') = (u_i(\tau_i - \bar{\tau}_i), 0) M_{f_i}. \tag{9.12.8}$$

Comparing first components in the first relation gives

$$\lambda u_i \tau_i' = \bar{u}_i(\bar{\tau}_i - \tau_i) g_{i3}. \tag{9.12.9}$$

Comparing the second components in the second relation gives

$$\lambda \bar{u}_i \bar{\tau}_i' = u_i(\tau_i - \bar{\tau}_i) g_{i2}. \tag{9.12.10}$$

(These relations are similar to (9.8.6.1) and (9.7.5)). Eliminating $\bar{\tau}_i - \tau_i$ gives

$$\frac{\bar{\tau}_i'}{\tau_i'} = -\frac{g_{i2}}{g_{i3}}\left(\frac{u_i}{\bar{u}_i}\right)^2 = -\frac{g_{i2}}{g_{i3}}\bar{\eta}_i^2 = -\frac{g_{i2}}{g_{i3}}\frac{1}{\eta_i^2}. \tag{9.12.11}$$

The analytic continuation of η_i shows that modulo $pW_z'^{(0)}$ it may be represented by an element ξ of $\mathbb{F}_p(\lambda)$. From equation (9.8.5) we deduce that for suitable $\alpha, \beta, \gamma, \delta \in \mathbb{F}_p$, $\gamma \neq 0$ (the key point here is that by Lemma 8.8, $c^{(i)} - a^{(i)} \not\equiv 0 \not\equiv c^{(i)} - b^{(i)} \bmod p$)

$$\xi^2 \equiv \alpha\xi + \beta\lambda\xi' + \gamma\frac{\lambda}{\lambda - 1} + \delta\lambda\frac{\xi}{\lambda - 1} \quad \bmod p. \tag{9.12.11.1}$$

This relation shows that ξ has only simple poles, no pole at infinity, and dividing through by ξ^2, we obtain a similar Riccati equation for $1/\xi$ which shows that ξ has only simple zeros. We note too that ξ cannot have a pole at $\lambda = 0$ nor can it have a zero at $\lambda = 1$. Using $\sim$ to designate equality mod p except for a nonzero constant factor,

$$\xi \sim \frac{\lambda^r}{(\lambda - 1)^s}\frac{\prod(\lambda - z_i)}{\prod(\lambda - w_j)},$$

where $r, s = 0, 1$; $z_i \neq 0, 1$, $w_j \neq 0, 1$, the z_i, w_j are distinct. Since g_{i2}/g_{i3} is of the form $\lambda/\lambda - 1$, we deduce from (9.12.11) that

$$\frac{\bar{\tau}_i'}{\tau_i'} \sim \frac{\lambda}{\lambda - 1}\left(\frac{(\lambda - 1)^s}{\lambda^r}\frac{\prod(\lambda - w_j)}{\prod(\lambda - z_i)}\right)^2 \quad \bmod p. \tag{9.12.12}$$

On the other hand we know by means of (11.6.3.1) together with a characteristic zero form of Proposition 10.3 that $pX_{p^m}^{(i)}/p\bar{X}_{p^m}^{(i)}$ is a ratio of polynomials each of degree not greater than $p^m - 1$ and when reduced modulo p the zeros of these polynomials distinct from 0, 1 have order congruent to 2 modulo p. If now z_0 is a pole of C_1 mod p then it is a zero of order at least p^m of

$$\frac{\tau_i'}{\bar{\tau}_i'} \cdot \frac{pX_{pm}^{(i)}}{p\bar{X}_{pm}^{(i)}} \equiv \frac{1}{C_1^{\phi m}}, \tag{9.12.13}$$

but the first factor has no zero of order greater than 2 and so the numerator of $pX_{pm}^{(i)}$ mod p must have z_0 as a zero of order at least $p^m - 2$. By the bound on the degree of this numerator we conclude that there can be no more than one pole of C_1. Likewise C_1 can have no more than one zero. If z_0 is a pole of C_1 mod p, $z_0 \neq 0, 1$ then it is a zero of the first factor in the right side of (9.12.13) of order $\pm 2, 0$ and hence in fact it must be a zero of the first factor of order 2 and of the numerator of $pX_{pm}^{(i)}$ of order $p^m - 2$. This implies that

$$p^m - 2 \equiv 2 \mod p, \tag{9.12.14}$$

which is impossible. Thus the poles of C_1 are either 0 or 1 and likewise for the zeros.

Our object is to exclude the possibility that 0 or 1 are zeros or poles of C_1. We first show that if 0 is a pole then 1 is a zero.

Since $\lambda = 0$ is a zero of the first factor of (9.12.13) of order ± 1, it follows that if $\lambda = 0$ is a pole of C_1 then 0 must be a zero of the first factor of order $+1$ and that the numerator of $pX_{pm}^{(i)}$ mod p is $\lambda^{p^m - 1}$ and that λ is not a factor of the numerator of $p\bar{X}_{pm}^{(i)}$. It follows that the factors $\lambda - w_j$ in (9.12.12) do not occur and that $r = 1$. But the degree of $\bar{\tau}_i'/\tau_i'$ mod p as element of $\mathbb{F}_p(\lambda)$ is zero (by (9.12.11) since by (9.12.11.1), ξ is of degree zero) and so

$$s = r + (\text{number of } z_i \text{ in (9.12.12)}). \tag{9.12.15}$$

Since $s \leq 1$, we conclude that $s = 1$ and $\bar{\tau}_i'/\tau_i' \sim (\lambda - 1)/\lambda$. By inspection of (9.12.13) we conclude that the numerator of $p\bar{X}_{pm}^{(i)}$ is $(\lambda - 1)^{p^m - 1}$ mod p which shows that $C_1 \sim (\lambda - 1)/\lambda$ has a zero at 1 as asserted.

A similar analysis for the case in which $\lambda = 0$ is a zero of C_1 shows that $\lambda = 1$ is a pole. We have thus reduced the discussion to two cases.

Case I. numerator of $pX_{pm}^{(i)} \equiv \lambda^{p^m - 1}$,

numerator of $p\bar{X}_{pm}^{(i)} \equiv (\lambda - 1)^{p^m - 1}$.

Case II. numerator of $pX_{pm}^{(i)} \equiv (\lambda - 1)^{p^m - 1}$,

numerator of $p\bar{X}_{pm}^{(i)} \equiv \lambda^{p^m - 1}$.

Now we know that the numerator of $pX_{pm}^{(i)}$ has zero at 0, 1 of orders congruent modulo p to the difference of exponents of the equation for

$F(a^{(i)}, b^{(i)}, 1 + c^{(i)}, \lambda)$ at 0, 1 respectively. For $p\bar{X}^{(i)}_{p^m}$ we have the same result replacing $F(a^{(i)}, b^{(i)}, 1 + c^{(i)}, \lambda)$ by $F(a^{(i)}, b^{(i)}, c^{(i)}, \lambda)$. The numerator is of degree congruent mod p to -1 plus the exponent difference at ∞. Thus in Case I

(9.12.16)

$$1 + c^{(i)} \equiv a^{(i)} + b^{(i)} \quad \text{(order of 1 as zero of numerator of } pX^{(i)}_{p^m}\text{)},$$

$$c^{(i)} \equiv 1 \quad \text{(order of 0 as zero of numerator of } p\bar{X}^{(i)}_{p^m}\text{)},$$

$$\pm(a^{(i)} - b^{(i)}) - 1 \equiv -1 \quad \text{(degree of numerator of } pX^{(i)}_{p^m}\text{)}.$$

This shows that $c^{(i)} \equiv a^{(i)} \equiv b^{(i)} \equiv 1 \bmod p$ and hence $c^{(i)} = a^{(i)} = 1$ which is impossible.

For Case II we have

$$c^{(i)} \equiv 0 \quad \text{(order of 0 as zero of numerator of } pX^{(i)}_{p^m}\text{)},$$

$$\text{(9.12.17)} \quad c^{(i)} \equiv a^{(i)} + b^{(i)} \quad \text{(order of 1 as zero of numerator of } p\bar{X}^{(i)}_{p^m}\text{)},$$

$$a^{(i)} \equiv b^{(i)} \quad \text{(degree of numerator of } pX^{(i)}_{p^m}\text{)}.$$

This is impossible. This shows that γ_i defined modulo p by (9.12.1) is indeed independent of λ. Independence of z follows from (9.12.11). This completes the proof of the lemma. □

9.12.18 *Remark.* From equation (9.12.11.1) we may deduce that at ∞, the rational function ξ is either 1 or $(c^{(i)} - a^{(i)})/(c^{(i)} - b^{(i)})$. Equation (9.12.11) then shows that $\bar{\tau}'_i/\tau'_i \bmod p$ assumes the values $((c^{(i)} - a^{(i)})/c^{(i)} - b^{(i)})^{\pm 1}$ at infinity.

9.12.19 *Explanation.* $(\lambda(1 - \lambda))^{p^m - 1} \cdot X^{(i)}_{p^m}$ is a polynomial of degree bounded by $p^m - 1$ which in the proof of Lemma 9.12 is referred to as the *numerator* of $X^{(i)}_{p^m}$.

Chapter 10

Nilpotent Second-Order Linear Differential Equations with Fuchsian Singularities

This chapter is self-contained with notation independent of the remainder of this work.

Let K be a field of characteristic $p \neq 2$ say algebraically closed. Let L be a linear differential operator

(10.00) $$L = D^2 - aD - b \in K(x)[D],$$

with $D = d/dx$. Let $\{\gamma_1, \ldots, \gamma_m, \gamma_\infty = \infty\} = T$ be the set of singularities of L and let

(10.01) $$f(x) = \prod_{i=1}^{m} (x - \gamma_i).$$

We assume

(10.02) All the singularities of L are fuchsian.

(10.03) The exponents of L at each singularity lie in $\mathbb{F}_p$.

(10.04) L is nilpotent but does not have two solutions in $K(x)$ linearly independent over $K(x^p)$.

By "nilpotent" we mean that L has a nontrivial solution in $K(x)$ and that the equation for the wronskian,

(10.04.1) $$Dw = wa$$

has a nontrivial solution in $K(x)$. We may assume that the zeros and poles of w lie in T.

We use the word "exponent" to refer to a root of the indicial polynomial. For $i = 1, \ldots, m, \infty$ let e_i, e_i' be the exponents at γ_i.

We choose a solution u of L in $K[x]$, unique up to factor in K, by the condition that no zero of u is of order greater than $p - 1$.

We write

(10.05) $$u = g(x)\prod_{i=1}^{m}(x-\gamma_i)^{\tilde{e}_i},$$

where

$$g \in K[x], \qquad (g, f) = 1,$$

$$\tilde{e}_i \in [0, p-1].$$

We define $\tilde{e}_\infty$ by the condition that $\tilde{e}_\infty \in [0, p-1]$,

(10.06) $$\tilde{e}_\infty \equiv -\deg u \mod p.$$

Clearly the $\tilde{e}_i$ represent exponents of L. For all $s \in \mathbb{N}$ we write

(10.07) $$D^s = a_s D + b_s \mod K(x)[D]L,$$

with $a_s, b_s \in K(x)$. It is known that

$$a_s = 0 = b_s, \qquad \forall s \geq 2p.$$

An *ad hoc* proof is given in §10.4.5 below.

Having defined $\tilde{e}_i$ ($i = 1, \ldots, m, \infty$), we define e_i ($\in \mathbb{F}_p$) to be the class of $\tilde{e}_i$ and we define e_i' to be the other exponent at γ_i (of course we may have $e_i = e_i'$). Thus we have uniquely defined the difference, $e_i - e_i'$, of exponents at γ_i. We define $t_i \in [0, p-1]$

(10.08) $$t_i \bmod p = e_i - e_i', \qquad i = 1, \ldots, m, \infty.$$

The object of this section is to prove

10.1 **Lemma.**

(10.1.1) $$(p-1)(m-1) = 2\deg g + (t_1 + \cdots + t_m + t_\infty) + pt,$$

where $t \in \mathbb{N}$, $t \geq 0$.

10.2 **Lemma.**

(10.2.1) $$f(x)^{p-1}a_p = g(x)^2\prod_{i=1}^{m}(x-\gamma_i)^{t_i}\cdot\theta(x^p),$$

where

$\theta \in K[x]$, *g is given by* (10.5), *i.e.*,

$g \in K[x]$,

g is prime to f,

g has only simple zeros.

We commence our treatment with an elementary proposition.

10.3 **Proposition.** *For each* $s \in \mathbb{N}$, $a_s f(x)^{s-1} \in K[x]$.

$$\deg a_s f(x)^{s-1} \leq (s-1)(m-1). \tag{10.3.1}$$

Proof. By differentiating (10.07) and using L to reduce the D^2 on the right-hand side we obtain the recursion formula

$$\begin{pmatrix} a_{s+1} \\ b_{s+1} \end{pmatrix} = \begin{pmatrix} a_s' \\ b_s' \end{pmatrix} + \begin{pmatrix} a & 1 \\ b & 0 \end{pmatrix} \begin{pmatrix} a_s \\ b_s \end{pmatrix}. \tag{10.3.2}$$

On the other hand,

$$\begin{pmatrix} a_0 \\ b_0 \end{pmatrix} = \begin{pmatrix} 0 \\ 1 \end{pmatrix}, \quad \begin{pmatrix} a_1 \\ b_1 \end{pmatrix} = \begin{pmatrix} 1 \\ 0 \end{pmatrix}, \quad \begin{pmatrix} a_2 \\ b_2 \end{pmatrix} = \begin{pmatrix} a \\ b \end{pmatrix}. \tag{10.3.3}$$

By hypothesis for $1 \leq i \leq m$, $a_2 = a$ (resp. $b_2 = b$) has a pole at γ_i of order not greater than one (resp. two). By induction on s and the recursion formula we show that for $s \geq 1$

(10.3.4) a_s (resp. b_s) has a pole at γ_i of order not greater than $s - 1$ (resp. s).

This shows that $a_s f(x)^{s-1}$ is a polynomial.

The condition that L is fuchsian everywhere implies that we may write L in the form

$$L = D^2 + \sum_{i=1}^{m} \frac{A_i}{x - \gamma_i} D + \sum_{i=1}^{m} \left(\frac{B_i}{x - \gamma_i} + \frac{C_i}{(x - \gamma_i)^2} \right), \tag{10.3.5}$$

where $A_i, B_i, C_i \in K$ for $i = 1, 2, \ldots, m$. The condition at infinity implies that

$$\sum_{i=1}^{m} B_i = 0. \tag{10.3.6}$$

Thus deg a_2 (resp. deg b_2) ≤ -1 (resp. -2) and by (10.3.2) and induction we show that

$$\deg a_s \text{ (resp. } \deg b_s) \leq -(s-1) \text{ (resp. } -s). \tag{10.3.7}$$

This completes the proof of the proposition. □

10.4

We now commence the proofs of the lemmas. By hypothesis the $K(x^p)$-space of solutions of L in $K(x)$ has dimension one but in a suitable differential extension field F, the F^p space of solutions of L has dimension two. More explicitly, we choose F so as to contain, τ, a solution of

$$\tau' = \frac{w}{u^2} \tag{10.4.1}$$

and then by a well-known calculation using (10.04.1):

$$L(u\tau) = \tau L(u) + \frac{w}{u}\left(\tau' \frac{u^2}{w}\right)' = 0$$

while the wronskian

$$\begin{vmatrix} u\tau & (u\tau)' \\ u & u' \end{vmatrix} = -w, \tag{10.4.2}$$

which shows that u, $u\tau$ are linearly independent over the kernel of D in F.

We now apply (10.07) and conclude that

$$\begin{aligned} (\tau u)^{(s)} &= a_s(\tau u)' + b_s(\tau u), \\ u^{(s)} &= a_s u' + b_s u. \end{aligned} \tag{10.4.3}$$

Eliminating b_s we obtain

$$a_s = \sum_{\substack{i+j=s \\ i\geq 1}} \frac{\tau'^{(i-1)}}{\tau'} \frac{u^{(j)}}{u} \binom{s}{i}, \tag{10.4.4}$$

a formula involving u and τ' but not τ. We observe that this formula is independent of the characteristic.

10.4.5 *Remark.* Since u and τ' lie in $K(x)$ they are annihilated by D^p. For $s \geq 2p$ either $i - 1$ or j on the right side of (10.4.4) exceeds $p - 1$ which shows again that $a_s = 0$ for $s \geq 2p$.

In particular for $s = p$, the above formula gives

$$a_p = \frac{u^2}{w} D^{p-1} \frac{w}{u^2}. \tag{10.4.6}$$

Now $a_p \neq 0$ as otherwise w/u^2 would lie in the kernel of D^{p-1} in $K(x)$, i.e.,

$$\frac{w}{u^2} \in K(x^p) + K(x^p)x + \cdots + K(x^p)x^{p-2},$$

which would show that (10.4.1) has a solution τ in $K(x)$ contrary to hypothesis concerning the dimensionality of the kernel of L in $K(x)$ (as $K(x^p)$ space).

By the same argument since $1, x, \ldots, x^{p-1}$ is basis of $K(x)$ as $K(x^p)$ space we conclude that D^{p-1} maps $K(x)$ into $K(x^p)$. Hence

$$a_p \in \frac{u^2}{w} K(x^p). \tag{10.4.7}$$

Putting

$$Q_p = a_p f(x)^{p-1},$$

we have

$$Q_p \in \frac{u^2}{w} \frac{1}{f(x)} K(x^p). \tag{10.4.8}$$

We have defined g as the factor of u prime to $f(x)$. If x_0 is a zero of g then the indicial polynomial of L at x_0 has 0, 1 (mod p) as zeros and by definition u has no zero of order greater than $p - 1$. This shows that the zeros of g are simple.

10.5

We continue our proof of the lemmas. We will show

$$\frac{u^2}{w} \frac{1}{f(x)} \in g(x)^2 \prod_{i=1}^{m} (x - \gamma_i)^{t_i} \cdot K(x^p). \tag{10.5.1}$$

With this in mind we use (10.3.5) to deduce

(10.5.2)
$$e_i + e_i' = 1 - A_i, \qquad i = 1, \ldots, m,$$
$$e_\infty + e_\infty' = \sum_{i=1}^{m} A_i - 1,$$

while

(10.5.3)
$$w \in \prod_{i=1}^{m} (x - \gamma_i)^{-\bar{A}_i} K(x^p),$$

where $\bar{A}_i$ is a representative in $\mathbb{N}$ of A_i ($1 \le i \le m$). Thus the order of γ_i as zero of the left side of (10.5.1) is congruent mod p to $2e_i + A_i - 1 = e_i - e_i' \equiv t_i$. This together with our discussion of g, the factor of u prime to f, concludes the demonstration of (10.5.1).

We now estimate the degree of the left side of (10.5.1). By hypothesis $\deg u = -\tilde{e}_\infty \equiv -e_\infty$ and so

$$\deg \frac{u^2}{wf(x)} \equiv -2e_\infty - m + \sum_{i=1}^{m} A_i.$$

By (10.5.2) this is the same as $-m + 1 - t_\infty$. Thus from (10.5.1),

(10.5.4)
$$m + 2 \deg g + t_\infty + \sum_{i=1}^{m} t_i \equiv 1 \mod p.$$

By (10.4.8) and (10.5.1) we obtain (10.2.1) with $\theta \in K(x)$. We assert that θ is a polynomial. Indeed Q_p is a polynomial and $Q_p/\theta(x^p)$ is by (10.2.1) a polynomial with zeros of order bounded by $p - 1$. Thus θ must be a polynomial. This completes the proof of Lemma 10.2. We continue with the proof of Lemma 10.1. By Proposition 10.3,

(10.5.5)
$$(p - 1)(m - 1) \ge \text{degree } Q_p = 2 \text{ degree } g + \sum_{i=1}^{m} t_i + p \deg \theta.$$

Let

(10.5.6)
$$\rho = (p - 1)(m - 1) - 2 \text{ degree } g - \sum_{i=1}^{m} t_i.$$

Then

(10.5.7)
$$\rho \ge p \text{ degree } \theta \ge 0.$$

On the other hand, by (10.5.4) and (10.5.6),

$$\rho \equiv t_\infty \mod p.$$

And hence, by (10.5.7),

$$\rho = t_\infty + pt$$

for some $t \geq 0$. Substitution in (10.5.6) completes the proof of Lemma 10.1 □

10.6 *Remark.* We view the sum of the t_i as the analogue of the sum of the angles of the image of the upper half plane under a ratio of solutions of L if K were say the reals and the γ_i were all real.

10.7

In general we are given L but not u and so there are two choices of t_i for each i. Thus applying Lemma 10.1 there are 2^{m+1} choices for $(t_1, \ldots, t_m, t_\infty)$ and t is not known.

Corollary. *If* $m = 2$ *then under hypotheses* (10.02)–(10.04), *we have* $t = 0$. *In equation* (10.2.1) *we may put* $\theta = 1$ *and there is just one possible choice for* t_0, t_1, t_∞. *Equation* (10.1.1) *takes the form*

$$p - 1 = 2 \deg g + t_0 + t_1 + t_\infty.$$

Proof. It is clear from (10.1.1) that $t = 0$. Since $p \neq 2$, it follows that

$$\text{(10.7.1)} \qquad \begin{aligned} t_0 + t_1 + t_\infty &\equiv 0 \mod 2, \\ p - 1 &\geq t_0 + t_1 + t_\infty. \end{aligned}$$

Now each t_i is fixed by L up to the transformation

$$t_i \to p - t_i.$$

The condition of parity shows that such a transformation, if applied at all, must be applied to two of the t_i, say to t_0, t_1 and we would then have

$$\text{(10.7.2)} \qquad p - 1 \geq p - t_0 + p - t_1 + t_\infty.$$

This is inconsistent with (10.7.1) as the sum would give

$$2(p-1) \geq 2p + 2t_\infty \geq 2p. \quad \square$$

Remark. The degree of g is at most $(p-1)/2$ and this occurs precisely when $t_0 = t_1 = t_\infty = 0$, i.e., $(a, b, c) = (\frac{1}{2}, \frac{1}{2}, 1)$.

Chapter 11

Second-Order Linear Differential Equations Modulo Powers of p

None of the symbols concerning equation (2.4.5.1) will be used here. However, the notation of Chapter 5 will be of use.

We consider a second-order linear differential operator

$$L = D^2 + fD + g, \tag{11.0.1}$$

with coefficients f, g analytic elements on $\Delta_0(r, 1)$ for some $r \in (0, 1)$. Let K be a field of definition of f and g; i.e., K is a subfield of Ω such that f and g are uniform limits on $\Delta_0(r, 1)$ of elements of $K(x)$ having no poles in the annulus. We may with no loss in generality assume the existence of $t \in \Omega$, $|t| = 1$ such that the image of t in the residue class field of Ω is transcendental over the residue class field of K. We shall be concerned with the action of L on elements of W_t, the ring of functions analytic on the generic disk, $D(t, 1^-)$. We observe that by virtue of recent work of Robba [32, Exposé 5], we could assume merely that f and g are elements of W'_0.

We will assume in the following that the value group of K is discrete. This would mean that if $u \in K(t)[[x - t]]$ lies in W_t then the reciprocal, $1/u$, is analytic on $\Delta_t(r, 1)$ for some r and hence lies in W'_t.

We assume that

$$\dim_\Omega W_t \cap \operatorname{Ker} L = 1, \tag{11.0.2}$$

$$\dim_\Omega W_t \cap \operatorname{Ker}(D + f) = 1. \tag{11.0.3}$$

Thus we have essentially unique u, $w \in W_t$ such that

$$\begin{aligned} Lu &= 0, \\ \frac{w'}{w} &= -f, \end{aligned} \tag{11.0.4}$$

and we normalize by the condition

$$|u(t)| = 1 = |w(t)|. \tag{11.0.5}$$

It is known (e.g., Theorem 2.6, Robba [30] or §11.10 below) that $\eta = u'/u$ is an analytic element on $D(t, 1^-)$ and hence u has no zeros on the disk. Thus $|u(x)| = 1$ everywhere on $D(t, 1^-)$.

We define τ by the differential equation

$$\tau' = \frac{w}{u^2} \tag{11.0.5.1}$$

but shall never use τ itself.

11.0.6 **Definition** (cf. §9.11). Let $s \in \mathbb{N}$. We say that L has *two solutions* mod p^s if there exists $v \in W_t'^{(0)}$ such that

$$Lv \equiv 0 \quad \text{mod } p^s W_t'^{(0)}$$

and

$$\|\text{wronskian}(u, v)\|_t = 1$$

(and then we say that u, v are two solutions mod p^s).

11.1 **Lemma.** *There exist two solutions* mod p^s *of L if and only if there exists $\xi \in W_t'^{(0)}$ such that*

$$\tau' \equiv \xi' \quad \text{mod} \quad p^s W_t'^{(0)}. \tag{11.1.1}$$

Proof. For $z \in W_t'$, we have,

$$L(uz) = zLu + \frac{w}{u}\left(\frac{z'}{\tau'}\right)'$$
$$= \frac{w}{u}\left(\frac{z'}{\tau'}\right)'.$$

Hence u and uz are two solutions mod p^s if and only if

$$\|u^2 z'\|_t = 1, \tag{11.1.2}$$

and

$$z' = \tau'\rho \tag{11.1.3}$$

with

$$\rho' \in p^s W_t'^{(0)}.$$

Condition (11.1.2) may be replaced by

(11.1.4) $$\|\rho\|_t = 1.$$

Thus

$$\tau' = \frac{z'}{\rho} = \left(\frac{z}{\rho}\right)' + \frac{z\rho'}{\rho^2},$$

which proves the lemma in one direction with $\xi = z/\rho$. Conversely if we assume (11.1.1) then let $v = \xi u \in W_t'^{(0)}$. We have

$$\|\text{wronskian}(u, v)\|_t = \|u^2\xi'\|_t = \|\xi'\|_t = \|\tau'\|_t = 1,$$

while

$$Lv = \frac{w}{u}\left(\frac{\xi'}{\tau'}\right)' \equiv 0 \mod p^s W_t'^{(0)}.$$

This completes the proof of the lemma. □

Notation. For each $s \in \mathbb{N}$, let

(11.2) $$\frac{1}{s!} D^s \equiv X_s D + Y_s \mod W_t'[D]L,$$

where X_s, Y_s lie in the differential ring generated by f and g (over $\mathbb{Q}$, not $\mathbb{N}$ as we divide by $s!$).

11.3 **Lemma.**

$$X_s = \sum_{\substack{i+j=s \\ i \geq 1}} \frac{1}{i} \frac{\tau'^{(i-1)}}{(i-1)!\tau'} \frac{u^{(j)}}{j!u}.$$

Proof. The proof is precisely the same as the proof of (10.4.4) except that we may now divide by $s!$ since the characteristic is zero. □

11.4 **Corollary.**

$$\|X_s\|_t \leq \operatorname{Sup}_{1 \leq i \leq s} \frac{1}{|i|}.$$

Proof. $(1/j!)D^j$ is bounded by 1 as operator on W_t'. □

11.5 **Lemma.** *Let $m \geq 1$, $m \in \mathbb{N}$. If L has two solutions* mod p^{m-1} *then*

$$X_{p^{m-1}} \in W_t^{\prime(0)}.$$

Proof. By Lemma 11.1

$$\tau' = \xi' + p^{m-1}\Lambda, \tag{11.5.1}$$

with $\xi, \Lambda \in W_t^{\prime(0)}$. Furthermore, $1 = \|\tau'\|_t$. Thus

$$\frac{1}{i}\frac{\tau'^{(i-1)}}{(i-1)!\tau'} \equiv \frac{p^{m-1}}{\tau'}\frac{\Lambda^{(i-1)}}{(i-1)!i} \quad \text{mod } W_t^{\prime(0)}. \tag{11.5.2}$$

Applying Lemma 11.3 with $s = p^{m-1}$, we find that we need consider only $i \leq p^{m-1}$ and so the lemma follows directly. □

11.6 **Lemma.** *If L has two solutions* mod p^{m-1} *but not* mod p^m *and if the field of definition, K, is unramified over $\mathbb{Q}_p$ then*

$$pX_{p^m} \in W_t^{(0)}, \tag{11.6.1}$$

$$\|pX_{p^m}\|_t = 1. \tag{11.6.2}$$

Proof. We apply (11.5.1) to the evaluation of pX_{p^m} by means of Lemma 11.3. We obtain

$$pX_{p^m} \equiv \frac{1}{\tau'}\frac{D^{p^m-1}}{(p^m-1)!}\Lambda \quad \text{mod } pW_t^{\prime(0)}$$

and by Lemma 5.14 this last may be written ($\delta = x(d/dx)$)

$$pX_{p^m} \equiv \frac{1}{\tau'}\phi^m \circ \frac{1}{x} \circ \psi^m x\Lambda \quad \text{mod } pW_t^{\prime(0)}; \tag{11.6.3.1}$$

$$pX_{p^m} \equiv \frac{1}{\tau'}\phi^m \circ \frac{1}{x} \circ \frac{1}{p^{m-1}}\psi^m \delta\tau \quad \text{mod } pW_t^{\prime(0)}. \tag{11.6.3.2}$$

It follows from (11.5.1) that

$$\psi^i x\tau' \equiv 0 \quad \text{mod } p^i W_t^{\prime(0)}, \qquad i = 0, 1, \ldots, m-1.$$

If this held also for $i = m$ then by Lemma 5.10, τ' would be congruent modulo p^m to the derivative of an element of $W_t^{\prime(0)}$, which by Lemma 11.1 would contradict the hypothesis that L does not have two solutions mod p^m.

This then shows that

$$1 \geq \left\| \frac{1}{p^{m-1}} \psi^m \delta\tau \right\|_t > |p|. \tag{11.6.4}$$

Now we know (§11.10) that $\eta = u'/u$ lies in the completion under the gauss norm of $K(x)$ and hence the same holds for

$$\frac{\tau''}{\tau'} = -\jmath - 2\eta.$$

This means that

$$\frac{\tau'(x)}{\tau'(t)} \in K_1[[x - t]],$$

where K_1 denotes the completion in Ω of $K(t)$. The same then holds for $\psi^m\delta\tau/(\tau'(t))$. Since $|\tau'(t)| = 1$ by (11.0.5), we conclude that $\|\psi^m\delta\tau\|_t$ lies in the value group of K_1, i.e., in the value group of $K(t)$. We now use the hypothesis that K is unramified over $\mathbb{Q}_p$ to deduce (11.6.2) from (11.6.4). □

11.7

From now on we will assume that L satisfies the hypotheses of Lemma 11.6. Equation (11.6.3.2) then gives pX_{p^m} mod p in terms of w/u^2 where u is a solution. We ask if a similar formula holds with u replaced by an approximate solution.

Theorem. *Let L satisfy the hypotheses of Lemma 11.6. Let $v \in W_t'^{(0)}$, $\|v\|_t = 1$*

$$Lv \equiv 0 \mod p^{3m-2}W_t'^{(0)},$$

then

$$pX_{p^m} \equiv \frac{1}{p^{m-1}} \frac{v^2}{w} \phi^m \frac{1}{x} \psi^m \frac{xw}{v^2} \mod pW_t'^{(0)}. \tag{11.7.1}$$

We need a preliminary lemma:

11.8 Lemma. *Under the hypotheses of Lemma 11.6, let $T \in \mathbb{N}$, $z \in W_t'^{(0)}$ such that*

$$\|z\|_t = 1, \tag{11.8.1}$$

$$L(zu) \equiv 0 \mod p^T W_t'^{(0)}; \tag{11.8.2}$$

(hence by the proof of Lemma 11.1,

$$z' = \tau'\rho, \tag{11.8.3}$$

$$\rho' \equiv 0 \mod p^T W_t'^{(0)}, \tag{11.8.4}$$

but we do not assume that $\|\rho\|_t = 1$). *Let*

$$l = \operatorname{ord} z' = \operatorname{ord} \rho \tag{11.8.5}$$

(*i.e.*, $\|z'\|_t = \|\rho\|_t = |p^l|$). *We assert that if*

$$T \geq 2l + m - 1, \tag{11.8.6}$$

then there exists $\Lambda_0, \Lambda_1 \in W_t'^{(0)}$, *such that*

$$\|\Lambda_0\|_t = 1, \tag{11.8.7}$$

$$\frac{zp^l}{\rho} = \Lambda_0^{\phi^m} + p^l \Lambda_1 \tag{11.8.8}$$

and if $m > l$ *then*

$$\|\delta\Lambda_1\|_t = 1. \tag{11.8.9}$$

Proof. Put

$$Y = p^l \frac{z}{\rho} \tag{11.8.10}$$

so that

$$\|Y\|_t = 1.$$

Then by (11.8.3) and (11.8.4),

$$\frac{1}{p^l}\delta Y = \delta z \cdot \frac{1}{\rho} + z\delta\left(\frac{1}{\rho}\right) \equiv \delta\tau \mod p^{T-2l} W_t'^{(0)} \tag{11.8.10.1}$$

and so by (11.5.1), we obtain

$$\frac{1}{p^l}\delta Y \equiv \delta\xi \mod(p^{m-1}, p^{T-2l}) W_t'^{(0)}. \tag{11.8.11}$$

Thus by (11.8.6), there exists $V \in W_t'^{(0)}$ such that

$$\frac{1}{p^l}\delta Y = \delta\xi + p^{m-1}V. \tag{11.8.12}$$

We assert that for $0 \le r \le m$ there exist

$$Y_{r,1} \in W_t'^{(0)}, \qquad Y_{r,0} \in W_{tp^r}'^{(0)},$$

such that

$$Y = Y_{r,0}^{\phi^r} + p^l Y_{r,1}. \tag{11.8.13$_r$}$$

This is trivial for $r = 0$. We now assume $r < m$, assume validity of (11.8.13)$_r$ and we will verify (11.8.13)$_{r+1}$. Thus by differentiating

$$\delta Y = p^r(\delta Y_{r,0})^{\phi^r} + p^l\delta Y_{r,1} \tag{11.8.14}$$

and by (11.8.12),

$$\delta\xi + p^{m-1}V = \frac{1}{p^l}\delta Y = p^{r-l}(\delta Y_{r,0})^{\phi^r} + \delta Y_{r,1}. \tag{11.8.15}$$

Applying ψ^r and using $r \le m - 1$, we obtain

$$p^{r-l}\delta Y_{r,0} \equiv 0 \mod p^r W_{tp^r}'^{(0)},$$

i.e.,

$$\delta Y_{r,0} \equiv 0 \mod p^l W_{tp^r}'^{(0)}. \tag{11.8.16}$$

We deduce from Lemma 5.9 that there exists $Y_{r+1,0}$ (resp. q) in $W_{tp^{r+1}}'^{(0)}$ (resp. $W_{tp^r}'^{(0)}$) such that

$$Y_{r,0} = Y_{r+1,0}^{\phi} + p^l q. \tag{11.8.17}$$

Substituting in (11.8.13)$_r$ and putting

$$Y_{r+1,1} = q^{\phi^r} + Y_{r,1} \tag{11.8.18}$$

completes the proof of (11.8.13)$_{r+1}$.

We may now put $r = m$ and let $\Lambda_0 = Y_{m,0}$, $\Lambda_1 = Y_{m,1}$. This verifies (11.8.8), and (11.8.7) now follows from $\|Y\|_t = 1$. To verify (11.8.9), we write (11.8.10.1) in the form

$$\text{(11.8.19)} \qquad \delta Y = p^l \frac{\delta z}{\rho} - p^l z \frac{\delta \rho}{\rho^2}$$

$$\equiv p^l \frac{\delta z}{\rho} \quad \text{mod } p^{T-l} W_t'^{(0)}.$$

By (11.8.5), (11.8.6)

$$\text{(11.8.20)} \qquad \text{ord } p^l \frac{\delta z}{\rho} = l < T - l.$$

Thus

$$\text{(11.8.21)} \qquad \text{ord } \delta Y = l,$$

while by (18.8.8)

$$\text{(11.8.22)} \qquad \delta Y = p^m (\delta \Lambda_0)^{\phi m} + p^l \delta \Lambda_1.$$

Equation (11.8.9) is now clear. This completes the proof of the lemma. □

11.9 Proof of Theorem 11.7

We use the notation of Lemma 11.8. Let $v = zu$, ρ defined by (11.8.3) and l as in (11.8.5). We assume (11.8.4) with

$$\text{(11.9.1)} \qquad T = 3m - 2.$$

We shall use ξ, Λ as in (11.5.1).

Case 1. $l \geq m$

By the definition of u and equation (11.5.1)

$$\text{(11.9.2)} \qquad \frac{xw}{v^2} = \frac{1}{z^2}(\delta\xi + p^{m-1} x\Lambda) = \delta\left(\frac{\xi}{z^2}\right) + p^{m-1}\frac{x\Lambda}{z^2} + \frac{2\xi\delta z}{z^3}$$

and so by (11.8.5), since $l \geq m$,

$$\text{(11.9.3)} \qquad \frac{xw}{v^2} \equiv \delta\left(\frac{\xi}{z^2}\right) + p^{m-1}\frac{x\Lambda}{z^2} \quad \text{mod } p^m W_t'^{(0)}.$$

This shows that

(11.9.4) $$\psi^m \frac{xw}{v^2} \equiv p^{m-1}\psi^m\left(\frac{x\Lambda}{z^2}\right) \quad \operatorname{mod} p^m W'^{(0)}_{tp^m}.$$

By (11.8.5) and Lemma 5.9, there exists $z_0 \in W'^{(0)}_{tp^m}$

$$\|z_0\|_{tp^m} = \|z\|_t = 1,$$

such that

(11.9.5) $$z \equiv z_0^{\phi^m} \quad \operatorname{mod} pW'^{(0)}_t.$$

Substituting in (11.9.4) gives

(11.9.6) $$\psi^m \frac{xw}{v^2} \equiv p^{m-1} \frac{1}{z_0^2} \psi^m(x\Lambda) \quad \operatorname{mod} p^m W'^{(0)}_{tp^m}.$$

Thus the right side of (11.7.1) is mod $pW'^{(0)}_t$ equal to

$$\frac{v^2}{w} \phi^m \frac{1}{xz_0^2} \psi^m(x\Lambda) = u^2 \frac{z^2}{w} \left(\frac{1}{z_0^2}\right)^{\phi^m} \phi^m \frac{1}{x} \psi^m x\Lambda$$

and by (11.9.5) this reduces mod $pW'^{(0)}_t$ to

$$\frac{u^2}{w} \phi^m \frac{1}{x} \psi^m x\Lambda,$$

which by (11.6.3.1) coincides mod p with pX_{p^m}.

Case 2. $m > l$.

Then by (11.9.1), condition (11.8.6) is satisfied. By (11.8.3) and (11.8.10),

(11.9.7) $$\begin{aligned}\delta\tau = \frac{\delta z}{\rho} = \frac{1}{\rho}\,\delta\,\frac{Y\rho}{p^l} &= \frac{1}{p^l}\,\delta Y + \frac{Y}{p^l\rho}\,\delta\rho \\ &\equiv \frac{1}{p^l}\,\delta Y \quad \operatorname{mod} p^{T-2l},\end{aligned}$$

i.e.,

(11.9.8) $$\delta\tau \equiv \frac{1}{p^l}\,\delta Y \quad \operatorname{mod} p^m.$$

Hence by (11.6.3.2),

$$pX_{p^m} \equiv \frac{p^l}{\delta Y} x\phi^m \frac{1}{x} \psi^m \frac{\delta Y}{p^l p^{m-1}} \quad \text{mod } p.$$

Thus

(11.9.9)
$$pX_{p^m} \equiv \frac{p^l}{\delta Y} x\phi^m \frac{1}{x} p\delta \frac{\psi^m Y}{p^l} \quad \text{mod } p.$$

By (11.8.8)

(11.9.10)
$$\psi^m Y = \Lambda_0 + p^l \psi^m \Lambda_1.$$

By (11.8.21),

$$\left\| \frac{p^l}{\delta Y} \right\|_t = 1$$

$$\frac{p}{p^l} \delta(p^l \psi^m \Lambda_1) \equiv 0 \quad \text{mod } p.$$

Thus substituting (11.9.10) in (11.9.9),

(11.9.11)
$$pX_{p^m} \equiv \frac{p^l}{\delta Y} x\phi^m \frac{1}{x} p \frac{\delta \Lambda_0}{p^l} \quad \text{mod } p$$

Equations (11.9.11) and (11.6.2) imply

(11.9.12)
$$\left\| \frac{\delta \Lambda_0}{p^{l-1}} \right\|_t = 1.$$

Thus

(11.9.12′)
$$pX_{p^m} \equiv \frac{1}{\delta \Lambda_1} x\phi^m \frac{1}{x} p \frac{\delta \Lambda_0}{p^l} \quad \text{mod } p,$$

since

$$\frac{p^l}{\delta Y} = \frac{p^l}{p^m (\delta \Lambda_0)^{\phi^m} + p^l \delta \Lambda_1} \equiv \frac{1}{\delta \Lambda_1} \quad \text{mod } p$$

by (11.8.9), using $m > l$.

By (11.8.1) and (11.9.8)

(11.9.13) $$\frac{wx}{v^2} = \frac{1}{z^2}\,\delta\tau \equiv \frac{1}{(Y\rho/p^l)^2}\frac{\delta Y}{p^l} \mod p^m.$$

We assert that there exists $\rho_0 \in W_t^{\prime(0)}$, $\|\rho_0\|_t = 1$ such that

(11.9.14) $$\left(\frac{\rho}{p^l}\right)^2 \equiv \rho_0^{\phi^m} \mod p^m.$$

Indeed by (11.8.4) and (11.9.1),

$$\delta\left(\left(\frac{\rho}{p^l}\right)^2\right) = \frac{2\rho}{p^l}\frac{\delta\rho}{p^l} \equiv 0 \mod p^{T-l}$$
$$\equiv 0 \mod p^{2m-1},$$

and so there exist $q_0, q_1, \ldots, q_{2m-1}$

$$\|q_0\| = 1, \qquad q_j \in W'^{(0)}_{tp^{2m-1-j}},$$

such that

(11.9.15) $$\left(\frac{\rho}{p^l}\right)^2 = q_0^{\phi^{2m-1}} + pq_1^{\phi^{2m-2}} + \cdots + p^{2m-1}q_{2m-1}.$$

Putting

$$\rho_0 = q_0^{\phi^{m-1}} + pq_1^{\phi^{m-2}} + \cdots + p^{m-1}q_{m-1},$$

we see that (11.9.15) implies (11.9.14).

Applying (11.9.14) to (11.9.13), we obtain

(11.9.16) $$\frac{xw}{v^2} \equiv \frac{\delta Y}{p^l}\frac{1}{Y^2\rho_0^{\phi^m}} \mod p^m.$$

Letting

(11.9.17) $$Q = \frac{v^2}{xw}\,x\phi^m\,\frac{1}{x}\,\psi^m\,\frac{xw}{v^2},$$

we now have

(11.9.18) $$Q \equiv \frac{Y^2p^l}{\delta Y}\,x\phi^m\,\frac{1}{x}\,\psi^m\,\frac{\delta Y}{p^l}\frac{1}{Y^2} \mod p^m.$$

Now

$$\psi^m \frac{1}{p^l} \frac{\delta Y}{Y^2} = -\psi^m \frac{1}{p^l}\, \delta\left(\frac{1}{Y}\right) = -p^{m-l}\, \delta\psi^m \frac{1}{Y},$$

while by (11.8.8)

$$\frac{1}{Y} \equiv \frac{1}{\Lambda_0^{\phi^m}} \mod p^l$$

and so

(11.9.19) $$\psi^m \frac{1}{p^l} \frac{\delta Y}{Y^2} \equiv -p^{m-l}\, \delta \frac{1}{\Lambda_0} \mod p^m.$$

Thus

(11.9.20) $$Q \equiv \frac{Y^2 p^l}{\delta Y}\, x\phi^m \frac{1}{x}\left(-p^{m-l}\delta \frac{1}{\Lambda_0}\right) \mod p^m.$$

Again by (11.8.8)

(11.9.21) $$\frac{1}{p^l}\, \delta Y = p^{m-l}(\delta\Lambda_0)^{\phi^m} + \delta\Lambda_1 \equiv \delta\Lambda_1 \mod p$$

and by (11.9.12)

(11.9.22) $$p^{m-l}\delta\left(\frac{1}{\Lambda_0}\right) \equiv 0 \mod p^{m-1}.$$

From (11.9.21) and (11.8.8),

$$\frac{Y^2 p^l}{\delta Y} \equiv \frac{\Lambda_0^{2\phi^m}}{\delta\Lambda_1} \mod p.$$

This together with (11.9.22) gives

$$\begin{aligned} Q &\equiv \frac{\Lambda_0^{2\phi^m}}{\delta\Lambda_1}\, x\phi^m \frac{1}{x}\left(-p^{m-l}\,\delta \frac{1}{\Lambda_0}\right) \mod p^m \\ &= \frac{1}{\delta\Lambda_1}\, x\phi^m \frac{1}{x}\, p^{m-l}\, \delta\Lambda_0 \end{aligned}$$

and so

$$\frac{Q}{p^{m-1}} \equiv \frac{1}{\delta\Lambda_1} x\phi^m \frac{1}{x} \frac{1}{p^{l-1}} \delta\Lambda_0 \mod p.$$

By comparison with (11.9.12′)

$$\frac{Q}{p^{m-1}} \equiv pX_{p^m} \mod p.$$

This completes the proof of the theorem. □

11.10 *Ad hoc* treatment of u'/u

Let L be given by (11.0.1) and again assume that L satisfies conditions (11.0.2) and (11.0.3). Let u be as in (11.0.4). We wish to explain briefly why $\eta = u'/u$ is an analytic element on $D(t, 1^-)$.

It follows from (11.0.2) that (in the notation of (11.2)) the elements $\{X_s, Y_s\}_{s=1,2,\ldots}$ cannot have a uniformly bounded gauss norm. For any real $r \in (0, 1)$, $s \in \mathbb{N}$, we have

$$\left|\frac{u^{(s)}}{s!u}\right|_t(r) \leq \frac{1}{r^s},$$

(11.10.1) $$|X_s|_t(r) \leq |X_s|_{\text{gauss}},$$

$$|Y_s|_t(r) \leq |Y_s|_{\text{gauss}}.$$

But by (11.2) we have

(11.10.2) $$\frac{u^{(s)}}{s!u} = X_s\frac{u'}{u} + Y_s.$$

Applying $|\ |_t(r)$ and letting $r \to 1$, we obtain

(11.10.3) $$|Y_s|_{\text{gauss}} \leq \text{Max}(1, |X_s|_{\text{gauss}}).$$

This shows that $\{|X_s|_{\text{gauss}}\}_{s=1,2,\ldots}$ is an unbounded set. We rewrite (11.0.2) in the form

$$\frac{1}{s!}\frac{u^{(s)}}{u}\frac{1}{X_s} = \frac{u'}{u} + \frac{Y_s}{X_s}.$$

As elements of W_t' we have

$$\left\| \frac{u'}{u} + \frac{Y_s}{X_s} \right\|_t \le \frac{1}{|X_s|_{\text{gauss}}}. \tag{11.10.4}$$

Restricting s to a sequence such that $|X_s|_{\text{gauss}} \to \infty$, we conclude that u'/u is an analytic element on $D(t, 1^-)$. (The article of P. Robba [30] referred to in §11.0 gives the generalization of this result to equations of higher order.) The analytic continuation of η is the subject of speculation.

Chapter 12

Dieudonné Theory

Let K be a complete unramified extension of $\mathbb{Q}_p$ with algebraically closed residue class field. Let σ denote the Frobenius automorphism of K over $\mathbb{Q}_p$. Let K_0 be a complete subfield of K, and let $\mathbf{0}$ (resp. $\mathbf{0}_0$) be the ring of integers of K (resp. K_0)

12.1 **Lemma.** *If* $\alpha, \beta, \gamma, \delta \in \mathbf{0}_0$, α *a unit and*

$$A = \begin{pmatrix} \alpha & \beta \\ p\gamma & p\delta \end{pmatrix}$$

is a matrix with

$$n = \text{ord det } A,$$

then there exists a matrix $X \in \mathrm{GL}(2, \mathbf{0})$ *such that*

$$X^\sigma A = \begin{pmatrix} 1 & 0 \\ 0 & p^n \end{pmatrix} X. \tag{12.1.1}$$

Proof. We give an *ad hoc* proof of this special case of the theory of semilinear transformations. We observe that

$$z \mapsto \frac{\beta + p\delta z^\sigma}{\alpha + p\gamma z^\sigma} \tag{12.1.2}$$

is a contractive map of $\mathbf{0}_0$ into itself and hence there exists a unique fixed point, $z_0 \in \mathbf{0}_0$. We may choose $x \in \mathbf{0}$, x a unit such that

$$x^{1-\sigma} = \alpha + p\gamma z_0^\sigma. \tag{12.1.3}$$

We then put

$$(x, y) = x(1, z_0) \tag{12.1.4}$$

and conclude that

$$(x, y)^\sigma A = (x, y). \tag{12.1.5}$$

We now put

$$A_1 = \begin{pmatrix} x & y \\ 0 & 1 \end{pmatrix}^\sigma A \begin{pmatrix} x & y \\ 0 & 1 \end{pmatrix}^{-1} \tag{12.1.6}$$

and compute

$$A_1 = \begin{pmatrix} 1 & 0 \\ p\gamma' & p^n\delta_1 \end{pmatrix}, \text{ where } \gamma' \text{ and } \delta_1 \text{ lie in } \mathbf{0},\ \delta_1 \text{ a unit.}$$

We may choose a unit $v \in \mathbf{0}$ such that

$$\begin{pmatrix} 1 & 0 \\ 0 & v \end{pmatrix}^\sigma A_1 \begin{pmatrix} 1 & 0 \\ 0 & v \end{pmatrix}^{-1} = \begin{pmatrix} 1 & 0 \\ p\gamma_1 & p^n \end{pmatrix} = A_2, \tag{12.1.7}$$

with $\gamma_1 \in \mathbf{0}$. We now observe that if we choose $w \in \mathbf{0}$ such that

$$w^\sigma + p\gamma_1 = p^n w, \tag{12.1.8}$$

then

$$\begin{pmatrix} 1 & 0 \\ w & 1 \end{pmatrix}^\sigma A_2 \begin{pmatrix} 1 & 0 \\ w & 1 \end{pmatrix}^{-1} = \begin{pmatrix} 1 & 0 \\ 0 & p^n \end{pmatrix}. \tag{12.1.9}$$

The lemma follows with

$$X = \begin{pmatrix} 1 & 0 \\ w & 1 \end{pmatrix} \begin{pmatrix} 1 & 0 \\ 0 & v \end{pmatrix} \begin{pmatrix} x & y \\ 0 & 1 \end{pmatrix}. \quad \square \tag{12.1.10}$$

12.2 **Proposition.** *Let*

$$A = \begin{pmatrix} p\alpha & \beta \\ p\gamma & p\delta \end{pmatrix},$$

with $\alpha, \beta, \gamma, \delta \in \mathbf{0}_0$, β and γ units. Then there exists $X_1 \in \mathrm{GL}(2, \mathbf{0})$ such that

$$X_1^\sigma A = \begin{pmatrix} 0 & 1 \\ p & 0 \end{pmatrix} X_1. \tag{12.2.1}$$

Proof. We will show the existence of $x, y \in \mathbf{0}$, x a unit such that

$$(x, y)^{\sigma^2} A^{\sigma} A = p(x, y). \tag{12.2.2}$$

If we then put $(u, v) = (x, y)^{\sigma} A$,

$$X_1 = \begin{pmatrix} x & y \\ u & v \end{pmatrix},$$

then clearly (12.2.1) would hold. But we also assert that under these conditions,

$$\det X_1 \neq 0, \tag{12.2.3}$$

i.e.,

$$\begin{vmatrix} p(\alpha x^{\sigma} + \gamma y^{\sigma}) & \beta x^{\sigma} + p\delta y^{\sigma} \\ x & y \end{vmatrix} \neq 0.$$

Since β is a unit, this is clear if x is a unit and so the proposition is reduced to the verification of (12.2.2).

By multiplication

$$A^{\sigma} A = p \begin{pmatrix} \gamma\beta^{\sigma} + p\alpha^{1+\sigma} & (\alpha^{\sigma}\beta + \beta^{\sigma}\gamma) \\ p(\gamma^{\sigma}\alpha + \delta^{\sigma}\gamma) & \gamma^{\sigma}\beta + p\delta^{1+\sigma} \end{pmatrix}.$$

Thus verification of (12.2.2) is reduced to the assertion:

Given

$$C = \begin{pmatrix} a & b \\ pc & d \end{pmatrix}$$

with a, d units and b, c elements of $\mathbf{0}$, there exists nontrivial (x, y) in $\mathbf{0}$ such that x is a unit and

$$(x, y)^{\sigma^2} C = (x, y). \tag{12.2.2$'$}$$

With no loss in generality we may put

$$C = \begin{pmatrix} 1 & b \\ pc & 1 \end{pmatrix}$$

with $b, c \in \mathbf{0}$.

Let $y/x = z$. Equation (12.2.2′) is satisfied if we choose x, z such that

$$z = \frac{z^{\sigma^2} + b}{pcz^{\sigma^2} + 1}, \tag{12.2.4}$$

$$x^{1-\sigma^2} = 1 + pcz^{\sigma^2}. \tag{12.2.5}$$

Thus we need only find a solution z of (12.2.4) in $\mathbf{0}$. Writing this equation in the form

$$z + pcz^{1+\sigma^2} = z^{\sigma^2} + b, \tag{12.2.4′}$$

it is clear that there exists z_0, a solution mod p; indeed we know that in the residue class field there exists a solution of the Artin–Schrier equation

$$\bar{z} = \bar{z}^{p^2} + \bar{b}$$

and a lifting to $\mathbf{0}$ of $\bar{z}$ gives the required solution mod p. Suppose now that z_m is a solution of (12.2.4′) mod p^{m+1}, we assert the existence of

$$z_{m+1} \equiv z_m \mod p^{m+1}$$

such that z_{m+1} is a solution mod p^{m+2}. Indeed setting

$$z_{m+1} = z_m + p^{m+1}u$$

and letting

$$z_m + pcz_m^{1+\sigma} - z_m^{\sigma^2} - b = p^{m+1}v, \qquad v \in \mathbf{0},$$

we reduce to choosing $u \in \mathbf{0}$ such that

$$u - u^{\sigma^2} + v \equiv 0 \mod p.$$

This completes the proof of the proposition. □

12.3 **Corollary.** *Let A be as in Proposition* 12.2. *If we extend σ to $K(\sqrt{p})$ by $\sigma\sqrt{p} = \sqrt{p}$ then there exists $X \in \mathrm{GL}(2, K(\sqrt{p}))$ such that*

$$X^{\sigma}A = \sqrt{p}X.$$

Proof. Choose z in K such that

$$z^{\sigma^2} = z,$$

$$z^{\sigma} \neq z,$$

(e.g., let z be a primitive $(p^2 - 1)$st root of unity). Let

$$X_2 = \begin{pmatrix} \sqrt{p} & 1 \\ z^\sigma \sqrt{p} & z \end{pmatrix},$$

then

$$X_2^\sigma \begin{pmatrix} 0 & 1 \\ p & 0 \end{pmatrix} = \sqrt{p}\, X_2$$

and $X_2 \in \mathrm{GL}(2, K(\sqrt{p}))$. The assertion thus follows by putting $X = X_2 X_1$, with X_1 as in (12.2.1). □

12.4. **Corollary.** *The matrix X of Corollary* 12.3 *is unique up to multiplication on the left by elements of* $\mathrm{GL}(2, \mathbb{Q}_p(\sqrt{p}))$ *but may be put in the form*

$$X = \begin{pmatrix} x_1 & x_1 z_1 \\ x_2 & x_2 z_2 \end{pmatrix},$$

with

(12.4.1) $$\operatorname{ord} z_1 = \operatorname{ord} z_2 = \operatorname{ord}(z_2 - z_1) = -\tfrac{1}{2}$$

and where

(12.4.2) x_2/x_1 *is a unit of* $\mathbf{0}[\sqrt{p}]$ *whose reduction* $\bmod \sqrt{p}$ *lies in* $\mathbb{F}_{p^2} - \mathbb{F}_p$.

Proof. If $x, y \in \mathbb{Q}_p(\sqrt{p})$, not both zero and

(12.4.3) $$(x, y)^\sigma A = \sqrt{p}(x, y),$$

then

(12.4.4) $$\sqrt{p}\, x = p(x^\sigma \alpha + y^\sigma \gamma),$$

which shows that x cannot be zero as otherwise y would also be zero.

Now clearly,

$$|\sqrt{p}\, x| > |p x^\sigma \alpha|$$

and hence

(12.4.5) $$|\sqrt{p}\, x| = |p y^\sigma \gamma| = |p||y|.$$

This shows that if we set $z = y/x$ then

(12.4.6) $$|z| = \sqrt{p}.$$

It follows from (12.4.3) that z must satisfy

(12.4.7) $$z = \frac{p\delta z^\sigma + \beta}{p\gamma z^\sigma + p\alpha},$$

a relation from which too we could deduce (12.4.6). (We first note that ord z must be negative as otherwise the numerator of the right side of (12.4.7) would be a unit in $\mathbf{0}[\sqrt{p}]$ while the denominator would be a nonunit. We would then observe that for ord $z < 0$, the denominator has ordinal $1 + \text{ord } z$ and so

$$1 + \text{ord } z > 1 + 2 \text{ ord } z = \text{ord}(\beta + p\delta z^\sigma),$$

from which we could again deduce $0 = 1 + 2 \text{ ord } z$).

If now z is a solution of (12.4.7) then we may construct a corresponding solution of (12.4.3) by setting $y = xz$ and choosing x (cf. (12.4.4)) such that

(12.4.8) $$x^{1-\sigma} = \sqrt{p}\alpha + (\sqrt{p}z)^\sigma\gamma,$$

which may be done since the right side would be a unit.

Thus we could construct X satisfying Corollary 12.3 by finding two distinct solutions z_1, z_2 of (12.4.7). We simplify that equation by choosing ε a unit in $\mathbf{0}$ such that

(12.4.9) $$\varepsilon^{1+\sigma} = \frac{\beta}{\gamma}$$

and putting

(12.4.10) $$z = \frac{\varepsilon\mu}{\sqrt{p}}$$

the corresponding condition on μ is

(12.4.11) $$\mu = \frac{1 + \sqrt{p}\varepsilon_1\mu^\sigma}{\sqrt{p}\varepsilon_2 + \mu^\sigma},$$

where $\varepsilon_1, \varepsilon_2$ are elements of $\mathbf{0}$,

$$\varepsilon_1 = \frac{\delta}{\gamma\varepsilon},$$

$$\varepsilon_2 = \frac{\alpha\varepsilon}{\beta}.$$

The proof of (12.4.1) now reduces to that of the existence of two solutions μ_1, μ_2 of (12.4.11) which are units of $\mathbf{0}[\sqrt{p}]$ lying in distinct residue classes mod $\sqrt{p}$.

Indeed if t_1 is a $(p+1)$st root of unity then $\mu = t_1$ is a solution mod $\sqrt{p}$ of (12.4.11). We assert that such an approximate solution may be refined to an exact solution in $K[\sqrt{p}]$. Let then t_s ($\equiv t_1$ mod $\sqrt{p}$) be a solution mod $\sqrt{p}^s$, i.e., there exists $\kappa_s \in \mathbf{0}[\sqrt{p}]$ such that

$$t_s + \sqrt{p}^s\kappa_s = \frac{1 + \sqrt{p}\varepsilon_1 t_s^\sigma}{\varepsilon_2\sqrt{p} + t_s^\sigma}. \tag{12.4.12}$$

Choose $T \in \mathbf{0}$ such that

$$T + T^\sigma \equiv \kappa_s/t_s \quad \text{mod } \sqrt{p}. \tag{12.4.13}$$

Let

$$t_{s+1} = t_s(1 + T\sqrt{p}^s) \tag{12.4.14}$$

and then mod $\sqrt{p}^{s+1}$

$$\begin{aligned} t_{s+1} - \frac{1 + \sqrt{p}\varepsilon_1 t_{s+1}^\sigma}{\varepsilon_2\sqrt{p} + t_{s+1}^\sigma} &\equiv t_{s+1} - \frac{1 + \sqrt{p}\varepsilon_1 t_s^\sigma}{\varepsilon_2\sqrt{p} + t_s^\sigma + t_s^\sigma T^\sigma\sqrt{p}^s} \\ &\equiv t_{s+1} - \frac{1 + \sqrt{p}\varepsilon_1 t_s^\sigma}{t_s^\sigma + \varepsilon_2\sqrt{p}} \frac{1}{1 + \sqrt{p}^s T^\sigma} \\ &\equiv t_{s+1} - (t_s + \sqrt{p}^s\kappa_s)\frac{1}{1 + \sqrt{p}^s T^\sigma} \\ &\equiv t_{s+1} - (t_s + \sqrt{p}^s(\kappa_s - T^\sigma t_s)) \\ &\equiv \sqrt{p}^s(t_s T - \kappa_s + T^\sigma t_s), \end{aligned}$$

which finally is congruent to zero mod $\sqrt{p}^{s+1}$ by (12.4.13).

Thus there exist solutions μ_1, μ_2 of (12.4.11) such that their reductions mod $\sqrt{p}$ are *distinct* elements of $\mathbb{F}_{p^2}$. Of course

$$\mu_i^{1+\sigma} \equiv 1 \mod \sqrt{p}, \qquad i = 1, 2. \tag{12.4.15}$$

Equation (12.4.8) may be written

$$x^{1-\sigma} = \mu^\sigma \frac{\beta}{\varepsilon} + \sqrt{p}\alpha \tag{12.4.16}$$

and so letting

$$x_i\left(1, \mu_i \frac{\varepsilon}{\sqrt{p}}\right), \qquad i = 1, 2$$

be the solutions of (12.4.3) corresponding to these two solutions of (12.4.11),

$$\left(\frac{x_2}{x_1}\right)^{1-\sigma} \equiv \left(\frac{\mu_2}{\mu_1}\right)^\sigma \mod \sqrt{p} \qquad (\not\equiv 1 \mod \sqrt{p}) \tag{12.4.17}$$

and so

$$\left(\frac{x_2}{x_1}\right)^{1-\sigma^2} \equiv 1 \mod \sqrt{p}. \tag{12.4.18}$$

Assertion (12.4.2) follows from these two relations, using the distinctness of the residue classes of μ_2 and μ_1. □

12.5 **Corollary.** *Under the hypothesis of* 12.2, *the equation*

$$c(x, y) = (x, y)^\sigma A$$

has no nontrivial solution with x, y, c in K.

Proof. Letting v_1, v_2 denote the two rows of the matrix X of Corollary 12.3, we may write

$$(x, y) = t_1 v_1 + t_2 v_2,$$

with $t_1, t_2 \in K(\sqrt{p})$. The condition on c is that

$$c(t_1 v_1 + t_2 v_2) = (t_1^\sigma v_1 + t_2^\sigma v_2)\sqrt{p}$$

and so

$$c(t_1, t_2) = \sqrt{p}(t_1^\sigma, t_2^\sigma).$$

If t_1, t_2 are not both zero then $|c| = |\sqrt{p}|$ which contradicts the hypothesis that (x, y) (and hence c) lie in K. □

Chapter 13

Canonical Liftings ($l \geq 1$)

13.1 **Definition.** A subset $\mathbf{T}$ of $\Omega \cup \{\infty\}$ is said to be superadmissible if the following conditions are satisfied:

(1) Under the residue class map, the image of $\mathbf{T}$ coincides with the image of Ω.
(2) For all but a finite set of residue classes $\mathbf{T}$ contains the entire residue class.
(3) For each residue class $\bar{z}$, $\mathbf{T}$ contains an annulus

$$\Delta_z(r, 1^-) = \{x \mid |x - z| \in (r, 1)\},$$

where z is a representative of $\bar{z}$.

In this chapter we use the notation of Chapter 9. In particular, hypotheses (9.0.1)–(9.0.6) will be retained.

We think of the mapping $\phi\colon \lambda \to \lambda^p$ as a lifting to the λ-line in characteristic p of the Frobenius mapping, Frob, on the λ-line in characteristic p given by the same formula.

13.2 **Definition.** Let $l \in \mathbb{N}$ be fixed. We shall say that θ is a *superadmissible* lifting of Frob^l if

(1) θ is an analytic element on a super admissible set
(2) $|\theta(\lambda) - \lambda^{p^l}| \leq |p|$ everywhere except on a finite number of residue classes.

Of course ϕ^l satisfies these conditions and the most general lifting of this type is of the form

$$\phi_l = \phi^l + q,$$

where q is an analytic element on a superadmissible set $\mathbf{T}$, dominated by $|p|$ on almost all residue classes.

13.3 **Theorem.** *Let* $\mathrm{Min}(a, b) > c$ (resp. $c > \mathrm{Max}(a, b)$). *Let* l *satisfy* (9.0.6). *Hypotheses* (9.0.1)–(9.0.6) *are to be understood. Then there exists* ϕ_l, *a unique* *superadmissible lifting of* Frob^l *such that if we put*

$$\tilde{G}_f = \frac{xf(x, \lambda)}{x^{p^l} f(x^{p^l}, \phi_l(\lambda))}, \tag{13.3.1}$$

then

$$\tilde{\alpha} = \gamma_- \circ \tilde{G}_f \circ \Phi^l$$

is a mapping of $\mathbf{K}_{f,\phi_l(\lambda)}$ *into* $\mathbf{K}_{f,\lambda}$ *whose matrix (relative to the bases* $\{1^*, (1/(1-x))^*\}$) *is of the form* $p^m \tilde{B}$ (m = *number of* α_i^* *of type* 4) *where*

$$\tilde{B} = \begin{pmatrix} \tilde{B}_1 & \tilde{B}_2 \\ 0 & \tilde{B}_4 \end{pmatrix} \quad \left(\text{resp.} \begin{pmatrix} \tilde{B}_1 & 0 \\ \tilde{B}_3 & \tilde{B}_4 \end{pmatrix}\right). \tag{13.3.2}$$

The $\tilde{B}_i$ *are analytic elements on superadmissible sets, which contain* S_H *(see* (9.6.1)). *Furthermore the* $\tilde{B}_i$ *are bounded by unity on* S_H *and* $\tilde{B}_4$ (resp. $\tilde{B}_1$) *is bounded by* $|p|$ *on* S_H.

Proof. We restrict our attention to the case $\text{Min}(a, b) > c$ as the other case presents no new difficulties. In (9.1.4) we discussed the matrix B of $p^{-m}\alpha_0^* \circ \cdots \circ \alpha_{l-1}^*$. Aside from the factor p^{-m} this is the mapping

$$\gamma_- \circ G_f \circ \Phi^l = \alpha^* \tag{13.3.3}$$

of $\mathbf{K}_{f,\lambda^{p^l}}$ into $\mathbf{K}_{f,\lambda}$, where

$$G_f = \frac{xf(x, \lambda)}{x^{p^l} f(x^{p^l}, \lambda^{p^l})}$$

is interpreted along the lines of (3.7.1). For any choice of ϕ_l approximately equal to ϕ^l, we have the mapping (§4.6), $T_{\lambda^{p^l}, \phi_l(\lambda)}$ of $\mathbf{K}_{f,\lambda^{p^l}}$ into $\mathbf{K}_{f,\phi_l(\lambda)}$ and the main point is that from the definitions

$$\tilde{\alpha} \circ T_{\lambda^{p^l}, \phi_l(\lambda)} = \alpha^*, \tag{13.3.4}$$

say for λ on the generic disk, subject only to the hypothesis that ϕ_l be defined on that disk and that

$$|\phi_l(\lambda) - \lambda^{p^l}| < 1, \qquad \forall \lambda \in D(t, 1^-). \tag{13.3.5}$$

Now as noted in 7.1, the matrix of $T_{z,\lambda}$ (subject to (4.6.1.1)) is given by $C_f(z, \lambda)$, which we may write in the form

$$C_f(z, \lambda) = \sum_{s=0}^{\infty} (\lambda - z)^s \frac{S}{s!} C_f^{(s)}(z, z), \tag{13.3.6}$$

where for $s \in \mathbb{N}$, $C_f^{(s)}(z, z)$ denotes the specialization at $\lambda = z$ of

$$\frac{\partial^s}{\partial \lambda^s} C_f(z, \lambda).$$

Fixing f, let us write M_1 for M_f in (7.1.3). Then by recursion we define M_s for $s \in \mathbb{N}$ such that

(13.3.7) $$\frac{1}{s!} \frac{\partial^s}{\partial \lambda^s} C_f(z, \lambda) = C_f(z, \lambda) M_s(\lambda).$$

(For $s = 0$ put $M_0 = I$, M_1 as indicated as above and in general

$$\frac{\partial^{s+1}}{\partial \lambda^{s+1}} C_f(z, \lambda) = s!(C_f(z, \lambda) M_s'(\lambda) + C_f(z, \lambda) M_1(\lambda) M_s(\lambda));$$

i.e.,

(13.3.8) $$(s + 1) M_{s+1} = M_s' + M_1 M_s.)$$

Thus

(13.3.9) $$C_f(z, \lambda) = \sum_{s=0}^{\infty} M_s(z)(z - \lambda)^s$$

and each M_s is a two by two matrix with coefficients in $Q(\lambda)$ and with singularities only at $\lambda = 0, 1$. Furthermore $s!M_s$ has coefficients lying in

$$\mathbb{Z}_p\left[\lambda, \frac{1}{\lambda}, \frac{1}{(1 - \lambda)}\right].$$

In particular the gauss norm of the coefficients of M_s are bounded by $1/|s!|$, a statement which we write in the form

(13.3.10) $$|M_s|_{\text{gauss}} \leq \frac{1}{|s!|}.$$

We show (for use in Chapter 15) that we may replace this trivial estimate by

(13.3.11) $$|M_s|_{\text{gauss}} \leq \operatorname{Sup}_{1 \leq i \leq s} \frac{1}{|i|}.$$

In the notation (9.9.4), let us write X_m, Y_m (resp. $\bar{X}_m$, $\bar{Y}_m$) rational functions associated with $(2.4.5.2)_{a,b,c}$ (resp. $(2.4.5.3)_{a,b,c}$). Letting $(u, \bar{u})$ denote a row of $C_f(t, \lambda)$ then by (13.3.7) we have

$$(u, \bar{u})M_s = \frac{1}{s!}(u, \bar{u})^{(s)} = (Y_s u + X_s u', \bar{Y}_s \bar{u} + \bar{X}_s \bar{u}'). \tag{13.3.12}$$

On the other hand, putting

$$M_1 = \begin{pmatrix} M_{11} & M_{12} \\ M_{13} & M_{14} \end{pmatrix}, \tag{13.3.12.1}$$

then by (7.1.4):

$$(u, \bar{u})' = (u, \bar{u})M_1,$$

by which we may eliminate u', $\bar{u}'$ from the right side of (13.3.12) and so

$$M_s = \begin{pmatrix} X_s M_{11} + Y_s & \bar{X}_s M_{12} \\ X_s M_{13} & \bar{Y}_s + \bar{X}_s M_{14} \end{pmatrix}. \tag{13.3.13}$$

Now we know by (9.4.1.3) that we may choose u, $\bar{u}$ bounded on $D(t, 1^-)$ such that

$$(u, \bar{u})' = (u, \bar{u})M_1 \tag{13.3.14}$$

and hence by Corollary 11.4, X_s and $\bar{X}_s$ have gauss norms bounded by (13.3.11). By relation (11.2), we have

$$\frac{1}{s!}\frac{u^{(s)}}{u} - X_s\frac{u'}{u} = Y_s, \tag{13.3.15}$$

a relation from which we may deduce the same estimate for Y_s. A similar argument holds for $\bar{Y}_s$. The assertion (13.3.11) now follows from (13.3.13).

To analyze equation (13.3.4) in greater detail we set

$$\phi_l(\lambda) = \lambda^{p^l} + q \tag{13.3.16}$$

and write the matrix $C_f(\lambda^{p^l}, \lambda^{p^l} + q)$ of $T_{\lambda^{p^l}, \phi_l(\lambda)}$ in the form

$$\begin{pmatrix} \chi_1 & \chi_2 \\ \chi_3 & \chi_4 \end{pmatrix}$$

with

$$\chi_i(\lambda^{p^l}, q) = \sum_{s=0}^{\infty} M_{s,i}(\lambda^{p^l})q^s, \tag{13.3.17}$$

where

$$M_s = \begin{pmatrix} M_{s,1} & M_{s,2} \\ M_{s,3} & M_{s,4} \end{pmatrix}, \tag{13.3.18}$$

M_s being defined by (13.3.7). Equations (13.3.4) and (13.3.2) now take the form

$$\begin{pmatrix} \chi_1 & \chi_2 \\ \chi_3 & \chi_4 \end{pmatrix}\begin{pmatrix} \tilde{B}_1 & \tilde{B}_2 \\ 0 & \tilde{B}_4 \end{pmatrix} = \begin{pmatrix} B_1 & B_2 \\ B_3 & B_4 \end{pmatrix}, \tag{13.3.19}$$

from which we deduce that

$$\begin{aligned} \chi_1 \tilde{B}_1 &= B_1, \\ \chi_3 \tilde{B}_1 &= B_3. \end{aligned} \tag{13.3.20}$$

We eliminate $\tilde{B}_1$ and obtain

$$\chi_1 B_3 - \chi_3 B_1 = 0,$$

which we write in the form $(\mathrm{Min}(a, b) > c)$

$$0 = \sum_{s=0}^{\infty} q^s\left(-M_{s,3}^{\phi^l} + M_{s,1}^{\phi^l}\frac{B_3}{B_1}\right). \tag{13.3.21}$$

It is clear that this relation implies that

$$\begin{pmatrix} \chi_1 & \chi_2 \\ \chi_3 & \chi_4 \end{pmatrix}^{-1}\begin{pmatrix} B_1 & B_2 \\ B_3 & B_4 \end{pmatrix} = \begin{pmatrix} * & * \\ 0 & * \end{pmatrix}. \tag{13.3.22}$$

(*Remark.* For $\mathrm{Max}(a, b) < c$, the equation is:

$$0 = \sum_{s=0}^{\infty} q^s\left(-M_{s,2}^{\phi^l} + M_{s,4}^{\phi^l}\frac{B_2}{B_4}\right).) \tag{13.3.21$'$}$$

In particular we observe that $M_0 = I$ and so

$$0 = \frac{B_3}{B_1} + q\left(-M_{1,3}^{\phi^l} + M_{1,1}^{\phi^l}\frac{B_3}{B_1}\right) + \sum_{s=2}^{\infty} q^s\left(-M_{s,3}^{\phi^l} + M_{s,1}^{\phi^l}\frac{B_3}{B_1}\right). \tag{13.3.21$''$}$$

We note that for each $s \in \mathbb{N}_1$ the coefficient of q^s is an analytic element on a superadmissible set. For $\lambda \in S_H$ we have $|B_1(\lambda)| = 1 = |\lambda| = |\lambda - 1|$ and hence

$$\left|\frac{B_3}{B_1}\right| \leq |p|, \tag{13.3.23.1}$$

$$\left|-M^{\phi^l}_{1,3} + M^{\phi^l}_{1,1}\frac{B_3}{B_1}\right| = 1, \tag{13.3.23.2}$$

$$\left|-M^{\phi^l}_{s,3} + M^{\phi^l}_{s,1}\frac{B_3}{B_1}\right| \leq 1, \qquad 1 < s < p, \tag{13.3.23.3}$$

$$\left|-M^{\phi^l}_{s,3} + M^{\phi^l}_{s,1}\frac{B_3}{B_1}\right| \leq s \qquad \text{for } s \geq p. \tag{13.3.23.4}$$

This means that the Newton polygon of equation (13.3.21) for $\lambda \in S_H$ has the form

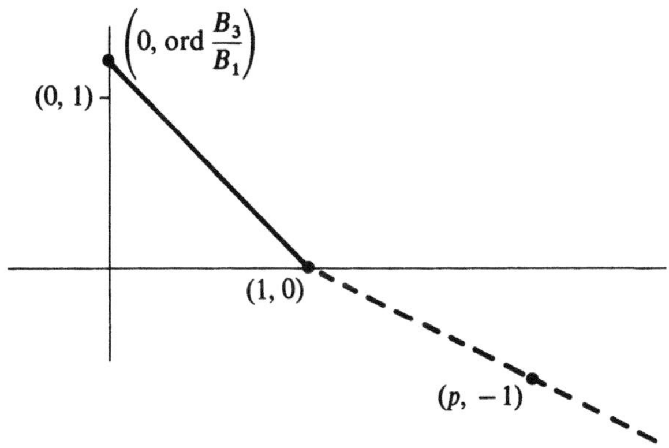

by which we mean that the first side has slope less than or equal to -1 while the succeeding sides have slope greater or equal to $-1/(p-1)$. Thus for $p \geq 3$ there is a unique solution, q_0, of (13.3.21) bounded by $|p|$. By Chapter 20 (or Newton's lemma) this defines a solution of (13.3.21) for q which is an analytic element of S_H. We observe that q_0 extends analytically to the region

$$\operatorname{Min}\left(|B_1(\lambda)|, |\lambda|, |\lambda - 1|, \left|\frac{1}{\lambda}\right|\right) > 1 - \varepsilon,$$

where ε is chosen so small that the Newton polygon for values of λ in the extended region differs only slightly from the indicated diagram. Thus in the extended region there will be a unique minimal solution and again by Chapter 20 below this minimal solution is an analytic element on the extended region. By replacing q by q_0 in (13.3.16) we obtain a superadmissible lifting of Frob^l as asserted. The remaining assertions follow from (13.3.19), the properties of the B_i (9.1.4), and the estimates for χ given by (13.3.17) for $|q| \leq |p|$. This completes the proof of the theorem. □

13.4 **Corollary.** *If* τ (resp. $\bar{\tau}$) *is the ratio of solutions of* (2.4.5.2), (resp. (2.4.5.3) *given by* (9.5.1) *with* $i = 0$, *then with* ϕ_l *as constructed in the theorem we have*

$$\tau^{\sigma^l} \circ \phi_l \text{ (resp. } \bar{\tau}^{\sigma^l} \circ \phi_l) = p^{l_0}\tau \text{ (resp. } p^{l_0}\bar{\tau}).$$

Proof. This follows from the calculation of Λ in (9.6.2) (see (9.6.7)). □

We introduce the following definition for future use:

13.5 **Definition.** If each of the sequences of maps α_i^* in (9.0.7) is unsplit then we say that every residue class is *ordinary* for (2.4.5.1). If there is at least one unsplit map in the sequence then we say that a residue class other than 0, 1, ∞ is *ordinary* if it lies in S_H for at least one choice of $(a^{(i)}, b^{(i)}, c^{(i)})$ for which $c^{(i)}$ fails to lie between $a^{(i)}$ and $b^{(i)}$.

Residue classes (other than (0, 1, ∞)) not ordinary are said to be supersingular.

As defined here ordinary and supersingular is the same for (a, b, c) as for (a', b', c').

Chapter 14

Abelian Differentials

The object of this chapter is to explain the meaning of the distinctions involving the relative magnitudes of a, b, and c. The notation here is that of Chapter 1 so the base field is $\Omega_2 = Q(\lambda)$. We define $\mathbf{V} = \mathbf{V}_f$ to be the Ω_2 space, $\mathbf{L}f\,dx$, of differentials of the form $\xi f\,dx$ with $\xi \in \mathbf{L}$. We say that $\omega \in \mathbf{V}_f$ is exact if

$$\omega = d(\xi f) \tag{14.0.1}$$

for some $\xi \in \mathbf{L}$.

14.1 **Proposition.** *Under the mapping* $\xi \mapsto \xi f\,dx$ *of* $\mathbf{L}$ *onto* $\mathbf{V}_f$, *the subspace* $D_f\mathbf{L}$ *corresponds to the subspace of exact elements of* $\mathbf{V}_f$.

Proof. This follows trivially from the definition of D_f. □

The Euler integral formula for $F(a, b, c)$),

$$\frac{\Gamma(b)\Gamma(c-b)}{\Gamma(c)} F(a, b, c, \lambda) = \int_0^1 \omega_{a,b,c} \qquad (\operatorname{Re} c > \operatorname{Re} b > 0)$$

involves the periods of

$$\omega_{a,b,c} = f\frac{1}{1-x}\,dx = x^{b-1}(1-x)^{c-b-1}(1-\lambda x)^{-a}\,dx. \tag{14.1.1}$$

It is clear that V_f (and hence dV_f) depends only upon a, b, c mod $\mathbb{Z}$. This is equivalent to the contiguity relations of Gauss.

We give the occurrence of differentials of the first kind in V_f. With no loss in generality we assume

$$\begin{gathered} 1 > a > 0, \\ 1 > b > 0, \\ 1 \geq c > 0. \end{gathered} \tag{14.1.2}$$

14.2 **Lemma.**

Type 1 $\operatorname{Min}(a, b) > c$: $\omega_{a,b,c+1}$ *is the unique element of* $\mathbf{V}_f$ *which is of the first kind.*

Type 2 $c > \operatorname{Max}(a, b)$: $\omega_{a,b,c}$ *is the unique element of* $\mathbf{V}_f$ *which is of the first kind.*

Type 3 $a > c > b$: *the* Ω_2 *span of* $\omega_{a,b,c}$ *and* $\omega_{a,b,c+1}$ *coincides with the set of all elements of* $\mathbf{V}_f$ *which are of the first kind (this set contains* $\omega_{a,b+1,c+1}$ *and* $\omega_{a-1,b,c}$*).*

Type 4 $b > c > a$: *No element of* $\mathbf{V}_f$ *is of the first kind.*

Proof. Let $\mathbf{p}_0$, $\mathbf{p}_1$, $\mathbf{p}_{1/\lambda}$, $\mathbf{p}_\infty$ be the divisors corresponding to the points 0, 1, $1/\lambda$, ∞, respectively. For $\xi \in \mathbf{L}$ we have

(14.2.1) $$\text{divisor } \xi = s_0\mathbf{p}_0 + s_1\mathbf{p}_1 + s_{1/\lambda}\mathbf{p}_{1/\lambda} + s_\infty\mathbf{p}_\infty + \text{positive divisor},$$

where $s_0, s_1, s_{1/\lambda}, s_\infty \in \mathbb{Z}$. The total degree of the divisor is zero and hence

(14.2.2) $$s_0 + s_1 + s_{1/\lambda} + s_\infty \leq 0.$$

Now $\omega_{a,b,c}\xi$ is of the first kind if for each $v \in S$

(14.2.3) $$\omega_{a,b,c}\xi \in \frac{dT_v}{T_v} T_v^{e_v}\Omega_2[[T_v]],$$

with $e_v > 0$. Thus the differential is of the first kind if and only if

(14.2.4) $$\begin{cases} b + s_0 > 0, \\ c - b + s_1 > 0, \\ 1 - a + s_{1/\lambda} > 0, \\ a + 1 - c + s_\infty > 0. \end{cases}$$

These inequalities imply

$$2 + s_0 + s_1 + s_{1/\lambda} + s_\infty > 0$$

and so by comparison with (14.2.2) we have

(14.2.5) $$s_0 + s_1 + s_{1/\lambda} + s_\infty = 0, -1.$$

On the other hand (14.2.4) together with (14.1.2) shows that

$$\text{(14.2.6)} \qquad \begin{cases} s_0 \geq 0, \\ s_1 \geq 0, \\ s_{1/\lambda} \geq 0, \\ s_\infty \geq -1. \end{cases}$$

These last two relations imply that there are at most five possible sets of values for the s_v. We list these together with the corresponding forms of the inequalities (14.2.4), omitting those inequalities already implied by (14.1.2). We also list the corresponding possibility for ξ (up to multiplication by a nonzero element of Ω_2).

s_0	s_1	$s_{1/\lambda}$	s_∞	(14.2.4)	ξ	$\xi\omega_{a,b,c}$
0	0	0	-1	$a > c > b$	$x - u, u \in \Omega_2 - \{0, 1, 1/\lambda\}$	$\omega_{a,b+1,c+1} - u\omega_{a,b,c}$
1	0	0	-1	$a > c > b$	x	$\omega_{a,b+1,c+1}$
0	1	0	-1	$a > c$	$1 - x$	$\omega_{a,b,c+1}$
0	0	1	-1	$a > c > b$	$1 - x\lambda$	$\omega_{a-1,b,c}$
0	0	0	0	$c > b$	1	$\omega_{a,b,c}$

We note that while $\mathbf{V}_f$ is an infinite-dimensional Ω_2 space, all the elements in each of the last two columns of the table span a two-dimensional Ω_2 space. The lemma follows directly from this table. □

Remark. Without (14.1.2), $\omega_{a,b,c}$ is of the first kind if and only if

$$2 > a + 1 > c > b > 0$$

a condition which appears in much of the classical literature. The purpose of the lemma is to explain that the different cases of Theorem 6.6 are consistent with the principle that $\alpha_{f,h}$ is a lifting of the Cartier operator and that the Cartier operator maps differentials of the second kind into differentials of the first kind.

Chapter 15

Canonical Lifting for $l = 1$

In this chapter we restrict the considerations of Chapter 13 to the case in which $l = 1$. Thus we assume

$$(p - 1)(a, b, c) \in \mathbb{Z}^3 \tag{15.0.1}$$

and we will continue to assume (9.0.2)–(9.0.5). We will concern ourselves with the situation in which c does not lie between a and b.

Theorem. *Under hypothesis* (15.0.1), *if c does not lie between a and b, then the canonical lifting, ϕ_1, described in Theorem* 13.3 *extends* in *each supersingular disk to the region*

$$\left\{\lambda \,|\, 0 < \operatorname{ord}(\lambda - z) < 1 - \frac{1}{p+1}\right\}, \tag{15.1}$$

where z denotes a nonramified representative of the supersingular disk.

Proof. To fix ideas let $\operatorname{Min}(a, b) > c$. (The case $c > \operatorname{Max}(a, b)$ involves minor modification of the argument.)

The proof is based upon a closer examination of equation (13.3.21″). We first observe that since $l = 1$, the matrix B coincides with the matrix

$$\mathbf{A} = \begin{pmatrix} A_1 & A_2 \\ pA_3 & pA_4 \end{pmatrix}$$

of Theorem 6.6 and so in particular by that theorem and Lemma 7.2, $(B_1, B_2) \bmod p\mathbf{Q}_0'$ is the unique polynomial solution of (2.4.5.1) of degree bounded by $p - 1$. In particular $B_1 \bmod p\mathbf{Q}_0'$ is a solution mod p of (2.4.5.2) of degree less than p and so the zeros of B_1 other than 0, 1 are *simple.* We conclude that in each supersingular disk, B_1 has a unique zero z, unramified over $\mathbf{Q}_p$ and this zero is a simple zero of B_1. From these remarks we conclude that for $\lambda \in D(z, 1^-)$ we have

$$\operatorname{ord} B_1(\lambda) = \operatorname{ord}(\lambda - z), \tag{15.2}$$

an estimate completely unknown in the general case studied in Chapter 13. Again in the region $D(z, 1^-)$ we have

$$\text{ord } B_3(\lambda) \geq 1 \tag{15.3}$$

and in fact by Theorem 6.6(iii), $(1/p)(B_1 B_4 - B_3 B_2)$ assumes unit values and B_4/p assumes values bounded by unity on Ω_1. Hence for $\lambda \in D(z, 1^-)$

$$\text{ord } B_3(\lambda) = 1. \tag{15.3'}$$

By means of the trivial estimate (13.3.10), we conclude that on $D(z, 1^-)$

(15.4)

$$\text{ord}\left(-M^{\phi}_{s,3} + M^{\phi}_{s,1}\frac{B_3}{B_1}\right) \geq -\frac{s}{p-1} + \text{Min}(0, 1 - \text{ord}(\lambda - z)), \qquad s \geq 2,$$

(15.4′)

$$\text{ord}\left(-M^{\phi}_{1,3} + M^{\phi}_{1,1}\frac{B_3}{B_1}\right) = 0.$$

From this we conclude that the Newton polygon of (13.3.21″) is of the form shown with a break at (1, 0) provided

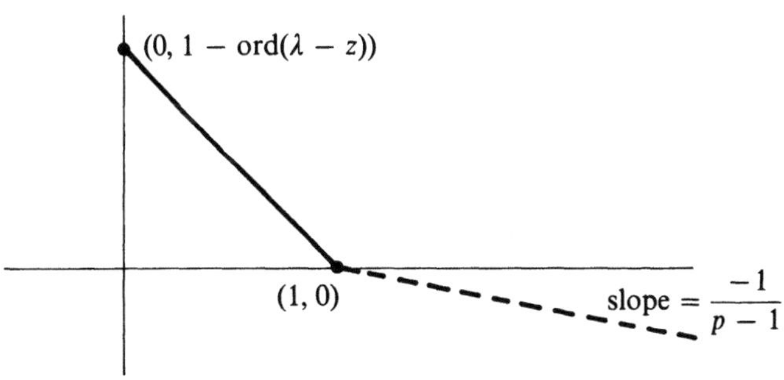

$$1 - \text{ord}(\lambda - z) > \frac{1}{p-1}. \tag{15.5}$$

This shows that ϕ_1 may be extended to the region

$$0 < \text{ord}(\lambda - z) < 1 - \frac{1}{p-1}, \tag{15.5'}$$

a somewhat weaker result than (15.1).

Our object is to show that a break occurs at $(1, 0)$ for λ in the region (15.1) and for this we propose to show that for $s \geq 2$, λ subject to (15.1) we have

$$1 - \operatorname{ord}(\lambda - z) > \frac{-1}{s-1} \operatorname{ord} M_{s,3}(\lambda^p), \tag{15.6}$$

$$1 - \operatorname{ord}(\lambda - z) > \frac{-1}{s-1}(1 + \operatorname{ord} M_{s,1}(\lambda^p) - \operatorname{ord}(\lambda - z)). \tag{15.7}$$

By the nontrivial estimate (13.3.11)

$$\operatorname{Min}(\operatorname{ord} M_{s,3}(\lambda^p), \operatorname{ord} M_{s,1}(\lambda^p)) \geq -\frac{\log s}{\log p} \tag{15.8}$$

and since $1 - \operatorname{ord}(\lambda - z) > 1/(p + 1)$ (by 15.1)), conditions (15.6) and (15.7) are certainly satisfied for those s such that

$$\frac{1}{p+1} \geq \frac{\log s}{\log p} \frac{1}{s-1}. \tag{15.9}$$

This inequality holds for $s \geq 2p + 3$ since $p^2 \geq 2p + 3$ for $p \geq 3$ and $1/(p + 1) \geq 1/s \log p$ for $s \geq p + 1$ if $p \geq 3$.

We now restrict our attention to $s \leq 2p + 2$ and for such s, by (13.3.11), we have

$$\operatorname{Min}(\operatorname{ord} M_{s,3}(\lambda^p), \operatorname{ord} M_{s,1}(\lambda^p)) \geq -1 \tag{15.10}$$

and so conditions (15.6) and (15.7) certainly hold if

$$\frac{1}{p+1} \geq \frac{1}{s-1}, \tag{15.11}$$

i.e., if $s \geq p + 2$. Again for $2 \leq s \leq p - 1$, estimate (13.3.11) gives

$$\operatorname{Min}(\operatorname{ord} M_{s,3}(\lambda^p), \operatorname{ord} M_{s,1}(\lambda^p)) \geq 0,$$

so that conditions (15.6) and (15.7) are satisfied trivially for such small s.

Summarizing: For λ satisfying (15.1), conditions (15.6) and (15.7) are now verified for all s except for $s = p, p + 1$.

Now for $s = p + 1$, with (15.10) we verify (15.7) for

$$1 - \operatorname{ord}(\lambda - z) > \frac{1}{p} \operatorname{ord}(\lambda - z),$$

which is precisely condition (15.1). It remains to show that

$$(15.12.1)\quad 1 - \operatorname{ord}(\lambda - z) > -\frac{1}{p-1}\operatorname{ord} M_{p,3}(\lambda^p),$$

$$(15.12.2)\quad 1 - \operatorname{ord}(\lambda - z) > -\frac{1}{p}\operatorname{ord} M_{p+1,3}(\lambda^p),$$

$$(15.12.3)\quad 1 - \operatorname{ord}(\lambda - z) > -\frac{1}{p-1}(1 - \operatorname{ord}(\lambda - z) + \operatorname{ord} M_{p,1}(\lambda^p))$$

subject to (15.1). But in fact these three follow from (15.10) if $\operatorname{ord}(\lambda - z) < 1 - 1/(p-1)$ and so with no loss in generality we may assume

$$(15.13)\qquad 1 - \frac{1}{p+1} > \operatorname{ord}(\lambda - z) \geq \frac{1}{p-1}.$$

We will in fact show: If

$$(15.14)\qquad \operatorname{ord}(\lambda - z) \geq \frac{1}{p},$$

then

$$\operatorname{Min}(\operatorname{ord} M_{p,3}(\lambda^p), \operatorname{ord} M_{p,1}(\lambda^p), \operatorname{ord} M_{p+1,3}(\lambda^p)) \geq 0.$$

This clearly implies (15.12).

With this object in mind we examine relationships between the zeros of $pM_{p,3}$ and B_1. Indeed by (13.3.13)

$$(15.15.1)\qquad M_{p,3} = X_p M_{1,3},$$

$$(15.15.2)\qquad M_{p,1} = X_p M_{1,1} + Y_p$$

and by the recursion relationship (13.3.8),

$$(p+1)M_{p+1,3} = M'_{p,3} + M_{1,3}M_{p,1} + M_{1,4}M_{p,3},$$

$$(p+1)M_{p+1,1} = M'_{p,1} + M_{1,1}M_{p,1} + M_{1,2}M_{p,3}$$

and so by the two preceding relations,

$$(15.15.3)\qquad \begin{aligned}(p+1)M_{p+1,3} &= M_{1,3}(X'_p + Y_p) + 2X_p M_{2,3},\\ (p+1)M_{p+1,1} &= M_{1,1}(X'_p + Y_p) + Y'_p + 2X_p M_{2,1}.\end{aligned}$$

We assert that the class $\bar{z}$ of z is related to pX_p and pY_p in the following way:

(15.16.0) $\bar{z}$ is a simple zero of B_1 mod p (as already noted),

(15.16.1) $\bar{z}$ is a zero of pX_p mod p of order two,

(15.16.2) $\bar{z}$ is a simple zero of pY_p mod p,

(15.16.3) $\bar{z}$ is a simple zero of $pX'_p + pY_p$ mod p.

B_1 mod p is a polynomial of degree bounded by $p - 1$ and is solution of (2.4.5.2) mod p. It follows that in (10.2.1) we may replace g by B_1 mod p up to powers of x and $1 - x$ and a_p by $p!X_p$. Assertion (15.16.1) follows from Corollary 10.7 and from (15.16.0). Again by (10.4.3)

$$0 \equiv B_1^{(p)} \equiv (p!X_p)B'_1 + (p!Y_p)B_1 \mod p. \tag{15.17}$$

This equation together with (15.16.0) and (15.16.1) imply (15.16.2). From (15.17) we compute

$$pX'_p + pY_p \equiv B_1\left(\frac{pX_p}{B_1}\right)' \mod p. \tag{15.18}$$

Assertion (15.16.3) follows directly.

From (15.15) and (15.16) we deduce

(15.19.1) $\bar{z}$ is a zero of order two of $pM_{p,3}$ mod p,

(15.19.2) $\bar{z}$ is a simple zero of $pM_{p,1}$ mod p,

(15.19.3) $\bar{z}$ is a simple zero of $pM_{p+1,3}$ mod p,

(15.19.4) $\bar{z}$ is not a zero of $pM_{p+1,1}$ mod p.

Thus $pM_{p,1}$ has a zero $z_1 \in D(z, 1^-)$ unramified over $\mathbb{Q}_p$ and likewise $pM_{p+1,3}$ has a zero $z_2 \in D(z, 1^-)$ unramified over $\mathbb{Q}_p$. Finally $pM_{p,3}$ has a numerator in $\mathbb{Z}_p[\lambda]$ with exactly two roots z_3, z_4 in $D(z, 1^-)$. The set $\{z_3, z_4\}$ is clearly invariant under the galois group of $\Omega/\mathbb{Q}_p(z)$ and hence the polynomial $(\lambda - z_3)(\lambda - z_4)$ has coefficients in $\mathbb{Q}_p(z)$, i.e., in Ω_0. (We do not exclude the possibility that $z_3 = z_4$.) Hence

$$\operatorname{ord}((z_3 - z) + (z_4 - z)) \geq 1,$$

$$\operatorname{ord}(z^2 - z_3 z_4) \geq 1$$

and so the numerator of $pM_{p,3}$ has a quadratic factor

$$\mu(\lambda) = (\lambda - z)^2 + p(c_0 + c_1\lambda), \tag{15.20}$$

with c_0, c_1 integral in Ω_0. Letting σ denote as usual the Frobenius automorphism of Ω_0 over $\mathbb{Q}_p$, we see that

$$\begin{cases} z_1^\sigma \text{ is a zero of } pM_{p,1}, \\ z_2^\sigma \text{ is a zero of } pM_{p+1,1}, \\ (\lambda - z^\sigma)^2 + p(c_0^\sigma + c_1^\sigma\lambda) \text{ is a factor of } pM_{p,3}. \end{cases} \tag{15.21}$$

Clearly $\operatorname{ord}(z - z_i) \geq 1$ $(i = 1, 2)$ and hence for $i = 1, 2$

$$\begin{aligned} \operatorname{ord}(\lambda^p - z_i^\sigma) &= \operatorname{ord}(\lambda^p - z^p + z^p - z_i^\sigma) \\ &\geq \operatorname{Min}(\operatorname{ord}(\lambda^p - z^p), 1) \\ &\geq \operatorname{Min}(1 + \operatorname{ord}(\lambda - z), p \operatorname{ord}(\lambda - z), 1). \end{aligned}$$

Thus under the hypothesis of (15.14), for $i = 1, 2$

$$\operatorname{ord}(\lambda^p - z_i^\sigma) \geq 1,$$

i.e.,

$$\operatorname{Min}(\operatorname{ord} M_{p,1}(\lambda^p), \operatorname{ord} M_{p+1,3}(\lambda^p)) \geq 0.$$

This proves part of (15.14). As noted

$$\mu^\sigma(\lambda) = (\lambda - z^\sigma)^2 + p(c_0^\sigma + c_1^\sigma\lambda)$$

is a factor of $pM_{p,3}$ and under the hypothesis of (15.14)

$$\operatorname{ord} \mu^\sigma(\lambda^p) \geq 1.$$

We conclude that $M_{p,3}$ also satisfies assertion (15.14). This completes the proof of (15.14) and hence of the theorem. □

Corollary. *Under the hypothesis of the theorem if (say)* $\operatorname{Min}(a, b) > c$ *then there exists* $\varepsilon > 0$ *such that* ϕ_1 *is analytic in the region defined by the simultaneous conditions,*

$$\operatorname{ord} B_1 < 1 - \frac{1}{p + 1},$$

$$\operatorname{Max}(\operatorname{ord} \lambda, \operatorname{ord}(1 - \lambda), -\operatorname{ord} \lambda) < \varepsilon.$$

Chapter 16

Supersingular Disks

We proceed to examine more carefully the canonical lifting ϕ_1 of Chapter 15. The hypotheses are those of that section and in particular $l = 1$. We will restrict our attention to the case $\operatorname{Min}(a, b) > c$ but the extension to the case $c > \operatorname{Max}(a, b)$ causes no difficulty. As before we write

$$\phi_1(\lambda) = \lambda^p + q. \tag{16.0.1}$$

Our object is to study q and $\tilde{B}$ (cf. (13.3.2)) on supersingular disks. In such a disk, B_1 has a unique zero, z. We know that $z \in \Omega_0$ and so is an unramified representative of the residue class. This representative need not be the most convenient one. In the following let z_0 be an element of Ω_0 representing the class such that

$$\phi(z_0) = z_0^p = \sigma z_0. \tag{16.0.2}$$

Of course

$$\operatorname{ord}(z - z_0) \geq 1.$$

Furthermore the support of ϕ_1 as analytic function contains the annulus

$$\left\{\lambda \,|\, \operatorname{ord}(\lambda - z_0) \in \left(0, 1 - \frac{1}{p+1}\right)\right\}. \tag{16.0.3}$$

16.1 **Lemma.** *In the annulus* (16.0.3),

$$\operatorname{ord} q(\lambda) = 1 - \operatorname{ord}(\lambda - z_0).$$

Proof. This follows from equations (13.3.21″), (15.4′), (15.2) and (15.3′). This completes the proof. □

16.2 **Corollary.** *In the annulus* (16.0.3), *q has the representation*

$$q(\lambda) = \sum_{j=0}^{\infty} k_j(\lambda - z_0)^j + \sum_{j=1}^{\infty} \frac{m_j}{(\lambda - z_0)^j}, \tag{16.2.0}$$

$$\text{ord } k_j \geq 1, \qquad j \geq 0, \tag{16.2.1}$$

$$\text{ord } m_j \geq 1 + (j-1)\frac{p}{p+1}, \qquad j \geq 1, \tag{16.2.2}$$

$$\text{ord } m_1 = 1. \tag{16.2.3}$$

Proof. The Laurent series representation follows from analyticity. By Cauchy's inequality and the lemma, for λ satisfying (16.0.3),

$$\text{ord } k_j + j\,\text{ord}(\lambda - z_0) \geq 1 - \text{ord}(\lambda - z_0), \qquad \forall j \geq 0, \tag{16.2.4}$$

$$\text{ord } m_j - j\,\text{ord}(\lambda - z_0) \geq 1 - \text{ord}(\lambda - z_0), \qquad \forall j \geq 1. \tag{16.2.5}$$

Letting $\text{ord}(\lambda - z_0) \to 0$ in (16.2.4) we obtain (16.2.1), letting $\text{ord}(\lambda - z_0) \to p/(p+1)$ in (16.2.5) we obtain (16.2.2).

Now again the equality part of Cauchy's inequality implies that for each λ in the annulus (16.0.3) there exists j (possibly depending upon λ) such that either

$$\text{ord } k_j + j\,\text{ord}(\lambda - z_0) = 1 - \text{ord}(\lambda - z_0) \tag{16.2.6}$$

or

$$\text{ord } m_j - j\,\text{ord}(\lambda - z_0) = 1 - \text{ord}(\lambda - z_0). \tag{16.2.7}$$

The suggested (16.2.6) is ruled out by (16.2.1), and by equation (16.2.2) we conclude that (16.2.7) can occur only for $j = 1$. This completes the proof of the corollary. □

16.3 **Lemma.** *In the annulus* (16.0.3)

$$\text{ord } \frac{dq}{d\lambda} = 1 - 2\,\text{ord}(\lambda - z_0).$$

Proof. By (16.2.0) we have

$$\frac{dq}{d\lambda} = \sum_{j=1}^{\infty} jk_j(\lambda - z_0)^{j-1} - \sum_{j=1}^{\infty} \frac{jm_j}{(\lambda - z_0)^{j+1}}.$$

Now for $j \geq 1$ by (16.2.1),

$$\operatorname{ord}(jk_j(\lambda - z_0)^{j-1}) \geq 1 > 1 - 2\operatorname{ord}(\lambda - z_0)$$

and for $j > 1$ by (16.2.2),

$$\begin{aligned}\operatorname{ord}\frac{jm_j}{(\lambda - z_0)^{j+1}} &\geq 1 + (j-1)\frac{p}{p+1} - (j+1)\operatorname{ord}(\lambda - z_0)\\ &= 1 - 2\operatorname{ord}(\lambda - z_0) + (j-1)\left(\frac{p}{p+1} - \operatorname{ord}(\lambda - z_0)\right)\\ &> 1 - 2\operatorname{ord}(\lambda - z_0),\end{aligned}$$

while for $j = 1$ by (16.2.3)

$$\operatorname{ord}\frac{m_1}{(\lambda - z_0)^2} = 1 - 2\operatorname{ord}(\lambda - z_0).$$

This completes the verification of the lemma. □

We now consider the values of $\tilde{B}_1$ (cf. (13.3.2)) in the annulus (16.0.3).

16.4 **Lemma.** *In the annulus* (16.0.3),

$$\operatorname{ord}\tilde{B}_1(\lambda) = \operatorname{ord}(\lambda - z_0).$$

Proof. By Corollary 13.4, on the generic disk (in fact on any ordinary residue class)

$$\tau^{\sigma}(\phi_1(\lambda)) = p\tau(\lambda). \tag{16.4.1}$$

Hence

$$\frac{d\phi_1}{d\lambda}\left(\frac{d\tau}{d\lambda}\right)^{\sigma\phi_1} = p\frac{d\tau}{d\lambda}. \tag{16.4.2}$$

By (9.7.5) ($g_3 = M_{1,3}$, cf. (13.3.12.1))

$$\frac{d\tau}{d\lambda} = g_3\frac{w}{u^2}, \tag{16.4.3}$$

where $(u, \bar{u})$ is the first row and w is the determinant of the normalized solution matrix $\mathbf{Y}_0$ of Theorem 9.4. Hence

$$(u, \bar{u})^{\sigma\phi_1}\begin{pmatrix}\tilde{B}_1 & \tilde{B}_2\\ 0 & \tilde{B}_4\end{pmatrix} = (u, \bar{u}), \tag{16.4.4}$$

and so

$$u^{1-\sigma\phi_1} = \tilde{B}_1. \tag{16.4.5}$$

From this we conclude that

$$\frac{d\phi_1}{d\lambda} = \frac{p(g_3 w)^{1-\sigma\phi_1}}{\tilde{B}_1^2}. \tag{16.4.6}$$

We recall that w is a branch of

$$\lambda^{-c}(1-\lambda)^{c-(a+b)} \tag{16.4.7}$$

normalized so as to be a unit at t (but by (9.4.2), w is unique up to a unit in $\mathbb{Z}_p$). It is clear that while equation (16.4.2) contains terms defined only in ordinary disks, the right side of (16.4.6) is defined on the support of ϕ_1. Furthermore $(g_3 w)^{1-\sigma\phi_1}$ assume unit values on the intersection of Ω_1 with the support of ϕ_1. Thus on the annulus (16.0.3) we have

$$1 - 2\operatorname{ord}(\lambda - z_0) = \operatorname{ord}\left(\frac{dq}{d\lambda}\right) = \operatorname{ord}\left(\frac{d\phi_1}{d\lambda}\right) = 1 - 2\operatorname{ord}\tilde{B}_1(\lambda). \tag{16.4.8}$$

The lemma follows directly. □

We extend σ, the Frobenius automorphism of Ω_0 over $\mathbb{Q}_p$, to $\Omega_0(\sqrt{p})$ by setting

$$\sigma\sqrt{p} = \sqrt{p}.$$

16.5 **Lemma.** *There exists a solution matrix* $\mathbf{Y}$ *of* (2.4.5.1) *on* $D(z_0, 1^-)$, *defined over* $\Omega_0(\sqrt{p})$ *such that*

$$\mathbf{Y}^{\sigma\phi_1}\tilde{B} = \sqrt{p}\mathbf{Y} = \mathbf{Y}^{\sigma\phi}B \tag{16.5.1}$$

and such that

$$\mathbf{Y} = \begin{pmatrix} u_1 & u_2 \\ v_1 & v_2 \end{pmatrix},$$

with

$$\operatorname{ord}\left(\frac{u_2}{u_1}\right)(z_0) = \operatorname{ord}\left(\frac{v_2}{v_1}\right)(z_0) = \operatorname{ord}\left(\frac{v_2}{v_1}(z_0) - \frac{u_2}{u_1}(z_0)\right) = -\frac{1}{2}, \tag{16.5.2}$$

$$\frac{v_1}{u_1}(z_0) \bmod \sqrt{p} \quad \textit{lies in } \mathbb{F}_{p^2} - \mathbb{F}_p. \tag{16.5.3}$$

Remark. We refer to such a pair of eigenvectors as a *polarized pair* of "eigenvectors."

Proof. The mapping

$$(u_1, u_2) \mapsto (u_1, u_2)^{\sigma\phi_1}\tilde{B} \tag{16.5.4}$$

is by construction the same as the mapping

$$(u_1, u_2) \mapsto (u_1, u_2)^{\sigma\phi}B \tag{16.5.5}$$

and by (9.4.5) the matrix of this mapping as semilinear transformation is $B(z_0)$ (by choice (16.0.2) of z_0). We are now in the supersingular situation with $B_1(z_0) \equiv 0 \bmod p$, $B_2(z_0)$, $p^{-1}B_3(z_0)$ units, $B_4(z_0) \equiv 0 \bmod p$. The assertion now follows from Corollaries 12.3 and 12.4. □

16.6 **Lemma.**

$$\operatorname{ord}(\phi_1(\lambda) - z_0^\sigma) = \begin{cases} p\operatorname{ord}(\lambda - z_0) & \textit{if}\operatorname{ord}(\lambda - z_0) < \dfrac{1}{p+1}, \\ 1 - \operatorname{ord}(\lambda - z_0) & \textit{if}\operatorname{ord}(\lambda - z_0) \in \left(\dfrac{1}{p+1}, \dfrac{p}{p+1}\right), \\ \geq \dfrac{p}{p+1} & \textit{if}\operatorname{ord}(\lambda - z_0) = \dfrac{1}{p+1}. \end{cases}$$

Proof. By Lemma 16.1:

$$\operatorname{ord}(\phi_1(\lambda) - \lambda^p) = 1 - \operatorname{ord}(\lambda - z_0) \quad \text{if } 0 < \operatorname{ord}(\lambda - z_0) < \frac{p}{p+1},$$

but as is well known,

$$\operatorname{ord}(\lambda^p - z_0^\sigma) = \begin{cases} p\operatorname{ord}(\lambda - z_0) & \text{if } \operatorname{ord}(\lambda - z_0) < \dfrac{1}{p-1}, \\ 1 + \operatorname{ord}(\lambda - z_0) & \text{if } \operatorname{ord}(\lambda - z_0) > \dfrac{1}{p-1}. \end{cases}$$

The graphs of these two quantities as function of $\operatorname{ord}(\lambda - z_0)$ cross at $\operatorname{ord}(\lambda - z_0) = 1/(p+1)$.

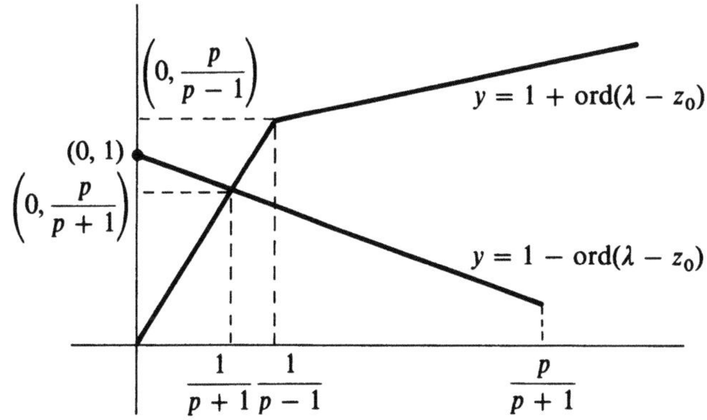

The lemma may now be deduced with the aid of the diagram. Note that while $\operatorname{ord}(\lambda^p - z_0^\sigma)$ is not uniquely determined when $\operatorname{ord}(\lambda - z_0) = 1/(p-1)$, this causes no difficulty since $\operatorname{ord}(\phi_1(\lambda) - \lambda^p)$ controls in that range. □

16.7 **Proposition.**

$$\phi_1(\lambda) - z_0^\sigma = (\lambda - z_0)^p + \sum_{j=0}^{\infty} k_j'(\lambda - z_0)^j + \sum_{j=1}^{\infty} \frac{m_j}{(\lambda - z_0)^j},$$

where m_j is as in (16.2.0) *and*

$$k_j' \equiv k_j \mod p, \qquad \forall j \geq 0.$$

Proof. This follows trivially from (16.0.1) and (16.2.0). □

From this we may construct the Newton polygon of $\phi_1(\lambda) - z_0^\sigma$:

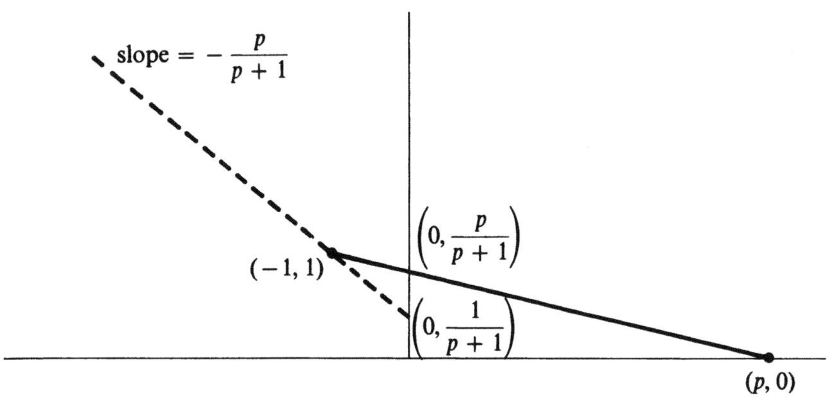

16.8 **Corollary.** *The equation(for* λ*),*

$$\phi_1(\lambda) = z_0^\sigma + y$$

has in the annulus (16.0.3)

$$p + 1 \textit{ solutions if } \operatorname{ord} y > \frac{1}{p+1},$$

$$p \qquad\qquad \textit{if } 0 < \operatorname{ord} y \le \frac{1}{p+1}.$$

The distribution of $\operatorname{ord}(\lambda - z_0)$ *as* λ *runs over the solutions is as follows: (The rows correspond to different ranges of* ord y*; the columns to values of* $\operatorname{ord}(\lambda - z_0)$. *The entries give the number of roots.)*

ord y \ $\operatorname{ord}(\lambda - z_0)$	$(1/p)\operatorname{ord} y$	$1 - \operatorname{ord} y$	$1/(p+1)$
$\left(\frac{1}{p+1}, \frac{p}{p+1}\right)$	p	1	0
$\ge \frac{p}{p+1}$	—	—	$p+1$
$\le \frac{1}{p+1}$	p	0	0

We now examine the rows of the matrix **Y** of Lemma 16.5.

16.9 **Lemma.** *If* $(u, \bar{u})$ *is row of* **Y** *then as* $|\lambda - z_0| = r \to 1$.

(16.9.1) $$0 < c_2 < |u|_{z_0}(r) \Big/ \sqrt{\frac{1}{\log(1/r)}} < c_1$$

for suitable constants c_1, c_2 *independent of* r.

Proof. By hypotheses

(16.9.2) $$u^{\sigma\phi_1}\tilde{B}_1 = \sqrt{p}u.$$

Fix

$$\operatorname{ord}(\lambda_0 - z_0) \in \left(\frac{1}{p+1}, \frac{p}{p+1}\right).$$

Then we may choose an infinite sequence $\lambda_1, \lambda_2, \ldots$ such that

(16.9.3) $$\lambda_s \in D(z_0^{\sigma^{-s}}, 1^-), \qquad s \geq 0,$$

(16.9.4) $$\phi_1(\lambda_s) = \lambda_{s-1}, \qquad s \geq 1,$$

(16.9.5) $$p \operatorname{ord}(\lambda_s - z_0^{\sigma^{-s}}) = \operatorname{ord}(\lambda_{s-1} - z_0^{\sigma^{-s+1}}), \qquad s \geq 1.$$

Now by (16.9.2)

(16.9.6) $$u^{\sigma^{-s}}(\lambda_s) = \frac{1}{p^{s/2}} \tilde{B}_1^{1+\phi_1+\cdots+\phi_1^{s-1}}(\lambda_s) u(\phi_1^s(\lambda_s))$$

and so by Lemma 16.4

(16.9.7) $$\operatorname{ord} u^{\sigma^{-s}}(\lambda_s) = -\frac{s}{2} + \sum_{i=0}^{s-1} \operatorname{ord}(\phi^i(\lambda_s) - z_0^{\sigma^{i-s}}) + \operatorname{ord} u(\lambda_0).$$

But

(16.9.8) $$\operatorname{ord}(\phi^i(\lambda_s) - z_0^{\sigma^{i-s}}) = \operatorname{ord}(\lambda_{s-i} - z_0^{\sigma^{i-s}}) = \frac{1}{p^{s-i}} \operatorname{ord}(\lambda_0 - z_0),$$

and so

(16.9.9) $$\operatorname{ord} u^{\sigma^{-s}}(\lambda_s) = -\frac{s}{2} + \frac{1-(1/p^s)}{p-1} \operatorname{ord}(\lambda_0 - z_0) + \operatorname{ord} u(\lambda_0).$$

To compute

(16.9.10) $$|u|_{z_0}(r) = |u^{\sigma^{-s}}|_{z_0^{\sigma^{-s}}}(r),$$

we compute $s \in \mathbb{N}$ such that

(16.9.11) $$-p^s \frac{\log r}{\log p} \in \left(\frac{1}{p+1}, \frac{p}{p+1}\right).$$

Let r_0 real be defined by

$$p^s \frac{\log r}{\log p} = \frac{\log r_0}{\log p}, \tag{16.9.12}$$

so

$$\left(\frac{1}{p}\right)^{1/(p+1)} > r_0 > \left(\frac{1}{p}\right)^{p/(p+1)}. \tag{16.9.13}$$

Then, by (16.9.9),

$$|u|_{z_0}(r) = p^{s/2}|u|_{z_0}(r_0) \cdot r_0^{(1-(1/p^s))/(p-1)}. \tag{16.9.14}$$

The lemma now follows from (16.9.11) and (16.9.13). □

16.10 **Corollary.** *No solution of* (2.4.5.1) *has asymptotic growth on* $D(z_0, 1^-)$ *slower than that represented by* (16.9.1). *In particular no solution is bounded on the disk. (Hence each solution has an infinite set of zeros in the disk.)*

Proof. Any solution, (u_1, u_2), of slower growth would be characterized up to constant multiples by this property. It follows that the space $\Omega(u_1, u_2)$ would be stable under the Galois group of Ω over $\mathbb{Q}_p(z_0)$. This would mean that $u_2(z_0)/u_1(z_0)$ lies in $\mathbb{Q}_p(z_0)$ and hence we may assume u_1, u_2 defined over $\mathbb{Q}_p(z_0)$. However order of growth is not increased by the mapping (16.5.5) and hence we would have for some scalar, c,

$$(u_1, u_2)^{\sigma\phi}B = c(u_1, u_2).$$

Specializing at z_0, putting $\mathbf{u} = (u_1(z_0), u_2(z_0))$, we would have

$$\mathbf{u}^\sigma B(z_0) = c\mathbf{u},$$

contradicting Corollary 12.5. □

16.11

We recall the function η_0 of (9.5.1). In the present context since $l = 1$, condition (9.6.2) may be written

$$\eta = \frac{B_2 + B_4\eta^\phi}{B_1 + B_3\eta^\phi} \tag{16.11.1}$$

and η lies in $H(S_H)$. This function coincides on each ordinary residue class with the ratio u_2/u_1 of the components of the unique bounded solution of (2.4.5.1) on that class.

On supersingular residue classes there are no bounded solutions and so one may suspect that η cannot be continued analytically outside S_H. We propose to show that this is indeed the case.

16.11.2 **Definition.** An open subset $\mathbf{C}$ of $\Omega \cup \{\infty\}$ is said to be an analytic set if $H(\mathbf{C})$ satisfies the principal of analytic continuation, i.e., if ξ is the uniform limit on $\mathbf{C}$ of elements of $\Omega(x)$ having no pole in $\mathbf{C}$ and if ξ coincides with the zero function on a disk contained by $\mathbf{C}$ then ξ is identically zero on $\mathbf{C}$.

Theorem. *Let $\mathbf{C}$ be an analytic set, $\mathbf{C} \supset S_H$, and let $\mathbf{C} \cap D(z_0, 1^-)$ contain an open subset $\mathbf{C}_0$. Then η does not lie in $H(\mathbf{C})$.*

Proof. Suppose otherwise. Let $z_1 \in \mathbf{C}_0$ and let u_1 be a nontrivial solution at z_1 of the linear first-order scalar equation,

$$u_1' = \left(-\frac{c}{\lambda} + \frac{c-b}{\lambda}\eta\right)u_1. \tag{16.11.3}$$

Then $u = (u_1, u_1\eta)$ is a solution at z_1 of (2.4.5.1) and hence must be analytic on $D(z_0, 1^-)$. This gives the extension of u_1 to a function analytic on $D(z_0, 1^-)$ and the extension of η to a function meromorphic on that set. It follows from the uniqueness of η as solution of (16.11.1) in $H_0(S_H)$ that η is invariant under all valuation preserving automorphisms of Ω over $\mathbb{Q}_p$. Hence $\eta(z_0)$ is invariant under all valuation preserving automorphisms of Ω over $\mathbb{Q}_p(z_0)$. We may with no loss in generality assume Ω to be maximally complete and hence (e.g., §8 of Dwork and Robba [16]) conclude that $\eta(z_0)$ is either infinite or an element of $\mathbb{Q}_p(z_0)$.

In either case we may normalize u so that $\mathbf{u}$ $(= u(z_0))$ is defined over $\mathbb{Q}_p(z_0)$ and hence in particular

$$u^\sigma(z_0^p) = (u(z_0))^\sigma. \tag{16.11.4}$$

If we put

$$v = u^{\sigma\phi}B, \tag{16.11.5}$$

then by (16.11.1), v_1, the first component of v, is again a solution of (16.11.3) and hence

$$v_1 = eu_1$$

for some constant e. Hence $v = eu$ and so

(16.11.6) $$eu = u^{\sigma\phi}B.$$

Specializing at z_0, and using (16.11.4), we have

$$e\mathbf{u} = \mathbf{u}^{\sigma}B(z_0),$$

which contradicts Corollary 12.5. □

16.12

We have already shown (Lemma 16.10) that equation (2.4.5.1) has no solution bounded on $D(z_0, 1^-)$. A second demonstration follows from the theorem of §16.11 together with part (ii) of Theorem 2.4 of Dwork and Robba [16].

Chapter 17

The Function τ on Supersingular Disks ($l = 1$)

In this section we continue the discussion of the matrix **Y** of Lemma 16.5. We again restrict our attention to the case $l = 1$, $\mathrm{Min}(a, b) > c$. By minor modifications, the case $c > \mathrm{Max}(a, b)$ may also be treated. As before $z_0^\sigma = z_0^p$, z_0 represents a supersingular disk.

We put $\tau = v_1/u_1$ in the notation of Lemma 16.5. Thus we have

$$\tau^{\sigma\phi_1} = \tau, \tag{17.1}$$

$$\tau(z_0) \bmod \sqrt{p} \in \mathbb{F}_{p^2} - \mathbb{F}_p. \tag{17.2}$$

As preparation for an examination of function theoretic properties of τ, we need estimates on the magnitude of the coefficients of the matrices M_s (cf. 13.3.7).

17.3 **Lemma.** *Under the hypotheses of Chapter* 15, *in particular* $\mathrm{Min}(a, b) > c$,

$$\operatorname{ord} M_{s,3}(z_0) \geq -\frac{s-1}{p+1} \quad \textit{for } s \geq 0, \tag{17.3.1}$$

$$\operatorname{ord} M_{s,1}(z_0) \geq -\frac{s-1}{p+1} \quad \textit{for } s \geq 1 \text{ except } \textit{for } s = p + 1, \tag{17.3.2}$$

$$\operatorname{ord} M_{p+1,1}(z_0) = -1. \tag{17.3.3}$$

Proof. The argument is similar to that of the proof of the theorem in Chapter 15, the distinction being that in Chapter 15 we considered the value of $M_{s,1}$ and $M_{s,3}$ at various points of $D(z_0^p, 1^-)$ and here we consider the value only at the center, z_0. The inequality (15.9) is known to hold for $s \geq 2p + 3$. This together with the nontrivial estimate (13.3.11), shows that (17.3.1) and (17.3.2) hold for $s \geq 2p + 3$. For $s < p$, the assertions are trivial. For $p \leq s < 2p + 3$, (13.3.11) shows that (17.3.1) and (17.3.2) hold provided

$$-\frac{s-1}{p+1} \leq -1.$$

This leaves only the cases $s = p, p + 1$.

(*Note.* Up to this point we have not used the hypothesis $\mathrm{Min}(a, b) > c$.)

Now by (15.19), we know that $M_{p,1}(z_0)$, $M_{p,3}(z_0)$, $M_{p+1,3}(z_0)$ are all bounded by 1 since z_0 is unramified over $\mathbb{Q}_p$. Again by (15.19), we know that $pM_{p+1,1}(z_0)$ is a unit. This completes the proof of the lemma. □

17.4 **Theorem.** *Let τ be the ratio v_1/u_1 of the first components of a polarized pair of eigenvectors (cf. Lemma 16.5). Let $\gamma = \tau(z_0)$. We assert that*

$$\lambda \mapsto \tau(\lambda) = y$$

is a one-to-one biholomorphic map of $D(z_0, |p|^{(1/(p+1))^-})$ onto

$$\mathscr{T}_\gamma = \left\{ y \,|\, \mathrm{ord}\left(\frac{y}{\gamma^\sigma} - 1\right) < \frac{1}{2} - \frac{1}{p+1} \right\}$$

and in particular for $y \in \mathscr{T}_\gamma$, the equation

$$\tau(\lambda) = y \tag{17.4.0.0}$$

has exactly one solution with $\mathrm{ord}(\lambda - z_0) > 1/(p+1)$.

Explicitly, (see equations (17.4.1)–(17.4.3) *for definition of* $\mu_1, \mu_2, \varepsilon$),

$$\mathrm{ord}(\lambda - z_0) = \tfrac{1}{2} + \mathrm{ord}\left(\frac{(y/\gamma) - 1}{(y/\gamma^\sigma) - 1}\right) \tag{17.4.0.1}$$

and furthermore, designating by e_y the right side of this equation,

(17.4.0.2)

$$\lambda - z_0 \in -\frac{(y/\gamma) - 1}{((y/\gamma) - 1)M_{1,1}(z_0) + (\mu_2/\sqrt{p})((y/\gamma)(\mu_1/\mu_2) - 1)\varepsilon M_{1,3}(z_0)} \times (1 + (p)^{e_y - 1/(p+1)}).$$

Proof. By Lemma 16.5 and Corollary 12.4, as elements of $\Omega_0(\sqrt{p})$,

$$\frac{u_2}{u_1}(z_0) = \varepsilon\mu_1 \frac{1}{\sqrt{p}}, \tag{17.4.1}$$

$$\frac{v_2}{v_1}(z_0) = \varepsilon\mu_2 \frac{1}{\sqrt{p}}, \tag{17.4.2}$$

$$\gamma^{1-\sigma} \equiv \frac{\mu_1}{\mu_2} \mod \sqrt{p}, \qquad \varepsilon^{1+\sigma} = p\left(\frac{B_2}{B_3}\right)(z_0), \tag{17.4.3}$$

$$|\mu_1 - \mu_2| = 1 = |\mu_1| = |\mu_2|, \tag{17.4.4}$$

$$\mu_1^{1+\sigma} \equiv \mu_2^{1+\sigma} \equiv 1 \mod \sqrt{p}. \tag{17.4.5}$$

By (13.3.9) for $\lambda \in D(z_0, 1^-)$,

$$(17.4.6)\quad u_1(\lambda) = \sum_{s=0}^{\infty} (\lambda - z_0)^s [u_1(z_0) M_{s,1}(z_0) + u_2(z_0) M_{s,3}(z_0)]$$

$$= u_1(z_0) \sum_{s=0}^{\infty} (\lambda - z_0)^s \left[M_{s,1}(z_0) + \frac{\mu_1 \varepsilon}{\sqrt{p}} M_{s,3}(z_0) \right];$$

likewise,

$$(17.4.7)\quad v_1(\lambda) = v_1(z_0) \sum_{s=0}^{\infty} (\lambda - z_0)^s \left[M_{s,1}(z_0) + \frac{\mu_2 \varepsilon}{\sqrt{p}} M_{s,3}(z_0) \right].$$

We now compute

(17.4.8)

$$\tau(\lambda) - \gamma^\sigma = \frac{(\gamma - \gamma^\sigma)\{1 + \sum_{s=1}^{\infty} (\lambda - z_0)^s M_{s,1}(z_0)\} + ((\gamma\mu_2 - \gamma^\sigma \mu_1)/\sqrt{p})\varepsilon \sum_{s=1}^{\infty} (\lambda - z_0)^s M_{s,3}(z_0)}{1 + \sum_{s=1}^{\infty} (\lambda - z_0)^s K_s}$$

where

$$(17.4.9)\quad K_s = M_{s,1}(z_0) + \frac{\varepsilon \mu_1}{\sqrt{p}} M_{s,3}(z_0).$$

Let ρ_1 denote the denominator and ρ_2 the numerator of the right side of (17.4.8). We will show that the image under τ lies in $\mathscr{T}_\gamma$ by showing

$$(17.4.10)\quad \operatorname{ord}(\lambda - z_0) > \frac{1}{p+1} \quad \text{implies } \operatorname{ord} \rho_1(\lambda) > -\frac{1}{2} + \frac{1}{p+1},$$

$$(17.4.11)\quad \operatorname{ord}(\lambda - z_0) > \frac{1}{p+1} \quad \text{implies } \rho_2(\lambda) \text{ is a unit.}$$

For (17.4.10) we estimate by (17.3.1)

(17.4.12)

$$\operatorname{ord}\left((\lambda - z_0)^s \frac{\mu_1 \varepsilon}{\sqrt{p}} M_{s,3}(z_0) \right) > -\frac{1}{2} + \frac{s}{p+1} - \frac{s-1}{p+1} = \frac{1}{p+1} - \frac{1}{2}, \quad s \geq 1.$$

From (17.3.2) we have

$$\operatorname{ord}(\lambda - z_0)^s M_{s,1}(z_0) > \frac{s}{p+1} + \begin{cases} -\dfrac{s-1}{p+1} & \text{if } s \neq 0,\ p+1, \\ -1 & \text{if } s = p+1. \end{cases}$$

Thus

$$\left.\begin{array}{l} \operatorname{ord}(\lambda - z_0)^s M_{s,1}(z_0) > 0 \\ \operatorname{ord}(\lambda - z_0)^s M_{s,3}(z_0) > 0 \end{array}\right\}, \qquad \forall s \geq 1, \operatorname{ord}(\lambda - z_0) > \frac{1}{p+1}. \tag{17.4.13}$$

This completes the verification of (17.4.10).

To verify (17.4.11) we note that by (17.2) or equivalently by (17.4.3) and (17.4.4), $\gamma^\sigma - \gamma$ is a unit. It is enough to show that this unit dominates the remaining terms of ρ_2. By (17.4.3),

$$\left|\frac{\gamma\mu_2 - \gamma^\sigma\mu_1}{\sqrt{p}}\right| \leq 1. \tag{17.4.14}$$

This together with (17.4.13) completes the verification of (17.4.11).

This completes the demonstration that τ maps $D(0, |p|^{(1/(p-1))^-})$ into $\mathscr{T}_\gamma$. We now consider equation (17.4.0.0) with $y \in \mathscr{T}_\gamma$. This equation may be rewritten:

$$0 = h_0 + h_1(\lambda - z_0) + h_2(\lambda - z_0)^2 + \cdots, \tag{17.4.15}$$

where

$$h_0 = \frac{y}{\gamma} - 1, \tag{17.4.16.1}$$

$$h_1 = \left(\frac{y}{\gamma} - 1\right) M_{1,1}(z_0) + \frac{\mu_2}{\sqrt{p}}\left(\frac{y}{\gamma}\frac{\mu_1}{\mu_2} - 1\right)\varepsilon M_{1,3}(z_0) \tag{17.4.16.2}$$

and in general

(17.4.16.3)

$$h_s = \left(\frac{y}{\gamma} - 1\right) M_{s,1}(z_0) + \frac{\mu_2}{\sqrt{p}}\left(\frac{y}{\gamma}\frac{\mu_1}{\mu_2} - 1\right)\varepsilon M_{s,3}(z_0), \qquad s \geq 1.$$

Now

$$\frac{y}{\gamma}\frac{\mu_1}{\mu_2} - 1 = \left(\frac{y}{\gamma^\sigma} - 1\right) + \frac{y}{\gamma}\left(\frac{\mu_1}{\mu_2} - \gamma^{1-\sigma}\right)$$

and hence by (17.4.3)

$$\frac{y}{\gamma}\frac{\mu_1}{\mu_2} - 1 \in \left(\frac{y}{\gamma^\sigma} - 1\right) + y(\sqrt{p}).$$

This shows that for $y \in \mathscr{T}_\gamma$ (and in particular for $|y| > 1$),

(17.4.17) $$\operatorname{ord}\left(\frac{y}{\gamma}\frac{\mu_1}{\mu_2} - 1\right) = \operatorname{ord}\left(\frac{y}{\gamma^\sigma} - 1\right).$$

In the following table we give ord h_s. For simplicity we treat the case ord $y \geq 0$ separately from the case ord $y < 0$. We assert that

(17.4.18)

	$\operatorname{ord} y < 0$	$0 \leq \operatorname{ord}\left(\frac{y}{\gamma^\sigma} - 1\right) < \frac{1}{2} - \frac{1}{p+1}$
ord h_0	$\operatorname{ord} y$	$\operatorname{ord}\left(\frac{y}{\gamma} - 1\right)$
ord h_1	$\operatorname{ord} y - \frac{1}{2}$	$-\frac{1}{2} + \operatorname{ord}\left(\frac{y}{\gamma^\sigma} - 1\right)$
ord $h_s \geq$	$\operatorname{ord} y - \frac{1}{2} - \frac{s-1}{p+1}$	$-\frac{1}{2} + \operatorname{ord}\left(\frac{y}{\gamma^\sigma} - 1\right) - \frac{s-1}{p+1}$

The value of ord h_0 follows directly from (17.4.16.1). For ord $y < 0$, the assertion in the table follow directly from Lemma 17.3. For ord $y \geq 0$, it follows from (17.4.17) and the definition of $\mathscr{T}_\gamma$ that

(17.4.19) $$\left|\frac{1}{\sqrt{p}}\left(\frac{y}{\gamma}\frac{\mu_1}{\mu_2} - 1\right)\right| > \left|\frac{y}{\gamma} - 1\right|.$$

The estimate for ord h_1 then follows from (17.4.17) and

$$|M_{1,3}(z_0)| = 1 = |M_{1,1}(z_0)|,$$

while the estimate for ord h_s, $s > 1$, follows from Lemma 17.3 except that for $s = p + 1$ we use the fact that

$$\operatorname{ord} M_{p+1,1}(z_0) = -1 > -\frac{1}{2} - \frac{p}{p+1} + \operatorname{ord}\left(\frac{y}{\gamma^\sigma} - 1\right).$$

This completes the verification of (17.4.18).

We now put

$$(\lambda - z_0) = \frac{h_0}{h_1} x. \tag{17.4.20}$$

Equation (17.4.15) now takes the form (after division by h_0),

$$0 = 1 + x + k_2 x^2 + \cdots, \tag{17.4.21}$$

where

$$k_s = \frac{h_s}{h_1}\left(\frac{h_0}{h_1}\right)^{s-1}.$$

It follows from (17.4.18) that for $s \geq 2$:

$$\operatorname{ord} k_s \geq (s-1)\left[\frac{1}{2} - \frac{1}{p+1} + \operatorname{ord}\left(\frac{(y/\gamma) - 1}{(y/\gamma^\sigma) - 1}\right)\right] = (s-1)\left(e_y - \frac{1}{p+1}\right).$$

If $y \in \mathcal{T}_\gamma$ then $e_y > 1/(p+1)$ and hence it follows from Newton's lemma that (17.4.21) has a unique solution x such that $\operatorname{ord} x > -e_y + 1/(p+1)$. Furthermore

$$x \in -1 + (p)^{e_y - 1/(p+1)}.$$

By means of (17.4.20) this may be translated into statements concerning solutions (17.4.15). This completes the proof of the theorem. (For the biholomorphic property see Chapter 20.) □

17.5 **Corollary.** *τ has just one zero in $D(z_0, |p|^{1/(p+1)^-})$,*

$$\frac{\lambda - z_0}{\sqrt{p}} \equiv \frac{-1}{\varepsilon\mu_2 M_{1,3}(z_0)} \mod p^{1/2}. \tag{17.5.1}$$

Proof. This follows by setting $y = 0$ in (17.4.0.2), the estimate $p^{1/2 - 1/(p+1)}$ being replaced by $p^{1/2}$ by ramification arguments. □

It now follows that if (u_1, u_2) is any "eigenvector" of (16.5.5) then u_1 has a unique zero in $D(z_0, |p|^{(1/(p+1))^-})$ and it is located at a point given by (17.5.1) with μ_2 replaced by a possibly different solution of (12.4.11) written in the form

$$\mu = \frac{1 + \sqrt{p}\,\varepsilon_1 \mu^\sigma}{\sqrt{p}\,\varepsilon_2 + \mu^\sigma}, \tag{17.5.2}$$

where

$$\varepsilon_1 = \frac{B_4(z_0)}{\varepsilon B_3(z_0)}, \qquad \varepsilon_2 = \frac{\varepsilon B_1(z_0)/p}{B_2(z_0)},$$

(17.5.3)
$$\varepsilon^{1+\sigma} = \frac{pB_2(z_0)}{B_3(z_0)}.$$

Chapter 18

The Defining Relation for the Canonical Lifting ($l = 1$)

We re-examine equation (13.3.21) for $q = \phi_1 - \phi$. The hypotheses are as in Theorem 13.3 but we restrict our attention to the case $\mathrm{Min}(a, b) > c$. (The extension to the case $c > \mathrm{Max}(a, b)$ causes no new difficulty.)

Let $\lambda \in \Omega_1$, $|q| < 1$,

(18.0.1) $$H(\lambda, q) = \sum_{s=0}^{\infty} M_{s,3}(\lambda^p)q^s,$$

(18.0.2) $$K(\lambda, q) = \sum_{s=0}^{\infty} M_{s,1}(\lambda^p)q^s.$$

Put

(18.0.3) $$N(\lambda, q) = B_3(\lambda)K(\lambda, q) - B_1(\lambda)H(\lambda, q).$$

Equation (13.3.21) may be written

(18.0.4) $$N(\lambda, q) = 0.$$

Let $z \in \Omega_1$; let (u_1, u_2) be any solution of (2.4.5.1) at z^p, defined over Ω and put

(18.0.5) $$(v_1, v_2) = (u_1, u_2)(\lambda^p)B,$$

a solution of (2.4.5.1) at z.

18.1 **Lemma.** *If* $\lambda \in D(z, 1^-)$, $|q| < 1$ *then*

(18.1.1) $$B_3(\lambda)u_1(\lambda^p + q) = H(\lambda, q)v_1(\lambda) + N(\lambda, q)u_1(\lambda^p),$$

(18.1.2) $$B_1(\lambda)u_1(\lambda^p + q) = K(\lambda, q)v_1(\lambda) - N(\lambda, q)u_2(\lambda^p).$$

Proof.

$$(u_1(\lambda^p + q), u_2(\lambda^p + q)) = \sum (u_1(\lambda^p), u_2(\lambda^p))M_s(\lambda^p) \cdot q^s.$$

Hence

$$u_1(\lambda^p + q) = u_1(\lambda^p)K(\lambda, q) + u_2(\lambda^p)H(\lambda, q). \tag{18.1.3}$$

Thus

$$u_1(\lambda^p + q) = \frac{u_1(\lambda^p)}{B_3(\lambda)}[N(\lambda, q) + B_1(\lambda)H(\lambda, q)] + u_2(\lambda^p)H(\lambda, q),$$

hence

$$B_3(\lambda)u_1(\lambda^p + q) = N(\lambda, q)u_1(\lambda^p) + H(\lambda, q)[u_1(\lambda^p)B_1(\lambda) + u_2(\lambda^p)B_3(\lambda)]$$

and (18.1.1) now follows from (18.0.5). On the other hand from (18.1.3) we have

$$u_1(\lambda^p + q) = u_1(\lambda^p)K(\lambda, q) + \frac{u_2(\lambda^p)}{B_1(\lambda)}[B_3(\lambda)K(\lambda, q) - N(\lambda, q)]$$

and so

$$B_1(\lambda)u_1(\lambda^p + q) = K(\lambda, q)[u_1(\lambda^p)B_1(\lambda) + u_2(\lambda^p)B_3(\lambda)] - u_2(\lambda^p)N(\lambda, q)$$

and now again (18.1.2) follows from (18.0.5). This completes the proof. □

18.2 **Corollary.** *For each specialization of* λ *in* Ω_1, *the roots of* (18.0.4) (*as equation for* q) *are all* simple.

Proof. Let q_1 be a multiple root at $\lambda = \lambda_1$, $|q_1| < 1$. Thus

$$\begin{aligned} N(\lambda_1, q_1) &= 0, \\ N_q(\lambda_1, q_1) &= 0. \end{aligned} \tag{18.2.1}$$

By differentiating (18.1.2), with respect to q,

$$B_1(\lambda)u_1'(\lambda^p + q) = K_q(\lambda, q)v_1(\lambda) - N_q(\lambda, q)u_2(\lambda^p). \tag{18.2.2}$$

We choose (v_1, v_2), a solution of (2.4.5.1) at λ_1 such that

$$v_1(\lambda_1) = 0$$

and let u_1, u_2 be defined at λ_1^p by (18.0.5). Then by (18.2.1), (18.1.2) and (18.2.2)

$$B_1(\lambda_1)u_1(\lambda_1^p + q_1) = 0,$$

$$B_1(\lambda_1)u_1'(\lambda_1^p + q_1) = 0.$$

Since u_1, u_1' cannot have a common zero except at 0, 1, ∞, we conclude that $B_1(\lambda_1) = 0$. Now by a similar calculation based upon (18.1.1) we may show $B_3(\lambda_1) = 0$, and so

$$\det B(\lambda_1) = 0.$$

This contradicts the invertibility of $\alpha^*_{f,f}$ as mapping of $K_{\lambda_1^p}$ into K_{λ_1}. □

18.3 **Corollary.** *Under the hypothesis of* (18.0.5), *if*

$$v_1(\lambda_1) = 0,$$

$$\lambda_1 \in D(z_0, 1^-), \qquad |q_1| < 1,$$

then

$$N(\lambda_1, q_1) = 0$$

if and only if

$$u_1(\lambda_1^p + q_1) = 0.$$

Proof. This follows from Lemma 18.1 and the fact that $u_1(\lambda_1^p)$, $u_2(\lambda_1^p)$ (resp. $B_1(\lambda_1)$, $B_3(\lambda_1)$) cannot be simultaneously zero. □

18.4

We now suppose that z_0 represents a supersingular disk and that

$$z_0^\sigma = z_0^p.$$

Theorem. *Let $\lambda_1 \in D(z_0, 1^-)$, and let (v_1, v_2) be a nontrivial solution of* (2.4.5.1) *on $D(z_0, 1^-)$ such that*

$$v_1(\lambda_1) = 0, \tag{18.4.1}$$

then the solutions of the equation

$$N(\lambda_1, q) = 0 \tag{18.4.2}$$

are precisely the values of

$$\phi_1(\xi) - \lambda_1^p \tag{18.4.3}$$

as ξ runs over all zeros of v_1 such that

$$0 < \operatorname{ord}(\xi - z_0) < \frac{p}{p+1}.$$

Proof. Equation (18.0.5) is equivalent to

$$u_1(\phi_1(\lambda))\tilde{B}_1 = v_1(\lambda). \tag{18.4.4}$$

If $N(\lambda_1, q_1) = 0$ then by Corollary 18.3

$$u_1(\lambda_1^p + q_1) = 0.$$

By Corollary 16.8 we may choose $\xi \in D(z_0, 1^-)$

$$\operatorname{ord}(\xi - z_0) < \frac{p}{p+1}$$

such that

$$\phi_1(\xi) = \lambda_1^p + q_1.$$

Hence by (18.4.4)

$$v_1(\xi) = u_1(\lambda_1^p + q_1)\tilde{B}_1(\xi) = 0,$$

which proves the assertion in one direction.

Conversely if

$$\operatorname{ord}(\xi - z_0) \in \left(0, \frac{p}{p+1}\right),$$

$$v_1(\xi) = 0,$$

$$\phi_1(\xi) = \lambda_1^p + q_1,$$

then

$$\tilde{B}_1(\xi)u_1(\lambda_1^p + q_1) = v_1(\xi) = 0$$

and so

$$u_1(\lambda_1^p + q_1) = 0,$$

which by Corollary 18.3 shows that $N(\lambda_1, q_1) = 0$. This completes the proof of the theorem. □

18.5 **Corollary.** *Let τ be a ratio, ρ_2/ρ_1, of independent solutions of* (2.4.5.2) *on a super-singular disk, $D(z_0, 1^-)$. For $\lambda_1 \in D(z_0, 1^-)$, $|q_1| < 1$, we have*

$$N(\lambda_1, q_1) = 0$$

if and only if there exists ξ,

$$\operatorname{ord}(\xi - z_0) \in \left(0, \frac{p}{p+1}\right),$$

such that

$$\tau(\lambda_1) = \tau(\xi),$$

$$\phi_1(\xi) = \lambda_1^p + q_1.$$

Proof. Let λ_1 be an element of $D(z_0, 1^-)$ and let (v_1, v_2) be a solution of (2.4.5.1) on $D(z_0, 1^-)$ satisfying condition (18.4.1). Thus

$$v_1 = (m + n\tau)\rho_1$$

for some choice of constants $m, n \in \Omega$. If $\rho_1(\lambda_1) = 0$ then v_1 is a constant multiple of ρ_1 and so $n = 0$. If $\rho_1(\lambda_1) \neq 0$ then $\tau(\lambda_1) = -m/n$. In the first case $v_1(\xi) = 0$ if and only if

$$\tau(\xi) = \infty = \tau(\lambda_1).$$

In the second case no zero of v_1 can coincide with a zero of ρ_1 and hence $v_1(\xi) = 0$ if and only if

$$\tau(\xi) = -\frac{m}{n} = \tau(\lambda_1).$$

The corollary now follows from Theorem 18.4. □

18.6

Our object is to study solutions of (18.0.4) by means of Corollary 18.5. For this purpose we shall choose τ which satisfies (17.1).

Theorem. *Let $\lambda_1 \in D(z_0, 1^-)$, a supersingular disk.*

Let τ be as in Chapter 17.

$$\mathbf{C}_{\lambda_1} = \left\{\xi \mid \tau(\xi) = \tau(\lambda_1), \operatorname{ord}(\xi - z_0) \in \left(0, \frac{p}{p+1}\right)\right\},$$

$\mathbf{D}_{\lambda_1}$ = *image of C_{λ_1} under ϕ_1 (viewed set-theoretically* without *multiplicities).*

We order the elements of $\hat{\xi}_1, \hat{\xi}_2, \ldots$ *of* $\mathbf{D}_{\lambda_1}$ *by the condition that*

(18.6.1) $$|\hat{\xi}_i - \lambda_1^p| \le |\hat{\xi}_{i+1} - \lambda_1^p|.$$

We assert that if

(18.6.2) $$\operatorname{ord}(\lambda_1 - z_0) > \frac{1}{p+1},$$

then

(18.6.3) $$|\hat{\xi}_{p+1} - \lambda_1^p| < |\hat{\xi}_{p+2} - \lambda_1^p|,$$

while if

(18.6.4) $$\operatorname{ord}(\lambda_1 - z_0) = \frac{1}{p+1},$$

then

(18.6.5) $$|\hat{\xi}_{p+1} - \lambda_1^p| = |\hat{\xi}_{p+2} - \lambda_1^p|.$$

Proof.

Case 1. λ_1 lines in the critical annulus about z_0 (i.e., $\operatorname{ord}(\lambda_1 - z_0) \in (1/(p+1), p/(p+1))$).

By Lemma 16.6, $\phi_1(\lambda_1)$ lies in the critical annulus about z_0^σ and so by recursion $\phi_1^s(\lambda)$ is defined for all $s \in \mathbb{N}$. Let $\xi \in \mathbf{C}_{\lambda_1}$. We choose $s \in \mathbb{N}$ such that

(18.6.6) $$\operatorname{ord}(\phi_1^s(\xi) - z_0^{\sigma^s}) > \frac{1}{p+1}.$$

Then by (17.1)

$$\tau^{\sigma^s}(\phi_1^s(\xi)) = \tau(\xi) = \tau(\lambda_1) = \tau^{\sigma^s}(\phi_1^s(\lambda_1)).$$

Now $\phi_1^s(\xi)$ and $\phi_1^s(\lambda_1)$ lie in $D(z_0^{\sigma^s}, |p|^{1/(p+1)^-})$ and hence by Theorem 17.4,

(18.6.7) $$\phi_1^s(\xi) = \phi_1^s(\lambda_1).$$

Conversely this condition implies that $\xi \in \mathbf{C}_{\lambda_1}$.

We have thus shown that

(18.6.8)
$$\mathbf{C}_{\lambda_1} = \bigcup_{s=0}^{\infty} H_s(\lambda_1),$$

where

$$H_0(\lambda_1) = \{\lambda_1\}$$

and for $s \geq 1$,

$$H_s(\lambda_1) = \{\xi \in D(z_0, 1^-) \mid \operatorname{ord}(\phi_1^{s-1}(\xi) - z_0^{\sigma^{s-1}}) \leq \frac{1}{p+1},$$

but (18.6.7) holds}. We make the following assertions

(18.6.9)
$$\xi \in H_s(\lambda_1) \quad \text{implies } \operatorname{ord}(\xi - z_0) = \begin{cases} \dfrac{1}{p^s} \operatorname{ord}(\lambda_1 - z_0) & \text{if } s \equiv 0 \mod 2, \\ \dfrac{1}{p^s}(1 - \operatorname{ord}(\lambda_1 - z_0)) & \text{if } s \equiv 1 \mod 2, \end{cases}$$

(18.6.10)
$$H_s(\lambda_1) \cap H_{s'}(\lambda) = \varnothing \quad \text{if } s \neq s',$$

(18.6.11) ϕ_1 maps $H_s(\lambda_1)$ onto $H_{s-1}(\phi_1(\lambda_1))$ for $s \geq 1$.

(18.6.11′) If $s \geq 2$ and $\phi_1(\xi) \in H_{s-1}(\phi_1(\lambda_1))$ then $\xi \in H_s(\lambda_1)$

(18.6.12)
$$\operatorname{Card} H_s(\lambda_1) = p^s,$$

(18.6.13)
$$D_{\lambda_1} = \bigcup_{s=0}^{\infty} H_s(\phi_1(\lambda_1)).$$

We commence the proof of these assertions. It follows from Lemma 16.6 that

$$\operatorname{ord}(\phi_1^s(\lambda_1) - z_0^{\sigma^s}) = \begin{cases} \operatorname{ord}(\lambda_1 - z_0) & \text{if } s \equiv 0 \mod 2, \\ 1 - \operatorname{ord}(\lambda_1 - z_0) & \text{if } s \equiv 1 \mod 2. \end{cases}$$

Assertion (18.6.9) now follows from Lemma 16.6 and the definitions. If ξ lies in $H_s(\lambda_1) \cap H_{s'}(\lambda_1)$ and $s \equiv s' \bmod 2$ then by (18.6.9),

$$\frac{1}{p^s} \operatorname{ord}(\lambda_1 - z_0) = \frac{1}{p^{s'}} \operatorname{ord}(\lambda_1 - z_0) \qquad \text{if } s \equiv 0 \mod 2,$$

$$\frac{1}{p^s}(1 - \operatorname{ord}(\lambda_1 - z_0)) = \frac{1}{p^{s'}}(1 - \operatorname{ord}(\lambda_1 - z_0)) \quad \text{if } s \equiv 1 \mod 2.$$

In either case $s = s'$. If $s \not\equiv s' \bmod 2$ then again

$$\frac{1}{p^s}\operatorname{ord}(\lambda_1 - z_0) = \frac{1}{p^{s'}}(1 - \operatorname{ord}(\lambda_1 - z_0)) \quad \text{if } s \equiv 0 \mod 2,$$

$$\frac{1}{p^{s'}}\operatorname{ord}(\lambda_1 - z_0) = \frac{1}{p^s}(1 - \operatorname{ord}(\lambda_1 - z_0)) \quad \text{if } s \equiv 1 \mod 2.$$

In either case we have a nontrivial intersection of the two open intervals

$$\left(\frac{1}{p^s}\frac{1}{p+1}, \frac{1}{p^s}\frac{p}{p+1}\right), \qquad \left(\frac{1}{p^{s'}}\frac{1}{p+1}, \frac{1}{p^{s'}}\frac{p}{p+1}\right),$$

which implies that $s = s'$. This completes the verification of (18.6.10).

If $s \geq 1$, $\xi \in H_s(\lambda_1)$ then $\phi_1^{s-1}(\phi_1(\xi)) = \phi_1^s(\xi) = \phi_1^s(\lambda_1) = \phi_1^{s-1}(\phi_1(\lambda_1))$. If $s = 1$ then $\phi_1(\xi) = \phi_1(\lambda_1) =$ unique element of $H_0(\phi_1(\lambda_1))$ (which proves (18.6.11) if $s = 1$). If $s \geq 2$ then $\xi \in H_s(\lambda_1)$ implies that $\phi_1^{s-2}(\phi_1(\xi)) = \phi_1^{s-1}(\xi)$ does *not* satisfy

$$\operatorname{ord}(\phi_1^{s-1}(\xi) - z_0^{\sigma^{s-1}}) > \frac{1}{p+1}.$$

This shows that ϕ_1 maps $H_s(\lambda_1)$ into $H_{s-1}(\phi_1(\lambda_1))$. Conversely given $y \in H_{s-1}(\phi_1(\lambda_1))$ ($s \geq 2$) then by (18.6.9)

$$\operatorname{ord}(y - z_0^\sigma) = \begin{cases} \dfrac{1}{p^{s-1}}\operatorname{ord}(\phi_1(\lambda_1) - z_0^\sigma) & \text{if } s - 1 \equiv 0 \mod 2, \\[2ex] \dfrac{1}{p^{s-1}}(1 - \operatorname{ord}(\phi_1(\lambda_1) - z_0^\sigma)) & \text{if } s - 1 \equiv 1 \mod 2, \end{cases}$$

Thus by Corollary 16.8 there exists ξ, such that

$$\phi_1(\xi) = y$$

$$\operatorname{ord}(\xi - z_0) = \begin{cases} \dfrac{1}{p^s}\operatorname{ord}(\phi_1(\lambda_1) - z_0^\sigma) & \text{if } s - 1 \equiv 0 \mod 2, \\[2ex] \dfrac{1}{p^s}(1 - \operatorname{ord}(\phi_1(\lambda_1) - z_0^\sigma)) & \text{if } s - 1 \equiv 1 \mod 2, \end{cases}$$

so in either case $\operatorname{ord}(\xi - z_0)$ lies in the interval

$$\left(\frac{1}{p^s}\frac{1}{p+1}, \frac{1}{p^s}\frac{p}{p+1}\right),$$

which by Lemma 16.6 shows that $\phi_1^{s-1}(\xi)$ does *not* satisfy

$$\operatorname{ord}(\phi_1^{s-1}(\xi) - \sigma^{s-1}z_0) > \frac{1}{p+1}.$$

Thus ξ lies in $H_s(\lambda_1)$ and so we conclude that ϕ_1 maps $H_s(\lambda_1)$ *onto* $H_{s-1}(\phi_1(\lambda_1))$. This completes the verification of assertion (18.6.11). Assertion (18.6.11′) is equally clear. Assertion (18.6.12) follows by induction on s, using (18.6.11), equation (18.6.9) and Corollary 16.8 which shows that for $m \in \mathbb{N}$, ϕ_1 is a p to 1 map of the annulus

$$\operatorname{ord}(\lambda - z_0) \in \left(\frac{1}{p^{m+1}}\frac{1}{p+1}, \frac{1}{p^{m+1}}\frac{p}{p+1}\right)$$

onto the annulus

$$\operatorname{ord}(\lambda - z_0^\sigma) \in \left(\frac{1}{p^m}\frac{1}{p+1}, \frac{1}{p^m}\frac{p}{p+1}\right).$$

Assertion (18.6.13) follows from (18.6.8) and (18.6.11) taking into account the fact that $H_0(\lambda_1) = \{\lambda_1\}$, the image of which under ϕ_1 coincides with that of $H_1(\lambda_1)$.

This completes the verification of assertions (18.6.9)–(18.6.13). Now for $s \geq 0$ there are p^s elements ξ in $H_s(\phi_1(\lambda_1))$ and these satisfy

$$(18.6.14) \qquad \operatorname{ord}(\xi - z_0^\sigma) = \frac{1}{p^s}\begin{cases} \operatorname{ord}(\phi_1(\lambda_1) - z_0^\sigma) & \text{if } s \equiv 0 \mod 2, \\ 1 - \operatorname{ord}(\phi_1(\lambda_1) - z_0^\sigma) & \text{if } s \equiv 1 \mod 2, \end{cases}$$

$$= \frac{1}{p^s}\begin{cases} 1 - \operatorname{ord}(\lambda_1 - z_0) & \text{if } s \equiv 0 \mod 2, \\ \operatorname{ord}(\lambda_1 - z_0) & \text{if } s \equiv 1 \mod 2. \end{cases}$$

On the other hand

$$\operatorname{ord}(\lambda_1^p - z_0^p) \geq \inf(p \operatorname{ord}(\lambda_1 - z_0); 1 + \operatorname{ord}(\lambda_1 - z_0)).$$

Thus for $\xi \in H_s(\phi_1(\lambda_1))$ we have

$$(18.6.15) \qquad \operatorname{ord}(\xi - \lambda_1^p) = \frac{1}{p^s}\begin{cases} 1 - \operatorname{ord}(\lambda_1 - z_0) & \text{if } s \equiv 0 \mod 2, \\ \operatorname{ord}(\lambda_1 - z_0) & \text{if } s \equiv 1 \mod 2, \end{cases}$$

since

$$1 + \operatorname{ord}(\lambda_1 - z_0) > \frac{1}{p^s}\operatorname{ord}(\lambda_1 - z_0) \qquad \text{for } s \geq 0,$$

$$1 + \operatorname{ord}(\lambda_1 - z_0) > \frac{1}{p^s}(1 - \operatorname{ord}(\lambda_1 - z_0)) \qquad \text{for } s \geq 0,$$

$$p\operatorname{ord}(\lambda_1 - z_0) > \frac{1}{p^s}\operatorname{ord}(\lambda_1 - z_0) \qquad \text{for } s \geq 0,$$

$$p\operatorname{ord}(\lambda_1 - z_0) > 1 - \operatorname{ord}(\lambda_1 - z_0) \geq \frac{1}{p^s}(1 - \operatorname{ord}(\lambda_1 - z_0)) \quad \text{for } s \geq 0,$$

the last holding since $\operatorname{ord}(\lambda_1 - z_0) > 1/(p+1)$.

It now follows from (18.6.13) and (18.6.15) that $(\hat{\xi}_1, \ldots, \hat{\xi}_{p+1}\} = H_0(\phi_1(\lambda_1)) \cup H_1(\phi_1(\lambda_1))$ and that the remaining elements, ξ, of $\mathbf{D}_{\lambda_1}$ satisfy the condition

(18.6.16)

$$\operatorname{ord}(\hat{\xi}_{p+1} - \lambda_1^p) = \frac{1}{p}\operatorname{ord}(\lambda_1 - z_0) > \frac{1}{p^2}(1 - \operatorname{ord}(\lambda_1 - z_0)) \geq \operatorname{ord}(\xi - \lambda_1^p).$$

This completes the treatment of Case 1.

Case 2. $(\operatorname{ord} \lambda_1 - z_0) \geq p/(p+1)$

Let $\xi \in \mathbf{C}_{\lambda_1}$ and let s be minimal such that

$$\operatorname{ord}(\phi_1^s(\xi) - z_0^{\sigma^s}) > \frac{1}{p+1}. \tag{18.6.17}$$

We assert

$$\begin{cases} s \equiv 0 \mod 2, \\ \operatorname{ord}(\phi_1^s(\xi) - z_0^{\sigma^s}) \geq \dfrac{p}{p+1}. \end{cases} \tag{18.6.18}$$

To demonstrate this, let $y = \tau(\lambda_1)$. By (17.1)

$$\tau^{\sigma^s}(\phi_1^s(\xi)) = \tau(\xi) = \tau(\lambda_1) = y.$$

By (17.4.0.1)

$$\operatorname{ord}\frac{(y/\gamma)-1}{(y/\gamma^\sigma)-1}=\operatorname{ord}(\lambda_1-z_0)-\frac{1}{2}\geq\frac{p}{p+1}-\frac{1}{2}.$$

Hence $|y|\leq 1$ and indeed

$$\operatorname{ord}\left(\frac{y}{\gamma}-1\right)\geq\frac{p}{p+1}-\frac{1}{2}. \tag{18.6.19}$$

Likewise

$$\operatorname{ord}(\phi_1^s(\xi)-z_0^{\sigma^s})=\tfrac{1}{2}+\operatorname{ord}\left(\frac{(y/\gamma^{\sigma^s})-1}{(y/\gamma^{\sigma^{s+1}})-1}\right). \tag{18.6.20}$$

By (17.4.3), $|\gamma^\sigma-\gamma|=1$ and hence by (18.6.19)

$$\left|\frac{y}{\gamma^{\sigma^c}}-1\right|=1\quad\text{if } c\equiv 1\mod 2. \tag{18.6.21}$$

By (17.4.3) and (17.4.5)

$$\gamma^{\sigma^2-1}\equiv 1\mod\sqrt{p}.$$

Hence

$$\operatorname{ord}\left(\frac{y}{\gamma}-\frac{y}{\gamma^{\sigma^c}}\right)\geq\frac{1}{2}\quad\text{if } c\equiv 0\mod 2$$

and so by (18.6.19)

$$\operatorname{ord}\left(\frac{y}{\gamma^{\sigma^c}}-1\right)\geq\frac{p}{p+1}-\frac{1}{2}\quad\text{if } c\equiv 0\mod 2. \tag{18.6.22}$$

This implies that

$$\operatorname{ord}\left(\frac{y}{\gamma^{\sigma^{s+1}}}-1\right)\geq\frac{p}{p+1}-\frac{1}{2}\quad\text{if } s\equiv 1\mod 2. \tag{18.6.23}$$

Equations (18.6.17), (18.6.20), (18.6.21) and (18.6.23) show for $s\equiv 1 \bmod 2$,

$$\frac{p}{p+1}-\frac{1}{2}\leq\operatorname{ord}\left(\frac{y}{\gamma^{\sigma^{s+1}}}-1\right)=\frac{1}{2}-\operatorname{ord}(\phi_1^s(\xi)-z_0^{\sigma^s})<\frac{1}{2}-\frac{1}{p+1},$$

a contradiction. Thus in fact $s \equiv 0 \bmod 2$ (which proves the first part of (18.6.18)). Hence

$$\text{(18.6.22}'\text{)} \qquad \operatorname{ord}\left(\frac{y}{\gamma^{\sigma^s}} - 1\right) \geq \frac{p}{p+1} - \frac{1}{2},$$

$$\text{(18.6.21}'\text{)} \qquad \operatorname{ord}\left(\frac{y}{\gamma^{\sigma^{s+1}}} - 1\right) = 0,$$

and so by (18.6.20) we verify the second part of assertion (18.6.18).

We now define for $s > 0$

$$K_s(\lambda_1) = \left\{\xi \,|\, \operatorname{ord}(\xi - z_0) \in \left(0, \frac{p}{p+1}\right), \tau(\xi) = \tau(\lambda_1),\right.$$

$$\left. s \text{ in minimal such that } \operatorname{ord}(\phi_1^{2s}(\xi) - \sigma^{2s} z_0) \geq \frac{p}{p+1}\right\}.$$

By (18.6.18),

$$\text{(18.6.24)} \qquad \mathbf{C}_{\lambda_1} = \bigcup_{s \geq 1} K_s(\lambda_1).$$

It follows from Corollary 16.8 that $\xi \in K_s(\lambda_1)$ implies

$$\text{(18.6.25)} \qquad \operatorname{ord}(\xi - z_0) = \frac{p}{p+1} \frac{1}{p^{2s}},$$

$$\text{(18.6.26)} \qquad \operatorname{ord}(\phi_1(\xi) - z_0^{\sigma}) = \frac{p}{p+1} \frac{1}{p^{2s-1}}.$$

Now as noted before $\xi \in K_s$ implies

$$\text{(18.6.27)} \qquad \tau^{\sigma^{2s}}(\phi_1^{2s}(\xi)) = \tau(\lambda_1),$$

while conversely (18.6.25) and (18.6.27) imply that $\xi \in K_s$. Thus with the aid of Corollary 16.8

$$\operatorname{Card} K_s(\lambda_1) = p^{2s-1}(p+1)$$

and so

$$\text{(18.6.28)} \qquad \operatorname{Card} \phi_1 K_s(\lambda_1) = p^{2s-2}(p+1).$$

We now have a decomposition

$$\mathbf{D}_{\lambda_1} = \bigcup_{s \geq 1} \phi_1 K_s(\lambda_1), \tag{18.6.29}$$

into subsets of cardinality given by (18.6.28) and at distance from z_0^σ given by (18.6.26). On the other hand

$$\operatorname{ord}(\lambda_1^p - z_0^\sigma) \geq 1 + \frac{p}{p+1}. \tag{18.6.30}$$

Thus equation (18.6.26) gives the distance from λ_1^p to each of the points of $\phi_1 K_s$. To summarize: there are $p + 1$ elements ξ of $\mathbf{D}_{\lambda_1}$ such that

$$\operatorname{ord}(\xi - \lambda_1^p) = \frac{1}{p+1} \tag{18.6.31}$$

and for all *other* $\xi \in \mathbf{D}_{\lambda_1}$ we have

$$\operatorname{ord}(\xi - \lambda_1^p) \leq \frac{1}{p^2(p+1)}. \tag{18.6.32}$$

This completes the treatment of Case 2.

Case 3. $\operatorname{ord}(\lambda_1 - z_0) = 1/(p + 1)$

Let $\xi \in \mathbf{C}_{\lambda_1}$. Let $s \geq 0$ be minimal such that

$$\operatorname{ord}(\phi_1^{s+1}(\xi) - z_0^{\sigma^{s+1}}) > \frac{1}{p+1}. \tag{18.6.33}$$

We assert

$$\begin{cases} s \equiv 0 \mod 2 \\ \operatorname{ord}(\phi_1^{s+1}(\xi) - z_0^{\sigma^{s+1}}) \geq \dfrac{p}{p+1}. \end{cases} \tag{18.6.34}$$

The proof is similar to that of (18.6.18) so we may be brief.

Let $y = \tau^\sigma(\phi(\lambda_1))$. By (17.1)

$$\tau^{\sigma^{s+1}}(\phi_1^{s+1}(\xi)) = \tau(\xi) = \tau(\lambda_1) = \tau^\sigma(\phi(\lambda_1)).$$

Since $\operatorname{ord} \phi(\lambda_1) \geq p/(p + 1)$, Theorem 17.4 applied to τ^σ gives

$$\operatorname{ord} \frac{(y/\gamma^\sigma) - 1}{(y/\gamma^{\sigma^2}) - 1} = \operatorname{ord}(\phi(\lambda_1) - z_0^\sigma) - \frac{1}{2} \geq \frac{p}{p+1} - \frac{1}{2},$$

from which we deduce

$$\text{ord}\,\frac{y}{\gamma^{\sigma}} - 1 \geq \frac{p}{p+1} - \frac{1}{2}. \tag{18.6.35}$$

Again by applying Theorem 17.4 to $\tau^{\sigma^{s+1}}$,

$$\text{ord}(\phi_1^{s+1}(\xi) - z_0^{\sigma^{s+1}}) = \tfrac{1}{2} + \text{ord}\left(\frac{(y/\gamma^{\sigma^{s+1}}) - 1}{(y/\gamma^{\sigma^{s+2}}) - 1}\right). \tag{18.6.36}$$

If s were odd then

$$\begin{cases} \text{ord}\left(\dfrac{y}{\gamma^{\sigma^{s+1}}} - 1\right) = 0, \\ \text{ord}\left(\dfrac{y}{\gamma^{\sigma^{s+2}}} - 1\right) \geq \dfrac{p}{p+1} - \dfrac{1}{2}, \end{cases} \tag{18.6.37}$$

and so

$$\text{ord}(\phi^{s+1}(\xi) - z_0^{\sigma\ \ +1}) \leq \frac{1}{2} - \left(\frac{p}{p+1} - \frac{1}{2}\right) = \frac{1}{p+1}, \tag{18.6.38}$$

which contradicts the definition of s. This completes the verification of the first part of (18.6.34).

By (18.6.35), since s is even,

$$\begin{cases} \text{ord}\left(\dfrac{y}{\gamma^{\sigma^{s+1}}} - 1\right) \geq \dfrac{p}{p+1} - \dfrac{1}{2}, \\ \text{ord}\left(\dfrac{y}{\gamma^{\sigma^{s+2}}} - 1\right) = 0. \end{cases} \tag{18.6.39}$$

The estimate (18.6.34) for $\text{ord}(\phi_1^{s+1}(\xi) - z_0^{s+1})$ now follows by means of (18.6.36).

This completes the verification of (18.6.34).

We now define for $s \geq 0$

$$K_s(\lambda_1)$$

$$= \left\{\xi \,|\, \tau(\xi) = \tau(\lambda_1),\ s \text{ minimal such that } \text{ord}(\phi_1^{1+2s}(\xi) - z_0^{\sigma^{1+2^s}}) \geq \frac{p}{p+1}\right\}.$$

Then

$$\mathbf{C}_{\lambda_1} = \bigcup_{s=0}^{\infty} K_s(\lambda_1),$$

$$\text{(18.6.40)} \qquad \operatorname{ord}(\xi - z_0) = \frac{p}{p+1}\frac{1}{p^{1+2s}}, \qquad \forall \xi \in K_s(\lambda_1).$$

This condition together with

$$\tau^{\sigma^{1+2s}}(\phi_1^{1+2s}(\xi)) = \tau(\lambda_1)$$

characterizes the elements of $K_s(\lambda_1)$, i.e., $K_s(\lambda_1)$ is the inverse image under ϕ_1^{1+2s} of a fixed element of

$$D(z_0^{\sigma^{1+2s}}, |p|^{p/(p+1)^+}).$$

Thus

$$\text{Card } K_s(\lambda_1) = (p+1)p^{2s}.$$

In particular, $K_0(\lambda_1)$ is a set of $p+1$ points with one element as image under ϕ_1, while for $s \geq 1$, $K_s(\lambda_1)$ is a set of $p^{2s}(p+1)$ points with $p^{2s-1}(p+1)$ points in the image under ϕ_1. Thus

$\phi_1(K_0(\lambda_1))$ = set of one point with distance from z_0^σ given by ordinal $\geq p/(p+1)$.

$(s \geq 1)$ $\phi_1(K_s(\lambda_1))$ = set of $p^{2s-1}(p+1)$ points with distance from z_0^σ given by ordinal $= (p^2/p^{2s+1}(p+1))$.

Finally

$$\operatorname{ord}(\lambda_1^p - z_0^\sigma) = \frac{p}{p+1}.$$

Thus in ordering the points of $\mathbf{D}_{\lambda_1}$ according to their distance from λ_1^p, there is

precisely 1 point whose distance has ordinal $\geq \dfrac{p}{p+1}$,

$p(p+1)$ points whose distance has ordinal $= \dfrac{1}{p(p+1)}$

and all other points have distance with strictly smaller ordinal. This completes the proof of the theorem. □

18.7

Let $\mathbf{N}$ be the quotient field of the ring $\mathbf{N}_0$, of power series in $\mathbb{Q}_p(z_0)[[\lambda - z_0]]$ which converge for

$$\operatorname{ord}(\lambda - z_0) > \frac{1}{p+1}.$$

The functions B_1, B_3 are defined over $\mathbb{Q}_p$ and are analytic on Ω_1, hence may be viewed as elements of $\mathbf{N}$. Thus N, the series of which $\phi_1 - \phi$ is a root, lies in $\mathbf{N}[[q]]$.

Theorem. *$\phi_1 - \phi$ is root of an element of $\mathbf{N}[q]$ of degree $p + 1$. Hence ϕ_1 is algebraic over $\mathbf{N}$ of degree bounded by $p + 1$.*

Proof. For each value of λ_1 such that

$$\operatorname{ord}(\lambda_1 - z_0) > \frac{1}{p+1}, \tag{18.7.1}$$

the specialization $N(\lambda_1, q)$ of N (at λ_1) has by Corollary 18.5 and Theorem 18.6 exactly $p + 1$ roots for q which are strictly smaller than all other roots. This means that the Newton polygon has a corner at $p + 1$ for all such λ_1. Letting K_{p+1} denote the coefficient of q^{p+1} in N, we see (Chapter 20) that K_{p+1} has no zero in the disk (18.7.1) and there is a factorization

$$\frac{1}{K_{p+1}} N(\lambda, q) = N_1(\lambda, q) N_2(\lambda, q), \tag{18.7.2}$$

where N_1 is a monic polynomial of degree $p + 1$ with coefficients in $\mathbf{N}_0$ (satisfied by the $p + 1$ minimal roots for q when λ lies in (18.7.1)) and N_2 being a power series in q with coefficients in $\mathbf{N}_0$. The series converge for λ in (18.7.1) and $|q| < 1$. Furthermore, for λ in the critical annulus, $q = \phi_1(\lambda) - \lambda^p$ is the minimal solution of $N(\lambda, q)$ and hence is a root of N_1. This shows that $\phi_1(\lambda)$ as function on the critical annulus satisfies a polynomial equation over $\mathbf{N}$. This completes the proof of the theorem. □

18.8 Theorem.

$$\deg \frac{\mathbf{N}(\phi_1)}{\mathbf{N}} = p + 1.$$

Proof. Let

$$g(\lambda, y) = \sum_{i=0}^{b} h_i(\lambda)(y - z_0^\sigma)^i \tag{18.8.1}$$

be the irreducible polynomial over $\mathbf{N}$ satisfied by ϕ_1. Thus $1 \le b \le p + 1$.

Clearly $g(\lambda, q + \lambda^p)$ is a factor of N_1 (cf. equation (18.7.2)) in $\mathbf{N}[q]$. In the proof of Theorem 18.6 we showed that for $\operatorname{ord}(\lambda_1 - z_0) \ge p/(p + 1)$, the minimal values of q are $p + 1$ in number and give $\operatorname{ord}(q + \lambda_1^p - z_0^\sigma) = 1/(p + 1)$. This means that for λ_1 so restricted the roots of $g(\lambda_1, y)$ all have the property

$$\operatorname{ord}(y - z_0^\sigma) = \frac{1}{p + 1}. \tag{18.8.2}$$

Thus $g(\lambda_1, y)$ as polynomial in $y - z_0^\sigma$ has a Newton polygon with just one slope, namely $1/(p + 1)$. Thus

$$\operatorname{ord} \frac{h_0(\lambda_1)}{h_b(\lambda_1)} = \frac{b}{p + 1} \tag{18.8.3}$$

for $\operatorname{ord}(\lambda_1 - z_0) \ge p/(p + 1)$. But we may specialize λ_1 in the nonramified field $\mathbb{Q}_p(z_0)$, showing that $b/(p + 1) \in \mathbb{Z}$. Thus $b = p + 1$ as asserted. □

18.9 **Corollary.** *For $\varepsilon > 0$ let $\mathbf{N}_\varepsilon$ be the quotient field of the ring of* Laurent *series with coefficients in $\mathbb{Q}_p(z_0)$ which converge in the annulus*

$$\left\{\lambda \,\middle|\, \frac{1}{p + 1} < \operatorname{ord}(\lambda - z_0) < \frac{p}{p + 1} + \varepsilon\right\}. \tag{18.9.1}$$

(Thus $\mathbf{N} \hookrightarrow \mathbf{N}_\varepsilon$). Then

$$\deg \frac{\mathbf{N}_\varepsilon(\phi_1)}{\mathbf{N}_\varepsilon} = p + 1.$$

Proof. Again let g given by (18.8.1) designate the irreducible polynomial for ϕ_1 over N_ε. We now specialize λ_1 so that $p(p + 1)^{-1} < \operatorname{ord}(\lambda - z_0) < \varepsilon + p(p + 1)^{-1}$ and again (18.8.3) is valid. We can no longer insist that λ_1 be unramified over $\mathbb{Q}_p(z_0)$, but we can specialize so that the ramification of $\mathbb{Q}_p(z_0, \lambda_1)$ is prime to $p + 1$. This then implies that $b/(p + 1) \in \mathbb{Z}$ and hence once again $b = p + 1$. This completes the proof. □

18.10 **Theorem.** *The canonical lifting ϕ_1 cannot be extended analytically to the region* (18.9.1).

This follows from the above corollary. □

We may pose the following question. We know that ϕ_1 is analytic on the set

$$0 < \operatorname{ord}(\lambda - z_0) < \frac{p}{p+1}$$

but not on the set

$$0 < \operatorname{ord}(\lambda - z_0) < \frac{p}{p+1} + \varepsilon.$$

We ask if ϕ_1 can be extended to a function analytic on the set

$$0 < \operatorname{ord}(\lambda - z_0) \leq \frac{p}{p+1}?$$

The ramification arguments of 18.8–18.10 do not seem to be capable of resolving this question. However, the answer is negative. Indeed we can show that ϕ_1 cannot be extended to an analytic element on a set of the form

$$\left\{\lambda \,\middle|\, \operatorname{ord}(\lambda - \zeta_1) > \frac{p}{p+1}\right\},$$

where ζ_1 is fixed with $\operatorname{ord}(\zeta_1 - z_0) = p/(p+1)$. The method is to start with Theorem 18.7, make a transformation of independent variable

$$x = \frac{\lambda - z_0}{\zeta_1 - z_0}$$

so that $\mathbf{N}_0$ becomes a ring of analytic functions on super admissible sets (see §13.1). It is known (Dwork and Robba [16, §3.7]) that a function algebraic over such a ring which is also an analytic element on a residue class is in fact also an analytic element on a superadmissible set. By Theorem 18.10 we know that ϕ_1 is not, in terms of x, an analytic element on a super admissible set. This completes our discussion of the prolongation of ϕ_1.

Chapter 19

Semisimplicity

We again restrict our attention to the supersingular disk $D(z_0, 1^-)$, we assume $l = 1$ and to fix ideas we take $c < \text{Min}(a, b)$. The case $c > \text{Max}(a, b)$ may be treated in the same way.

19.1 **Lemma.** *ϕ_1^2 is not defined on the set*

$$\left\{\lambda \mid \text{ord}(\lambda - z_0) = \frac{1}{p+1}\right\} \tag{19.1.1}$$

but it does extend from a function on the critical annulus to an analytic function mapping the disk

$$\left\{\lambda \mid \text{ord}(\lambda - z_0) > \frac{1}{p+1}\right\} \tag{19.1.2}$$

onto the disk

$$\left\{\lambda \mid \text{ord}(\lambda - z_0^{\sigma^2}) > \frac{1}{p+1}\right\}.$$

Proof. On the critical annulus about z_0 we have (with τ as in (17.1))

$$\tau^{\sigma^2} \circ \phi_1^2 = \tau, \tag{19.1.3}$$

which shows that

$$\phi_1^2 = (\tau^{\sigma^2})^{-1} \circ \tau \tag{19.1.4}$$

on that annulus. We assert that the right side extends to an analytic function on the set (19.1.2). Indeed τ is a one-to-one map of $D(z_0, |p|^{1/(p+1)-})$ onto $\mathcal{T}_\gamma$. τ^{σ^2} is a map of $D(z_0^{\sigma^2}, |p|^{1/(p+1)-})$ onto $\mathcal{T}_{\gamma^{\sigma^2}}$. Since

$$\text{ord}(\gamma^{\sigma^2 - 1} - 1) \geq \tfrac{1}{2},$$

we conclude that

$$\mathcal{T}_\gamma = \mathcal{T}_{\gamma\sigma^2}.$$

From this the assertion is clear. □

19.2 **Corollary.** *For $m \in \mathbb{N}$, ϕ_1^{2m} extends from a map of the critical annulus about z_0 into the critical annulus about $z_0^{\sigma^{2m}}$ to a mapping of $D(z_0, |p|^{1/(p+1)-})$ onto $D(z_0^{\sigma^{2m}}, |p|^{1/(p+1)-})$.*

19.3

We recall that

$$(u_1, u_2) \mapsto (u_1, u_2)(\lambda^p)B(\lambda) \tag{19.3.0}$$

is a linear transformation of solutions of (2.4.5.1) at z_0^p into solutions at z_0. If we choose m such that

$$z_0^{\sigma^m} = z_0, \tag{19.3.1}$$

then

$$(u_1, u_2) \mapsto (u_1, u_2)^{\phi^m} B^{\phi^{m-1}} B^{\phi^{m-2}} \cdots B \tag{19.3.2}$$

is a linear endomorphism of the $\mathbb{Q}_p(z_0)$ space of solutions of (2.4.5.1) at z_0 defined over $\mathbb{Q}_p(z_0)$. We will take m *even* so not necessarily the minimal value for which (19.3.1) holds.

Theorem. *The endomorphism* (19.3.2) *is semisimple.*

Proof. Suppose otherwise. Then there exists a unique eigenvector (u_1, u_2) with eigenvalue ε,

$$|\varepsilon| = |p|^{m/2}. \tag{19.3.3}$$

Thus $u = (u_1, u_2)$ is defined over $\mathbb{Q}_p(z_0, \varepsilon)$ and there exists another solution $v = (v_1, v_2)$ of (2.4.5.1) at z_0 defined over the same field such that

$$\begin{aligned} u^{\phi^m} B^{\phi^{m-1}} \cdots B &= \varepsilon u, \\ v^{\phi^m} B^{\phi^{m-1}} \cdots B &= \varepsilon v + u. \end{aligned} \tag{19.3.4}$$

Let now $C = B(z_0)$, a matrix with coefficients in $\mathbb{Q}_p(z_0)$, $\mathbf{u} = u(z_0) = (u_1(z_0), u_2(z_0))$. Then specializing (19.3.4) at z_0, we have

(19.3.5) $$\varepsilon\mathbf{u} = \mathbf{u}C^{\sigma^{m-1}}C^{\sigma^{m-2}} \cdots C.$$

Now ε is a double root of the (quadratic) characteristic polynomial of $C^{\sigma^{m-1}} \cdots C$. The polynomial is invariant under σ, as σ permutes the C^{σ^i} cyclically. Hence ε lies in $\mathbb{Q}_p$ and so $\mathbf{u}$ is rational over $\mathbb{Q}_p(z_0)$. Applying σ to (19.3.5) we have

$$\varepsilon\mathbf{u} = \mathbf{u}^\sigma C C^{\sigma^{m-1}} \cdots C^\sigma,$$

and multiplying by C on the right

(19.3.6) $$\varepsilon\mathbf{u}^\sigma C = (\mathbf{u}^\sigma C)C^{\sigma^{m-1}} \cdots C^\sigma C,$$

which shows that $\mathbf{u}^\sigma C$ is again an eigenvector of $C^{\sigma^{m-1}} \cdots C$, a matrix which has just one eigenvector up to constant multiples. Thus

(19.3.7) $$\mathbf{u}^\sigma C = e\mathbf{u}$$

for some constant e. This shows that $u_2(z_0)/u_1(z_0)$ satisfies the equation (12.4.7) and hence

(19.3.8) $$\operatorname{ord} \frac{u_2(z_0)}{u_1(z_0)} = -\frac{1}{2}.$$

This contradicts the rationality of $\mathbf{u}$ over $\mathbb{Q}_p(z_0)$, which completes the proof of the theorem. □

19.4

We again consider the linear transformation (19.3.2). Let $u = (u_1, u_2)$, $v = (v_1, v_2)$ denote independent eigenvectors (analytic on $D(z_0, 1^-)$) corresponding to eigenvalues $\varepsilon_1, \varepsilon_2$. We define recursively a pair of solutions $u^{(i)}, v^{(i)}$ of (2.4.5.1) on $D(z_0^{\sigma^{-i}}, 1^-)$ ($i = 1, 2, \ldots, m$) by setting

$$u^{(0)} = u = (u_1, u_2),$$

$$v^{(0)} = v = (v_1, v_2),$$

and putting

(19.4.0.1) $$\begin{aligned} u^{(i-1)}(\lambda^p)B &= u^{(i)}(\lambda), \\ v^{(i-1)}(\lambda^p)B &= v^{(i)}(\lambda). \end{aligned}$$

In particular, $u^{(m)}$, $v^{(m)}$ are solutions in $D(z_0, 1^-)$ and in fact

$$u^{(m)} = u^{\phi^m} B^{\phi^{m-1}} \cdots B,$$

$$v^{(m)} = v^{\phi^m} B^{\phi^{m-1}} \cdots B,$$

and so since u and v are eigenvectors of (19.3.2)

$$\text{(19.4.0.2)} \qquad \begin{aligned} u^{(m)} &= \varepsilon_1 u, \\ v^{(m)} &= \varepsilon_2 v. \end{aligned}$$

On the other hand (19.4.0.1) may also be written

$$\text{(19.4.0.3)} \qquad \begin{aligned} u^{(i-1)}(\phi_1(\lambda))\tilde{B} &= u^{(i)}(\lambda), \\ v^{(i-1)}(\phi_1(\lambda))\tilde{B} &= v^{(i)}(\lambda), \end{aligned}$$

for $\operatorname{ord}(\lambda - z_0^{\sigma^{-i}}) \in (0, p/(p+1))$. Thus putting

$$\tau^{(i)} = \frac{v_1^{(i)}}{u_1^{(i)}},$$

meromorphic on $D(z_0^{\sigma^{-i}}, 1^-)$, we have

$$\text{(19.4.0.4)} \qquad \tau^{(i-1)} \circ \phi_1 = \tau^{(i)}, \qquad i = 1, 2, \ldots, m$$

and by (19.4.0.2) we have

$$\tau^{(m)} = \frac{\varepsilon_1}{\varepsilon_2} \tau^{(0)}.$$

Now equation (19.4.0.4) is certainly valid for

$$\operatorname{ord}(\lambda - z_0^{\sigma^{-i}}) \in \left(0, \frac{p}{p+1}\right).$$

We restrict the relation to λ in the critical annulus about $z_0^{\sigma^{-1}}$ and by iteration obtain

$$\text{(19.4.1)} \qquad \tau^{(0)} \circ \phi_1^m = \frac{\varepsilon_1}{\varepsilon_2} \tau^{(0)},$$

a relation between functions meromorphic on the critical annulus about z_0. However $\tau^{(0)}$ is meromorphic on $\mathscr{D} = D(z_0, |p|^{1/(p+1)-})$ and the restriction of ϕ_1^m to the critical annulus about z_0 extends by Corollary 19.2 to an analytic map of $\mathscr{D}$ into itself. Thus (19.4.1) is an identity on $\mathscr{D}$.

19.4.2 *Remark.* $\tau^{(0)}$ need not coincide with the ratio τ in Theorem 17.4 which involved "eigenvectors" of a semilinear transformation.

19.4.3 If $\varepsilon_1 = \varepsilon_2$ then we may by extending the base field to $\Omega_0(\sqrt{p})$ choose u, v to be the rows of the matrix $\mathbf{Y}$ of Lemma 16.5. With this choice $\tau^{(0)}$ coincides with τ and by Theorem 17.4 we conclude that if $\varepsilon_1 = \varepsilon_2$ then

$$\phi_1^m = I \tag{19.4.3.1}$$

as analytic function on $D(z_0, |p|^{1/(p+1)})$.

19.4.4 We now restrict our attention to the case $\varepsilon_1 \neq \varepsilon_2$. These are eigenvalues of the matrix $C^{\sigma^{m-1}}C^{\sigma^{m-2}} \cdots C$ of (19.3.5). This matrix is defined over $\mathbb{Q}_p(z_0)$ but the characteristic polynomial has coefficients in $\mathbb{Q}_p$ and hence ε_1 either lies in $\mathbb{Q}_p$ or generates a quadratic extension field.

Lemma. *If $\varepsilon_1 \neq \varepsilon_2$ then $\varepsilon_1 \notin \mathbb{Q}_p$.*

Proof. Suppose otherwise, i.e., $\varepsilon_1 \in \mathbb{Q}_p$.

Using σ in the sense of the Frobenius automorphism of Ω_0 over $\mathbb{Q}_p$ (or of its restriction to $\mathbb{Q}_p(z_0)$), we have in the notation of (19.3.5)

$$\varepsilon_1 \mathbf{u} = \mathbf{u} C^{\sigma^{m-1}} \cdots C. \tag{19.4.4.1}$$

The vector $\mathbf{u} = (u_1(z_0), u_2(z_0))$ has coefficients in $\mathbb{Q}_p(z_0)$ and applying σ and multiplying on the right by C, we find that $\mathbf{u}^\sigma C$ is eigenvector of $C^{\sigma^{m-1}} \cdots C$ corresponding to eigenvalue $\varepsilon_1^\sigma = \varepsilon_1 \neq \varepsilon_2$. Hence we conclude that

$$\mathbf{u}^\sigma C = e\mathbf{u}$$

for some $e \in \mathbb{Q}_p(z_0)$. This implies that $u_2(z_0)/u_1(z_0) = z$ satisfies relation (12.4.7) (with $\beta = B_2(z_0)$, $p\gamma = B_3(z_0)$, $p\alpha = B_1(z_0)$, $p\delta = B_4(z_0)$). The conclusion is that ord $z = 1/2$, which contradicts the nonramification of $\mathbb{Q}_p(z_0)$ over $\mathbb{Q}_p$. This completes the proof of the lemma. $\square$

19.4.5 The field $\mathbb{Q}_p(\varepsilon_1)$ is either ramified or unramified over $\mathbb{Q}_p$. In the second case ε_1 lies in Ω_0 and the Frobenius σ of Ω_0 over $\mathbb{Q}_p$ maps ε_1 into ε_2. In either case we extend σ to a continuous automorphism $\hat{\sigma}$ of Ω over $\mathbb{Q}_p$ such that

$$\hat{\sigma}\varepsilon_1 = \varepsilon_2.$$

19.4.6 **Definition.** We say that two elements x_1, x_2 of $D(z_0, 1^-)$ are *equivalent* if there exist integers $n_1, n_2 \geq 0$ such that

$$n_1 \equiv n_2 \mod 2,$$

$$\hat{\sigma}^{-n_1}\phi_1^{n_1}(x_1) = \hat{\sigma}^{-n_2}\phi_1^{n_2}(x_2),$$

but such that none of the elements in either sequence

$$\{(\hat{\sigma}^{-i}\phi_1^i)(x_1)\}_{i<n_1}, \qquad \{(\hat{\sigma}^{-j}\phi_1^j)(x_2)\}_{j<n_2}$$

lies in $D(z_0, |p|^{1/(p+1)-})$.

19.4.7 Given x_1 equivalent to x_2 we may increase (if necessary) the integers n_1, n_2 without changing the parity of the difference such that $y = \hat{\sigma}^{-n_1}\phi_1^{n_1}(x_1)$ lies in $D(z_0, |p|^{1/(p+1)-})$. The element y is equivalent to x_1 if n_1 is even but not otherwise. We refer to y as the *indicator* of the class of x_1. We organize equivalence classes into types, depending upon the location of the indicator, y, and upon whether or not n_1 is even.

Type I	y lies in the critical annulus,	n even,
Type II	y lies in the critical annulus,	n odd,
Type III	y lies in the inner disk,	n even,
Type IV	y lies in the inner disk,	n odd.

We arrange the elements of each class into rings, the sth ring having N_s elements each having a common value, D_s for $\text{ord}(\lambda - z_0)$. The following table based upon Corollary 16.8 lists all the elements in each class except for the indicator, y, in Case III. We recall that y lies in the class in Cases I and III but not in Cases II and IV.

Type	I	II	III	IV
D_s	$\frac{1}{p^{2s}}\text{ord}(y - z_0)$	$\frac{1}{p^{2s+1}}\text{ord}(y - z_0)$	$\frac{1}{p^{2s}}\frac{p}{p+1}$	$\frac{1}{p^{2s}}\frac{1}{p+1}$
N_s	p^{2s}	p^{2s+1}	$p^{2s-1}(p+1)$	$(p+1)p^{2s}$
$s \geq$	0	0	1	0

19.5 **Theorem.** *If u, v are eigenvectors of* (19.3.2) *with distinct eigenvalues then there are just two possibilities for the zero sets of u_1 and of v_1.*

Case (a)

The zero set of u_1 is the union of two classes, one of Type I, the other of Type II and each with indicator y satisfying

$$\operatorname{ord}(y - z_0) = \tfrac{1}{2}.$$

This case cannot occur unless $\mathbb{Q}_0(\varepsilon_1)$ is ramified over $\mathbb{Q}_p$.

Case (b)

The zero set of u_1 consists of just one class which is either of Type III or Type IV. In this case the zero set of v_1 also consists of just one class which is either of Type IV or Type III, the types of the zero sets of u_1 and v_1 being distinct in Case (b). *If y is the indicator of the class of Type III or Type IV then*

$$\operatorname{ord}(y - z_0) \geq 1, \qquad 2\operatorname{ord}(y - z_0) \in \mathbb{N}.$$

Proof. We have chosen $\hat{\sigma}$ an extension of σ to Ω such that

$$\hat{\sigma}\varepsilon_1 = \varepsilon_2. \tag{19.5.1}$$

The construction (19.4.0.1) gives eigenvectors $u^{(i)}$, $v^{(i)}$ of (19.3.2) (viewed as linear transformations of the space of solutions of (2.4.5.1) at $z_0^{\sigma^{-i}}$) with eigenvalues ε_1, ε_2 respectively. On the other hand if we let $\tilde{C} = B(z_0^\sigma)$ and apply $\hat{\sigma}$ to (19.4.4.1) we obtain

$$\varepsilon_2 \hat{\sigma}\mathbf{u} = \hat{\sigma}\mathbf{u}\tilde{C}^{\sigma^{m-1}}\tilde{C}^{\sigma^{m-2}} \cdots \tilde{C}. \tag{19.5.2}$$

From this we deduce that there exists constants e_1, e_2 such that

$$\begin{aligned} u^{\hat{\sigma}^{-1}} &= e_1 v^{(1)}, \\ v^{\hat{\sigma}^{-1}} &= e_2 u^{(1)}. \end{aligned} \tag{19.5.3}$$

Again from (19.4.0.3) we deduce

$$\begin{aligned} u_1(\phi_1(\lambda))\tilde{B}_1 &= u_1^{(1)}(\lambda), \\ v_1(\phi_1(\lambda))\tilde{B}_1 &= v_1^{(1)}(\lambda). \end{aligned} \tag{19.5.4}$$

As before we use $u^{(0)}$ (resp. $v^{(0)}$) to denote u (resp. v). Let us use $S^{(i)}$ (resp. $T^{(i)}$) to denote the zero set of $u_1^{(i)}$ (resp. $v_1^{(i)}$). We deduce from (19.5.3)

(19.5.5)
$$\hat{\sigma}S^{(1)} = T^{(0)},$$
$$\hat{\sigma}T^{(1)} = S^{(0)}.$$

(19.5.6)
$$S^{(1)} \subset \phi_1^{-1}(S^{(0)}),$$
$$T^{(1)} \subset \phi_1^{-1}(T^{(0)}),$$

and also

(19.5.7)
$$\phi_1(S^{(1)} \cap \text{support } \phi_1) \subset S^{(0)},$$
$$\phi_1(T^{(1)} \cap \text{support } \phi_1) \subset T^{(0)}.$$

From these relations we deduce that each zero set is a union of equivalence classes.

If y is a zero of u_1 in the critical annulus (in which case $S^{(0)}$ would contain a class of Type I) then by (19.5.6), the set $S^{(1)}$ would contain an element y_1 ($\in \phi_1^{-1}(y)$) in the critical annulus about $z_0^{\sigma^{-1}}$ and hence by (19.5.5) $T^{(0)}$ would contain $\hat{\sigma}y_1 = y_2$, which lies in the critical annulus about z_0 (and so $T^{(0)}$ would also contain a class of Type I). Now y_2 cannot lie in $S^{(0)}$ but is the indicator of a class of Type II which by (19.5.5) and (19.5.6) lies in $S^{(0)}$.

This shows that if $S^{(0)}$ contains a class of Type I then it also contains a set of Type II. The converse is now clear. Furthermore,

(19.5.8)
$$\text{ord}(y - z_0) + \text{ord}(y_2 - z_0) = 1.$$

We now choose a rational number r close to but less than $p/(p + 1)$ such that u_1 has no zeros on the set

$$\{\lambda \mid \text{ord}(\lambda - z_0) = r\}.$$

In this case u_1 has no zeros on the circumference

(19.5.9)
$$\{\lambda \mid \text{ord}(\lambda - z_0) = r/p^{2s}\}.$$

If M_s denotes the common value of $\text{ord}(u_1(\lambda))$ for λ on this circumference, then by (16.9.1),

(19.5.10)
$$M_s = -s + 0(1).$$

If **C** is any class, we define the polynomial

$$L_s(\mathbf{C})(\lambda) = \prod_{\substack{z \in \mathbf{C} \\ \operatorname{ord}(z - z_0) > r/p^{2s}}} (1 - \lambda/z), \tag{19.5.11}$$

and let

$$M_s(\mathbf{C}) = \operatorname{Min} \operatorname{ord} L_s(\mathbf{C})(\lambda)$$

the minimum being over all λ on circumference (19.5.9). Clearly

$$M_s = \sum M_s(\mathbf{C}) + 0(1) \tag{19.5.12}$$

the sum being over all classes **C** in the zero set of u_1. By means of the previous table we establish

Type	I	II	III	IV
$M_s(\mathbf{C}) + 0(1)$	$-s \operatorname{ord}(y - z_0)$	$-s \operatorname{ord}(y - z_0)$	$-s$	$-s$

It follows from (19.5.12), (19.5.10), and (19.5.8) that $S^{(0)}$ either contains one class of Type III or one of Type IV or precisely two classes one of Type I and the other of Type II.

If $S^{(0)}$ contains a class of Type III then u_1 has a zero y in the inner disk about z_0. Hence by (19.5.6), $S^{(1)}$ has an element y_1, with $\operatorname{ord}(y_1 - z_0^{\sigma^{-1}}) = 1/(p + 1)$ and hence by (19.5.5), $T^{(0)}$ contains $y_2 = \hat{\sigma} y_1$ such that

$$\operatorname{ord}(y_2 - z_0) = 1/(p + 1).$$

In this way we show that $T^{(0)}$ is a class of Type IV. In the same way it is clear that if $S^{(0)}$ is of Type IV the $T^{(0)}$ is of Type III.

Suppose $S^{(0)}$ is the union of a class of Type I and a class of Type II. Let y be the indicator of the class of Type I. Then y is the unique element of $S^{(0)}$ in the disk $D(z_0 | p|^{1/(p+1)-})$. This means that $\operatorname{ord}(y - z_0)$ may be computed from the Newton polygon of u_1 as function of $(\lambda - z_0)$. Indeed we must have

$$\operatorname{ord}(y - z_0) = \operatorname{ord} \frac{u_1(z_0)}{u_1'(z_0)}.$$

Since $u_1'(z_0)$ lies in the field $\mathbb{Q}_p(z_0, u_1(z_0), u_2(z_0)) \subset \mathbb{Q}_p(z_0, \varepsilon_1)$, a field of ramification 1 or 2, we conclude that

$$\operatorname{ord}(y - z_0) \in \mathbb{Z} \cdot \tfrac{1}{2}.$$

On the other hand by hypothesis

$$\frac{1}{p+1} < \operatorname{ord}(y - z_0) < \frac{p}{p+1}.$$

This shows that if $S^{(0)}$ contains a class of Type I then the indicator, y, must satisfy

$$\operatorname{ord}(y - z_0) = \tfrac{1}{2}$$

and that the field $\mathbb{Q}_p(\varepsilon_1)$ is ramified over $\mathbb{Q}_p$.

Finally if $S^{(0)}$ is a class of Type III or IV then either u_1 or v_1 has a unique zero in the disk $D(0, |p|^{(1/p+1)^-})$ and hence by the same argument as above the indicator y satisfies the conditions

$$\operatorname{ord}(y - z_0) \geq \frac{p}{p+1},$$

$$\operatorname{ord}(y - z_0) \in \tfrac{1}{2}\mathbb{Z}.$$

This show that

$$\operatorname{ord}(y - z_0) \geq 1.$$

This completes the proof of the theorem. □

19.6.

We try to make the situation of Theorem 19.5 more explicit. Let us write

$$C = B(z_0) = \begin{pmatrix} p\alpha & \beta \\ p\gamma & p\delta \end{pmatrix}. \tag{19.6.1}$$

Then

$$\frac{1}{p} C^\sigma C = \begin{pmatrix} \gamma\beta^\sigma + p\alpha^{1+\sigma} & \alpha^\sigma\beta + \beta^\sigma\delta \\ p(\gamma^\sigma\alpha + \delta^\sigma\gamma) & \gamma^\sigma\beta + p\delta^{1+\sigma} \end{pmatrix}. \tag{19.6.2}$$

Letting $m = 2m_1$, we obtain modulo p,

(19.6.2′)

$$\frac{1}{p^{m_1}} C^{\sigma^{m-1}} C^{\sigma^{m-2}} \cdots C \equiv \begin{pmatrix} (\gamma\beta^\sigma)^{1+\sigma^2+\cdots+\sigma^{2(m_1-1)}} & * \\ 0 & (\beta\gamma^\sigma)^{1+\sigma^2+\cdots+\sigma^{2(m_1-1)}} \end{pmatrix},$$

where $*$ denotes a quantity which is without interest to us. We note that

$$-p^{-1}\det B(z_0) \equiv \beta\gamma. \tag{19.6.3}$$

We deduce from equations (9.4.1.1) and (2.4.5.1) since $l = 1$ that

$$\frac{1}{p}\det B(\lambda) = e\cdot\lambda^{(p-1)c}\left(\frac{1-\lambda}{1-\lambda^p}\right)^{c-(a+b)}, \tag{19.6.4}$$

where e is a constant unit, independent of the residue class of λ. By restricting λ to a residue class in the prime field, we conclude that $e \in \mathbb{Z}_p$. (cf. Chapter 25).

The quantity β may be computed mod p by substituting z_0 for λ in the formula for χ_2 given by equation (6.4.8). Alternately up to factors in $\mathbb{F}_p$ it may be computed from the unique solution mod p of degree less than p defined over $\mathbb{F}_p$ of $(2.4.5.3)_{a,b,c}$.

Let $\theta = (\gamma\beta^{\sigma})^{1+\sigma^2+\cdots+\sigma^{2(m_1-1)}}$ mod p. Clearly θ lies in $\mathbb{F}_{p^2}$. If θ does not lie in $\mathbb{F}_p$ then we must be in case b of Theorem 19.5. Since $\beta^{1+\sigma+\cdots+\sigma^{m-1}}$ lies in $\mathbb{F}_p$, we see that

$$\theta \in \left(\frac{\gamma}{\beta}\right)^{1+\sigma^2+\cdots+\sigma^{2(m_1-1)}}\mathbb{F}_p.$$

In particular if $m = 2$ then the condition becomes

$$\frac{\lambda^{(p-1)c}(1-\lambda)^{(1-p)(c-a-b)}}{\beta^2} \notin \mathbb{F}_p. \tag{19.6.5}$$

Example 1. Let $c = 1$, $a = b = \frac{1}{2}$, $p = 11$. Here we are in Case 2 of §6.4 and so the supersingular classes are given by the zeros of B_4, i.e., of the mod p "branch" of $F(\frac{1}{2}, \frac{1}{2}, 1, \lambda)$ of degree $(p-1)/2$. Here β must be replaced by B_3, the mod p branch of $F(\frac{1}{2}, \frac{1}{2}, 2, \lambda)$ of degree less than p. Thus the supersingular classes are given by the zeros of $F(-5, -5, 1, \lambda) \equiv (1 + 3\lambda + \lambda^2)(1 - \lambda + \lambda^2) \times (\lambda + 1)$. Except for the zeros of $\lambda^2 - \lambda + 1$, these zeros lie in $\mathbb{F}_p$. Thus condition (19.6.5) cannot hold here except possibly for the zeros of that quadratic factor. Condition (19.6.5) reduces to the question of whether $(\lambda F(-5, -5, 2, \lambda))^2$ fails to lie in $\mathbb{F}_p$ modulo the relation R, $\lambda^2 - \lambda + 1 = 0$. We compute

$$\begin{aligned}\lambda F(-5, -5, 2, \lambda) &= \lambda(1 - 4\lambda + 4\lambda^2 + 3\lambda^3 + 5\lambda^4 + 2\lambda^5)\\ &\equiv -\lambda(7\lambda - 7) \text{ modulo } R\\ &\equiv 7 \in \mathbb{F}_p.\end{aligned}$$

Thus condition (19.6.5) is not satisfied. Indeed it is known (see, for example, Dwork [11, §8]) that the eigenvalues are equal in this case and hence by

(19.4.3.1) ϕ_1^2 is mapping of $D(z_0, |p|^{1/(p+1)-})$ into itself which coincides with the identity mapping.

EXAMPLE 2. Let $c = 1$, $a = \frac{1}{5}$, $b = \frac{2}{5}$, $p = 11$. Here the supersingular classes are given by the zeros of

$$F(-2, -4, 1, \lambda) = 1 - 3\lambda - 5\lambda^2,$$

which lie in $\mathbb{F}_{p^2} - \mathbb{F}_p$. In this case β must be replaced by

$$F(-2, -4, 2, \lambda) \equiv 1 + 4\lambda + 2\lambda^2$$

and condition (19.6.5) reduces to

$$(1 - \lambda)^4\lambda^2(1 + 4\lambda + 2\lambda^2)^2 \notin \mathbb{F}_{11}$$

modulo the relation R, $1 - 3\lambda - 5\lambda^2 = 0$, i.e., $\lambda^2 = -2 + 6\lambda$. Modulo R

$$\begin{aligned}(1 - \lambda)^4 &= (1 - 2\lambda + \lambda^2)^2 \equiv (-1 + 4\lambda)^2 = 1 + 3\lambda + 5\lambda^2\\ &\equiv 2,\end{aligned}$$

$$\begin{aligned}(\lambda(1 + 4\lambda + 2\lambda^2))^2 &\equiv (\lambda(-3 + 5\lambda))^2 \equiv (1 + 5\lambda)^2\\ &\equiv (1 - \lambda + 3\lambda^2) \equiv -5(1 + \lambda).\end{aligned}$$

This shows that condition (19.6.5) holds. We conclude from (19.4.1) that ϕ_1^2 is a mapping of $D(z_0, |p|^{1/(p+1)-})$ into itself which does not coincide with the identity mapping. Our test fails for ϕ_1^6.

Chapter 20

Analytic Factors of Power Series

The object of this chapter is to justify the main step in the proof of Theorem 18.7. It may also be useful in explaining the analyticity of the canonical lifting in the proof of Theorem 13.3. The notation is independent of those chapters.

The results of this section are closely related to results of Lubin [24] first communicated to us in 1969.

Let K be a complete subfield of Ω and $\mathbf{C}$ an open subset of Ω. We define $H_K(\mathbf{C})$ to be the completion under the supremum norm on $\mathbf{C}$ of elements of $K(x)$ having no poles in $\mathbf{C}$.

We impose upon the pair $(\mathbf{C}, K)$ the following conditions:

(20.1.1) $H_K(\mathbf{C})$ is a ring of bounded functions, each nonzero element of which has only a finite number of zeros, and elements without zeros are units.

(20.1.2) $H_K(\mathbf{C})$ satisfies the principle of analytic continuation.

These conditions hold if for example $\mathbf{C}$ is the complement in a disk $D(0, R^+)$ of the finite union $\bigcup_{i=1}^{l} D(a_i, r_i^-)$ where

$$R \geq \sup(r_1, \ldots, r_l) \geq \operatorname{Inf}(r_1, \ldots, r_l) > 0.$$

Let y be a single indeterminate and let

$$F(x, y) = \sum_{s=0}^{\infty} \xi_s(x) y^s, \tag{20.2}$$

with each $\xi_s \in H_K(\mathbf{C})$.

Let n be a fixed integer $n \geq 1$. We will consider the hypothesis

(20.3) For each $x \in \mathbf{C}$ the newton polygon of F (as element of $\Omega[[y]]$) has a vertex at $(n, \operatorname{ord} \xi_n(x))$.

Subject to this hypothesis, we conclude that for each $x \in \mathbf{C}$ there exist n zeros $y_1, \ldots, y_n$ of the power series in y which either constitute a full set of zeros of F (for the given value of x) or are strictly smaller than all other zeros.

Thus we may uniquely define a polynomial (in y),

$$P(x, y) = \prod_{i=1}^{n} (y - y_i(x)) \tag{20.4}$$

for each $x \in \mathbf{C}$.

Our object is to show that the coefficients of P are not only well defined but are even elements of $H_K(\mathbf{C})$.

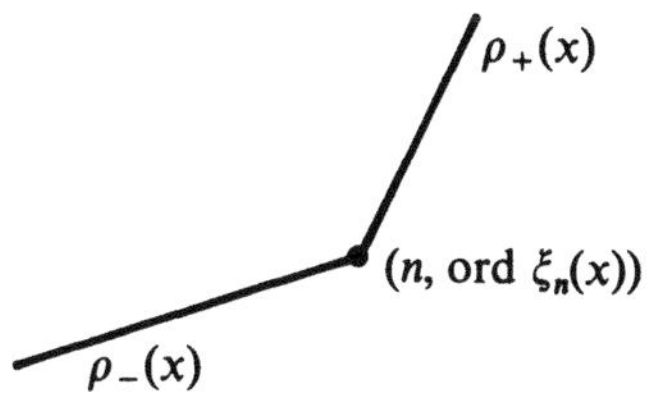

We attain this objective at the expense of strengthening the hypothesis (20.3). We formulate these conditions by defining $\rho_-(x)$ (resp. $\rho_+(x)$) to be the slope of the side of the polygon of F to the left (resp. right) of the vertex $(n, \text{ord } \xi_n(x))$. We consider the hypotheses

(20.5.1) There exists $\varepsilon_1 > 0$, ε_1 independent of x such that

$$\rho_+(x) - \rho_-(x) \geq \varepsilon_1, \qquad \forall x \in \mathbf{C}.$$

(20.5.2) There exists $\varepsilon_2 \in \mathbb{R}$, independent of x such that

$$\rho_+(x) \geq \varepsilon_2, \qquad \forall x \in \mathbf{C},$$

(ε_2 may be negative). This last condition implies that for all $x \in \mathbf{C}$, the series F converges for ord $y > -\varepsilon_2$.

Our procedure will be to first treat the case $n = 1$ with a slightly weaker hypothesis. We will then treat the case in which F is a polynomial by reducing this to the slightly stronger form of the case $n = 1$. The general case in which F is a power series will then be obtained by taking limits associated with polynomial approximations. Hypothesis (20.5.2) is not needed for $n = 1$, and will be seen to be automatic for F a polynomial. It will be used for the general case.

20.6 *Remark.* The main point in the construction of P is its uniqueness. Uniqueness could also be attained with a weaker form of condition (20.3):

(20.3′) For each $x_0 \in \mathbf{C}$ either the newton polygon of F has a vertex at $(n, \text{ord } \xi_n(x_0))$ or the roots $y_1, y_2, \ldots$ may be ordered so that $|y_1(x_0)| \leq |y_2(x_0)| \leq \cdots$ and such that

$$|y_i(x_0)| = |y_n(x_0)| \quad \text{implies } y_i(x_0) = y_n(x_0).$$

This weaker condition is analyzed (to some extent) for $n = 1$ in §§20.7 and 20.8.

If F is a polynomial then this question might be pursued by comparing the y_i with the solutions of F in the algebraic closure of $\Omega((x - x_0))$. This would be a method by which one could show that x_0 is not an isolated singularity.

20.7 The case $n = 1$

For $n = 1$ we do not need (20.5.1) on all of $\mathbf{C}$.

Let

$$M = \{x \in \mathbf{C} \mid \xi_0(x) = 0\},$$

a finite set.

Lemma. *Suppose that for each $x \in \mathbf{C} - M$, the newton polygon of F has a vertex at* $(1, \operatorname{ord} \xi_1(x))$ *and that there exists $\varepsilon_1 > 0$ such that*

$$\rho_+(x) - \rho_0(x) \geq \varepsilon_1, \qquad \forall x \in \mathbf{C} - M. \tag{20.7.1}$$

For $x \in \mathbf{C}$ let $y_1(x)$ be the minimal zero of $\mathbf{C}$, *hence zero for $x \in M$. Then $y_1 \in H_K(\mathbf{C})$.*

Note. We do not assume that (for $x \in M$) $y = 0$ is a simple zero of $F(y)$.

Proof. By hypothesis ξ_1 is never zero for $x \in \mathbf{C} - M$. On this set let us write $y = z \cdot \xi_0/\xi_1$ and then

$$\frac{1}{\xi_0} F(x, y) = 1 + z - \sum_{s=2}^{\infty} g_s(x) z^s, \tag{20.7.2}$$

where

$$-g_s = \frac{\xi_s \xi_0^{s-1}}{\xi_1^s}, \qquad s \geq 2. \tag{20.7.3}$$

For $x \in \mathbf{C} - M$,

$$\operatorname{ord} \frac{\xi_1(x)}{\xi_0(x)} = \rho_-,$$

$$\operatorname{ord} \frac{\xi_s(x)}{\xi_1(x)} \geq (s - 1)\rho_+.$$

Since $-g_s = (\xi_s/\xi_1)(\xi_0/\xi_1)^{s-1}$, we have for $x \in \mathbf{C} - M$

$$\text{(20.7.4)} \qquad \operatorname{ord} g_s(x) \geq (s-1)(\rho_+(x) - \rho(x)) \geq (s-1)\varepsilon_1.$$

Now g_s is meromorphic on $\mathbf{C}$ with possible poles on M, since ξ_1 cannot be zero on $\mathbf{C} - M$. But in fact (20.7.4) shows that g_s has no poles and so that relation must hold for all $x \in \mathbf{C}$.

The unique minimal zero for z (with $x \in \mathbf{C} - M$) is again by (20.7.4) a unit. Furthermore the mapping

$$\text{(20.7.5)} \qquad z \to G(x) = -1 + \sum_{s=2}^{\infty} g_s(x) z^s$$

is a contraction map of $H_0(\mathbf{C})(=\{z \in H_K(\mathbf{C}))\,|\,z$ bounded by 1 on $\mathbf{C}\})$ into itself since

$$\begin{aligned} |G(z_1) - G(z_2)| &= |z_1 - z_2| \left| \sum_{s=2}^{\infty} g_s(x) \sum_{i+j=s-1}^{\infty} z_1^i z_2^j \right| \\ &\leq |z_1 - z_2|\,|p|^{\varepsilon_1}. \end{aligned}$$

Thus there exists unique $z \in H_0(\mathbf{C})$ such that $z = G(z)$. This z coincides with $y\xi_1/\xi_0$ on $\mathbf{C} - M$ and since ξ_1 is never zero on $\mathbf{C} - M$, y coincides with $z\xi_0/\xi_1$ on that set. Now with no loss in generality we may assume for fixed $x_0 \in M$ that not all ξ_s vanish at x_0. Let m_0 (resp. m_1) be the order of zero of ξ_0 (resp. ξ_1) at x_0. By hypothesis $m_0 > 0$. If $m_1 \geq 1$ then there exists $s_1 \geq 2$ such that ξ_{s_1} has no zero, but g_{s_1} has no pole and hence

$$(s_1 - 1)m_0 \geq s_1 m_1.$$

This shows that $m_0 > m_1$. Thus $z\xi_0/\xi_1$ vanishes at x_0 and hence y coincides with $z\xi_0/\xi_1$ on M as well as on $\mathbf{C} - M$. Thus $y = z\xi_0/\xi_1$ and hence $y \in H_K(\mathbf{C})$. This completes the proof of the lemma. □

20.8 It is natural to ask whether it is possible for F to have a multiple minimal zero (other than $y = 0$) at some point $x_0 \in \mathbf{C}$ while (20.7.1) holds in a punctured neighborhood of x_0. We show that this cannot happen.

Corollary. *If $x_0 \in \mathbf{C}$, $\xi_0(x_0) \neq 0$ and in a punctured neighborhood of x_0, F has a unique minimal zero and*

$$\text{(20.8.1)} \qquad \rho_+(x) - \rho_-(x) \geq \varepsilon$$

for all x in the punctured neighborhood and if $F(x_0, y)$ has any zero (for y) then it has a unique (i.e., nonmultiple) minimal zero.

Proof. By hypothesis ξ_1 is never zero on the punctured neighborhood. We assert that $\xi_1(x_0) \neq 0$. To show this we again define g_s by (20.7.3) for $s \geq 2$ and by means of (20.8.1) obtain estimate (20.7.4) on the punctured neighborhood which shows that g_s has no pole at x_0. If $\xi_1(x_0) = 0$ then by (20.7.3), $\xi_s(x_0)\xi_0(x_0)^{s-1} = 0$ for all $s \geq 2$, i.e., $\xi_s(x_0) = 0$ for all such s. Hence

$$F(x_0, y) = \xi_0(x_0) \neq 0,$$

which contradicts the hypothesis that $F(x_0, y)$ has at least one zero. Now for $|x - x_0| = \mu$, close to but distinct from zero, we have

$$\frac{1}{s-1} \operatorname{ord} \frac{\xi_s(x)}{\xi_1(x)} \geq \rho_+(x) \geq \rho_-(x) + \varepsilon_1 = \varepsilon_1 + \operatorname{ord} \frac{\xi_1(x)}{\xi_0(x)}.$$

The two sides of this inequality depend only upon μ and vary continuously with μ. Letting $\mu \to 0$ we obtain

$$\frac{1}{s-1} \operatorname{ord} \frac{\xi_s(x_0)}{\xi_1(x_0)} \geq \varepsilon_1 + \operatorname{ord} \frac{\xi_1(x_0)}{\xi_0(x_0)}.$$

Since $\xi_1(x_0)$ and $\xi_0(x_0)$ are distinct from zero we conclude that the Newton polyton of $F(x_0, y)$ has a vertex at $(1, \operatorname{ord} \xi_1(x_0))$. This shows that there exists a unique minimal zero. □

20.9 F a Polynomial

In this section we take F to be a polynomial of degree $N > n$. We observe that subject to (20.3), ξ_n is never zero on **C** and so by (20.1.1) we may assume

(20.9.0) $$\xi_n = 1.$$

We now have

$$\rho_+(x) = \operatorname*{Inf}_{n < s \leq N} \frac{1}{s-n} \operatorname{ord} \xi_s(x).$$

For fixed s, by (20.1.1) there is a finite lower bound for ord $\xi_s(x)$ as x runs over **C**. This then shows that condition (20.5.2) holds.

Lemma. *If F is a polynomial of degree N satisfying* (20.5.1) *then the polynomial P has coefficients in $H(\mathbf{C})$.*

Proof. To avoid giving special treatment to the values of x which are zeros of ξ_N, we take $y = \infty$ to be a zero of $F(x_0, y)$ of order v if $N - v$ is maximal such that $\xi_{N-v}(x_0) \neq 0$.

The roots $y_1(x), \ldots, y_N(X)$ of $F(x, y)$ may be ordered for each $x \in \mathbf{C}$ such that

(20.9.1)
$$|y_1(x)| \le |y_{i+1}(x)|, \qquad 0 \le i < N,$$
$$|y_n(x)| < |y_{n+1}(x)| = |p|^{-\rho_+(x)}.$$

By (20.5.2) there exists $e > 0$ such that

$$|y_{n+1}(x)| \ge e, \qquad \forall x \in \mathbf{C}.$$

We fix $e' \in (0, e)$ and let $t \in \mathrm{D}(0, e') \cap K$.

For all $x \in \mathbf{C}$, $i \ge n + 1$

(20.9.2)
$$|t + y_i(x)| = |y_i(x)| \ge |p|^{-\rho_+(x)} \ge e.$$

We assert that

(20.9.3)
$$z = \prod_{i=1}^{n} (t + y_i(x)) \in H_K(\mathbf{C}).$$

(Note that z is defined by this formula.)

To prove this let $\mathbf{S}$ be the set of all subsets $S \subset \{1, 2, \ldots, N\}$ such that card $S = n$. There are $m = \binom{N}{n}$ such subsets. In particular let

$$S_0 = \{1, 2, \ldots, n\}.$$

For each $S \in \mathbf{S}$, let

$$z_S = \prod_{i \in S} (t + y_i(x))$$

and so

$$z = z_{S_0}.$$

Let

(20.9.4)
$$\mathbf{G}_t = \prod_{S \in \mathbf{S}} (y - z_S),$$

the monic polynomial of degree m satisfied by all z_s (if $\xi_N(x) \neq 0$). The coefficients of $\mathbf{G}_t$ are symmetric polynomials in $y_1, \ldots, y_N$ and hence lie in $\mathbb{Z}[t, \xi_0/\xi_N, \ldots, \xi_{N-1}/\xi_N]$. Multiplying by a suitable element of $H_K(\mathbf{C})$ we obtain, G_t, an element of $H_K(\mathbf{C})[y]$ of degree m whose zeros consists of the elements $\{z_S\}_{S \in \mathbf{S}}$. Let

$$M_t = \{x \mid z_S(x) = 0 \text{ for some } S \in \mathbf{S}\}. \tag{20.9.5}$$

If $x \in M_t$ then $t + y_i(x) = 0$ for some $i \in [1, N]$ and hence

$$F(x, -t) = 0. \tag{20.9.6}$$

Thus by (20.1.1), M_t is a finite set. It follows from the definitions that M_t coincides with the zeros of the coefficient of y^0 in G_t.

We assert that G_t satisfies the hypothesis of Lemma 20.7. To verify this we estimate z/z_S for $x \in \mathbf{C} - M_t$. Let $S \in \mathbf{S}$, $S \neq S_0$. Put $B = S \cap S_0$ and let us write as disjoint unions

$$S_0 = A \cup B, \qquad S = B \cup B'$$

and so $\mu = \text{card } B' = \text{card } A \geq 1$. Then for $x \notin M_t$.

$$\frac{z}{z_S} = \frac{\prod_{i \in A} (t + y_i)}{\prod_{j \in B'} (t + y_j)}. \tag{20.9.7}$$

For $i \in A$

$$|t + y_i(x)| \leq \operatorname{Sup}(|t|, |y_i(x)|) \leq \operatorname{Sup}(e', |p|^{-\rho_-(x)}).$$

Thus by (20.9.2)

$$\left|\frac{z}{z_S}\right| \leq \frac{\operatorname{Sup}(e', |p|^{-\rho_-(x)})^\mu}{|p|^{-\rho_+(x)\mu}} \leq \frac{\operatorname{Sup}(e', |p|^{-\rho_-(x)})^\mu}{e^\mu}.$$

Since $\mu \geq 1$, we obtain by (20.5.1)

$$\left|\frac{z}{z_S}\right| \leq \operatorname{Sup}\left(\frac{e'}{e}, |p|^{\varepsilon_1}\right), \tag{20.9.8}$$

an estimate independent of $x \in \mathbf{C} - M_t$. This estimate is equivalent to an estimate of type (20.7.1).

Hence by Lemma 20.7, assertion (20.9.3) is valid. Letting t assume $n + 1$ distinct values in K, we conclude that the symmetric polynomials in $y_1, \ldots, y_n$ lie in $H_K(\mathbf{C})$. This completes the proof of the lemma. □

20.10

We now extend the result to power series.

Theorem. *Let F be a power series in $H_K(\mathbf{C})[[Y]]$ satisfying* (20.5.1) *and* (20.5.2) *then the coefficients of P also lie in $H_K(C)$.*

Proof. We first observe that we may again assume $\xi_n = 1, n > 1$.

We restrict our attention to integers $N > n$ such that ξ_N is not identically zero. By truncation of F we obtain a polynomial of degree N,

$$F_N(x, y) = \sum_{s=0}^{N} \xi_s(x)y^s.$$

The polygon of F_N coincides with that of F up to the vertex at $(n, \text{ord } \xi_n)$ and to the right of that vertex the polygon of F_N lies on or above that of F. Thus by Lemma 20.9, F_N has a monic factor P_N of degree n in $H_K(\mathbf{C})[Y]$ whose Newton polygon coincides (for each $x \in \mathbf{C}$) with the first part of the Newton polygon of F_N.

It remains to show that the coefficients of P are uniform limits of the corresponding coefficients of P_N as $N \to \infty$.

By the definitions, for each $x \in \mathbf{C}$, $s \geq n$ we have

$$\text{ord } \xi_s(x) \geq (s - n)\rho_+(x). \tag{20.10.1}$$

For fixed $x \in \mathbf{C}$, let $-\mu$ be the slope of a side of the Newton polygon of F lying to the left of the vertex at $(n, 0)$. Then for $\text{ord } y = \mu$ we have

$$\text{ord}(F(y) - F_N(y)) \geq \operatorname*{Inf}_{s>N}[s\mu + (s - n)\rho_+(x)].$$

By hypothesis

$$-\mu \leq \rho_-(x)$$

and so

$$\rho_+(x) + \mu \geq \rho_+(x) - \rho_-(x) \geq \varepsilon_1.$$

This shows that

$$\operatorname*{Inf}_{\text{ord } y=\mu} \text{ord}(F(y) - F_N(y)) \geq (N - n)\varepsilon_1 + n\mu. \tag{20.10.2}$$

The line of support of the Newton polygon of F of slope $-\mu$ lies on or below the line of that slope passing through the vertex $(n, 0)$. This shows that

$$\operatorname*{Inf}_{\operatorname{ord} y = \mu} \operatorname{ord} F(y) \le n\mu.$$

This gives the estimate

$$\operatorname*{Inf}_{\mu} \operatorname{ord}(F(y) - F_N(y)) - \operatorname*{Inf}_{\mu} \operatorname{ord} F(y) \ge (N - n)\varepsilon_1, \tag{20.10.3}$$

where we use Inf_μ to indicate the minimum over all y such that $\operatorname{ord} y = \mu$. This estimate is independent of x and μ.

Let P_μ (resp. $P_{\mu, N}$) denote the monic factor of P (resp. P_n) whose roots consist of all roots of P (resp. P_N) of ordinal μ. Then by Lazards' [23*a*, Proposition 3] improvement of our earlier estimate (Dwork [9a, Lemma 1.7]),

$$\operatorname*{Inf}_{\mu} \operatorname{ord}(P_\mu(y) - P_{N,\mu}(y)) - \operatorname*{Inf}_{\mu} P_\mu \ge (N - n)\varepsilon_1. \tag{20.10.4}$$

We write

$$P_\mu(y) = \sum_{j=0}^{l(\mu)} y^j a_{\mu,j}(x),$$

$$P_{N,\mu}(y) = \sum_{j=0}^{l(\mu)} y^j a_{\mu,N,j}(x),$$

with

$$l(\mu) \le n,$$

$$1 = a_{\mu, l(\mu)} = a_{\mu, N, l(\mu)}.$$

In any case since P_μ is monic with Newton polygon of a single slope, $-\mu$, we conclude that

$$\operatorname{ord}_\mu P_\mu = ul(\mu).$$

We now conclude that from (20.10.4) that

$$\operatorname{ord}(a_{\mu,j}(x) - a_{\mu,N,j}(x)) \ge (N - n)\varepsilon_1 + (l(\mu) - j)\mu, \tag{20.10.5}$$

$$n \ge l(\mu) \ge l(\mu) - j \ge 0.$$

A lower bound for μ independent of x may be deduced from (20.1.1) together with

(20.10.6)
$$\mu \geq -\rho_{-}(x) \geq \operatorname*{Inf}_{0 \leq j \leq n-1} \frac{1}{n-j} \operatorname{ord} \xi_j(x).$$

Furthermore since

(20.10.7)
$$\operatorname{ord} a_{\mu,j}(x) \geq (n-j)\mu,$$
$$\operatorname{ord} a_{\mu,N,j}(x) \geq (n-j)\mu,$$

we may deduce upper bounds for $a_{\mu,j}$, $a_{\mu,N,J}$ independent of x, N and of j. Finally, we write

$$P - P_N = \prod_\mu P_\mu - \prod_\mu P_{N,\mu}$$

and deduce that the coefficients of P_N do indeed uniformly approximate those of P. This completes the proof of the theorem. □

Chapter 21

p-adic Gamma Functions

The object of this chapter is to introduce the p-adic gamma function. This will be used to explain the p-adic beta function and ultimately for our discussion of singular disks in Chapters 24–26. Our treatment is strongly influenced by Boyarsky [5]. We will be brief since similar material has been discussed by Lang [23] and by Katz [21].

21.1 **Proposition.** *Let* $F(x) = \exp \pi(x - x^p) = \sum_{j=0}^{\infty} c_j x^j$. *Then*

$$\operatorname{ord} c_j \geq j \frac{p-1}{p^2}, \qquad j \geq 0.$$

Proof. We start with the fact that the Artin–Hasse series

$$\theta(x) = \exp\left(\sum_{s=0}^{\infty} x^{p^s}/p^s \right)$$

satisfies the formal identity

$$\theta(x^p) = \theta(x)^p h(x),$$

where $h = \exp(-px) \in 1 + p\mathbb{Z}_p[[x]]$. Writing $\theta(x) = \prod_{n=1}^{\infty} (1 - a_n X^n)$ with all $a_i \in \mathbb{Q}$, it is clear that $a_1 \in \mathbb{Z}_p$ and then by induction the same holds for all a_n. Thus $\theta \in \mathbb{Z}_p[[x]]$ and so $\theta(\pi x) = \sum_{j=0}^{\infty} e_j x^j$ with $\operatorname{ord} e_j \geq j/(p-1)$. Since $\pi^p = -p\pi$, we know that

$$F(x) = \theta(\pi x) \cdot \prod_{s=2}^{\infty} \exp(-(\pi x)^{p^s}/p^s)$$

and the proposition follows from the fact that the function indexed by s is bounded by unity on the disk

$$p^s \operatorname{ord}(x\pi) > s + \frac{1}{p-1}. \quad \square$$

21.2

We pause to give uniform estimates for $\operatorname{ord}(a)_s$. (Cf. Clark [6].)

Lemma. *Let $a \in \mathbb{Q} \cap \mathbb{Z}_p$. If $a \in \mathbb{Z}$ let $a \geq 1$. Then for $s \in \mathbb{N}$,*

$$0 \geq \operatorname{ord} s! - \operatorname{ord}(a)_s \geq -\frac{\log d + \log(s + |a|)}{\log p}, \tag{21.2.1}$$

where d = denominator a and log *and* $|\ |$ *are used in the classical sense.*

Proof. The left inequality follows from the fact that

$$\sum \frac{(a)_s}{s!} x^s = (1 - x)^{-a} \in \mathbb{Z}_p[[x]].$$

The right inequality is more delicate. We first observe by the product formula that

$$\operatorname{ord}_p a \leq \frac{\log(d|a|)}{\log p}. \tag{21.2.2}$$

By Kummer's identity for the confluent hypergeometric function (Watson [36, p. 102]), ${}_1F_1(a, \rho, x) = e^x\, {}_1F_1(\rho - a, \rho, -x)$, we obtain in particular

$${}_1F_1(1, a; x) = e^x\, {}_1F_1(a - 1, a, -x). \tag{21.2.3}$$

Thus

$$\sum \frac{x^s}{(a)_s} = e^x \sum \frac{(-x)^s}{s!(a + s - 1)}.$$

Comparing coefficients of x^s and multiplying by $s!$, we obtain

$$\frac{s!}{(a)_s} = \sum_{i+j=s} (-1)^j \binom{s}{j} \frac{1}{a + j - 1}. \tag{21.2.4}$$

The lemma now follows by applying (21.2.2) to get an upper bound for $\operatorname{ord}(a + j - 1)$. □

21.3

We denote by $L_{0,\infty}$ the ring of Laurent series in x with coefficients in Ω which converge for $|T_\infty| > \varepsilon_\infty$, $|T_0| > \varepsilon_0$ for some unspecified real numbers $\varepsilon_0, \varepsilon_\infty$ in the interval (0, 1). For $a \in \mathbb{Q} \cap \mathbb{Z}_p - \mathbb{Z}$, we view x^a as an abstract symbol, designate by Ω_a^0 the set of formal products $\{x^a\xi \mid \xi \in L_{0,\infty}\}$. We identify Ω_{a+m}^0 with Ω_a^0 (for $m \in \mathbb{Z}$) by identifying $x^{a+m} \cdot \xi$ in the first space with $x^a \cdot x^m\xi$ which lies in the second. Thus Ω_a^0 depends only upon a mod $\mathbb{Z}$. We define a differential operator D stable on Ω_a^0 by setting for all $\xi \in L_{0,\infty}$,

$$Dx^a\xi = x^a(E + a + \pi x)\xi.$$

We put $\bar{\Omega}_a = \Omega_a^0/D\Omega_a^0$.

Lemma ($a \notin \mathbb{Z}$). *$\bar{\Omega}_a$ is a vector space of dimension* 1 *with x^a representing a basis. In particular* (*using the classical gamma function*),

$$x^{a+m} \equiv \frac{\Gamma(a+m)(-\pi)^{-m}x^a}{\Gamma(a)} \mod D\Omega_a^0. \tag{21.3.1}$$

Proof. For $m \in \mathbb{N}$, $m \geq 1$, let

$$v_m = \pi^{-1}\sum_{j=0}^{m-1}(-1)^j\binom{m+a-1}{j}\frac{j!}{\pi^j}x^{m-j-1},$$

$$u_m = \sum_{j=0}^{m-1}\frac{\pi^j(-1)^j}{(a-m)_{j+1}}\frac{1}{x^{m-j}}.$$

Then for $m \geq 1$,

$$Dx^a v_m + (-1)^m\frac{(a)_m}{\pi^m}x^a = x^{a+m},$$

$$Dx^a u_m + \frac{\pi^m}{(1-a)_m}x^a = \frac{1}{x^m}x^a.$$

This shows that for $\xi = \sum_{m=-\infty}^{+\infty} b_m x^m \in L_{0,\infty}$, we have formally

$$x^a\xi = \left(b_0 + \sum_{m=1}^{\infty} b_m(-1)^m\frac{(a)_m}{\pi^m} + \sum_{m=1}^{\infty} b_{-m}\frac{\pi^m}{(1-a)_m}\right)x^a$$

$$+ Dx^a\sum_{m=1}^{\infty}(b_m v_m + b_{-m}u_m).$$

With the aid of Proposition 21.2 this shows that the dimension of $\bar{\Omega}_a$ is at most unity.

To show the converse we fix a, replace D on Ω_0^a by $D_a = E + a + \pi x$ on $L_{0,\infty}$ and construct a space dual to $L_{0,\infty}/D_a L_{0,\infty}$ by finding the kernel of $D_a^* = \gamma_- \circ (-E - 1 + a + \pi x)$ in $R = R_0 \oplus R_\infty$. Indeed we obtain

$$1^* = \left(\sum_{s=0}^{\infty} \frac{\pi^{s+1} x^s}{(1-a)_{s+1}}, -\sum_{s=1}^{\infty} \frac{(a)_{s-1}}{(-\pi)^{s-1}} \frac{1}{x^s}\right),$$

from which it follows that if $x^a = Dx^a\xi$, $\xi \in L_{0,\infty}$ then $\langle 1^*, 1\rangle = 0$ which is impossible. To verify (21.3.1) we observe that $Dx^a = ax^a + \pi x^{a+1}$ and then use iteration. □

21.4

Let $a, b \in \mathbb{Z}_p \cap \mathbb{Q} - \mathbb{Z}$,

(21.4.1) $$pb \equiv a \mod \mathbb{Z}.$$

Define a map α of Ω_a^0 into Ω_b^0 by the formula, for $\xi \in L_{0,\infty}$,

$$\alpha x^a \xi = x^b \psi(\xi x^{a-pb} F).$$

Since F lies in $L_{0,\infty}$, ψ is used here in the sense of (3.5.6) as a mapping of $L_{0,\infty}$ into itself. Of course α is independent of the choice of representatives a, b in given classes mod $\mathbb{Z}$. It is clear that

$$\alpha \circ D = pD \circ \alpha$$

and hence α maps $D\Omega_a^0$ into $D\Omega_b^0$. For all a, b satisfying (21.4.1) there exists a unique element $\gamma_p(a, b)$, in $\mathbb{Q}_p(\pi)$ such that

(21.4.2) $$\alpha x^a \equiv \gamma_p(a, b) x^b \mod D\Omega_b^0.$$

This modified gamma function does vary as a and b are varied in a fixed pair of classes mod $\mathbb{Z}$. Indeed for $m, n \in \mathbb{Z}$ we have

(21.4.3) $$\gamma_p(a+m, b+n) = \gamma_p(a, b) \frac{\Gamma(a+m)}{\Gamma(a)} \frac{\Gamma(b)(-\pi)^{n-m}}{\Gamma(b+n)},$$

a direct consequence of (21.3.1) and (21.4.2). An explicit formula for γ_p is obtained by using the notation of Proposition 21.1 and the reduction formula (21.3.1) to compute (with $pb - a = t$)

(21.4.4) $$\gamma_p(a, b) = \sum_{pi+t \geq 0} c_{pi+t} \frac{\Gamma(b+i)}{\Gamma(b)} \Big/ (-\pi)^i$$

a series which converges rather rapidly. Indeed we deduce from Proposition 21.1 that with t fixed, the right side extends to a function of b analytic for ord $b > -e$, where

$$e = 1 - \frac{1}{p} - \frac{1}{p-1}.$$

We define the *first Boyarsky* function Γ^B, on the disjoint union ($\rho = |p|^{1-e}$)

$$Q = \bigcup_{t=0}^{p-1} D(-t, \rho^-) \supset \mathbb{Z}_p$$

by setting for $x = -t + py \in Q$, $t = 0, 1, \dots, p-1$

$$\pi^t \Gamma^B(-t + py) = \sum_{i=0}^{\infty} C_{pi+t}(y)_i/(-\pi)^i. \tag{21.4.5}$$

21.4.6 **Lemma.**

$$\frac{\Gamma^B(1+x)}{\Gamma^B(x)} = \begin{cases} -x & \textit{if } |x| = 1, \\ -1 & \textit{if } |x| < 1, \end{cases}$$

$$\Gamma^B(0) = 1;$$

i.e., the first Boyarsky function coincides with the function, Γ_p, commonly known as the Morita p-adic gamma function.

Proof. The value of Γ^B at zero may be deduced from (21.4.5). By analyticity it is enough to verify the functional relation for $y \in \mathbb{Q} \cap \mathbb{Z}_p - \mathbb{Z}$. Thus

$$\pi^t \Gamma^B(-t + py) = \gamma_p(-t + py, y),$$

and now we may use equation (21.4.3). If $t \neq 0$ (resp. $t = 0$) then $t - 1$ (resp. $p - 1$) is the representative of $-(x + 1)$ in $\{0, 1, \dots, p - 1\}$. Thus for $t \neq 0$ we compute

$$\pi^{t-1}\Gamma^B(x+1) = \gamma_p(py - (t-1), y) = \gamma_p(py - t, y)\frac{py - t}{-\pi}$$

$$= \pi^t \Gamma^B(x) \cdot \frac{x}{(-\pi)}.$$

Again for $t = 0$ we have $x + 1 = p(y + 1) - (p - 1)$ and so

$$\pi^{p-1}\Gamma^B(x + 1) = \gamma_p(p(y + 1) - (p - 1), y + 1) = \gamma_p(py + 1, y + 1)$$
$$= \frac{py}{y}\gamma_p(py, y) = -\pi^{p-1}\Gamma^B(x).$$

This completes the proof of the lemma. □

We have now shown that Γ^B may be characterized by the values at $n \in \mathbb{N}$,

(21.4.7) $$\Gamma_p(n) = \prod_{\substack{j=0\\ p \nmid j}}^{n-1} j \cdot (-1)^n.$$

In particular by continuity we have (t = representative of $-x$ in $\{0, 1, \ldots, p - 1\}$)

(21.4.8) $$\Gamma_p(x)\Gamma_p(1 - x) = -(-1)^t.$$

Thus letting $G = \Gamma_p'/\Gamma_p$, the logarithmic derivative of Γ_p, we have

(21.4.9) $$G(x) = G(1 - x).$$

21.4.10 Equations (21.4.2) and (21.4.4) and their relation to Gauss sums and to Stickelberger's characterization of gauss sums have been well known since 1961. Local analyticity of Γ_p was known to Morita but the improved estimate for domain of analyticity is due to Boyarsky (cf Barsky [4b]). Lemma 21.4.6 relating Γ^B with Γ_p is the content of the Gross–Koblitz formula (Boyarsky [5], Lang [23]). As noted in 1961 an infinite set of functions similar to Γ^B in their relation to Gauss sums may be deduced by the methods of this chapter by merely replacing the analytic "additive character" function F (Proposition 21.1) by other functions of similar type (Dwork [9a, §4a]). These successive Boyarsky functions should be locally analytic on larger disks than Γ_p.

21.5.

We use Γ_p to obtain formulae for ratios of binomial type coefficients.

Lemma. *Let* $\theta, \theta' \in \mathbb{Q} \cap \mathbb{Z}_p - \mathbb{Z}$,

$$p\theta' - \theta = \mu \in \{0, 1, \ldots, p - 1\}.$$

Let t also be an element of $\mathbb{N}$, $t \leq p - 1$. Then for all $m \in \mathbb{N}$, we have

$$\frac{(-1)^{pm+t}}{p^m}\frac{(\theta)_{t+pm}}{(\theta')_m} = \frac{\Gamma_p(\theta + t + pm)}{\Gamma_p(\theta)}\chi_{t,\mu}(p(\theta' + m)),$$

where

$$\chi_{t,\mu}(z) = \begin{cases} 1 & \text{if } t \leq \mu, \\ z & \text{if } t > \mu. \end{cases}$$

Proof. This lemma is deduced by writing $\prod_{j=0}^{pm+t-1}(\theta + j)$ as a product of units and a product of nonunits. The product of units is identified with $(-1)^{pm+t}\Gamma_p(\theta + t + pm)/\Gamma_p(\theta)$. The nonunits involve $j \equiv -\theta \bmod p$, i.e., $j = \mu + ps$ and here

$$0 \leq s < m + (t - \mu)p^{-1}.$$

This gives two possibilities. If $t \leq \mu$ then the second product is

$$\prod_{s=0}^{m-1}(\theta + \mu + ps) = p^m \prod_{s=0}^{m-1}(\theta' + s) = p^m(\theta')_m.$$

It only remains to observe that if $t > \mu$ then there is an additional factor, $\theta + \mu + pm = p(\theta' + m)$. This completes the proof. □

As a corollary we obtain a similar formula for

(21.5.1) $$A_\theta(s) = \frac{(\theta)_s}{s!} = \binom{-\theta}{s}(-1)^s.$$

Corollary. *Under the conditions of the lemma,*

(21.5.2) $$\frac{A_\theta(t + pm)}{A_{\theta'}(m)} = -\frac{\Gamma_p(\theta + t + pm)}{\Gamma_p(1 + t + pm)}\frac{1}{\Gamma_p(\theta)}\chi_{t,\mu}(p(\theta' + m)).$$

Proof. We apply the lemma again to the case $\theta = 1$ in which case $\mu = p - 1$ and so $\chi_{t,\mu}(x) = 1$. Taking ratios and using $\Gamma_p(1) = -1$, we obtain the assertion. □

21.6 Gauss Multiplication Formula

We shall need this relation among the gamma functions.

Theorem. *Let $n \in \mathbb{N}$, $p \nmid n$, $x = -t + py$ lie in the support of Γ_p then*

$$\prod_{i=0}^{n-1} \Gamma_p\left(\frac{x+i}{n}\right) = \Gamma_p(x) \cdot \left(\frac{1}{n}\right)^{-t+(p-1)y} k_n,$$

where k_n is a constant depending only on n.

Proof. Put

$$g_n(x) = \frac{1}{\Gamma_p(x)} \prod_{i=0}^{n-1} \Gamma_p\left(\frac{x+i}{n}\right).$$

Then by the functional equation

$$\frac{g_n(x+1)}{g_n(x)} = \frac{\Gamma_p((x/n)+1)}{\Gamma_p(x/n)} \frac{\Gamma_p(x)}{\Gamma_p(x+1)} = \begin{cases} 1 & \text{if } |x| < 1, \\ \dfrac{1}{n} & \text{if } |x| = 1. \end{cases}$$

We deduce that for $x \in \mathbb{N}$,

$$\frac{g_n(x)}{g_n(0)} = \left(\frac{1}{n}\right)^{x-1-[(x-1)/p]} = \left(\frac{1}{n}\right)^{-t+(p-1)y}.$$

Finally we set $k_n = g_n(0)$. We find from (21.4.8):

$$k_n = (-1)^A \Gamma_p(\tfrac{1}{2})^B,$$

$B = 0$ (resp. 1) if n is odd (resp. even), $A = \sum_{1 \le i < n/2}$ representative of $(-i/n)$ + number of such i. □

Chapter 22

p-adic Beta Functions

22.1 Algebraic Theory

Let a_1, a_2 be elements of $\mathbb{Q}$, $a_2 \notin \mathbb{Z}$. Let

$$\mathbf{L}_0 = \Omega\left[x, \frac{1}{x}, \frac{1}{1-x}\right] \subset \Omega(x).$$

We put

$$f_0 = x^{a_1-1}(1-x)^{a_2},$$

$$D_{f_0} = E + a_1 + a_2 - \frac{a_2}{1-x},$$

$$\mathbf{W}_{f_0} = \mathbf{L}_0 / D_{f_0}\mathbf{L}_0.$$

Lemma. $\mathbf{W}_{f_0}$ *is a vector space of dimension one. A basis is represented by* $1/(1-x)$ *if either*

$$\text{(22.1.1)} \qquad \begin{array}{ll} a_1, a_1 + a_2 & \textit{both outside } \mathbb{Z}, \\ a_1 + a_2 \in \mathbb{Z} & \textit{but } a_1 + a_2 > 0, \\ a_1 \in \mathbb{Z} & \textit{but } a_1 < 1. \end{array}$$

Proof. To compute the dimension we may assume that one of these three conditions are met and then use the reduction formulae as in Chapter 1. The lemma may be demonstrated in this way. We may also proceed by associating $1/(1-x)$ with the one-form

$$\tau_{a_1, a_2} = x^{a_1-1}(1-x)^{a_2-1}\, dx$$

and noting that if $a_1 + a_2 \in \mathbb{Z}$, $a_1 + a_2 \geq 1$ then this one-form has a residue at infinity while if $a_1 \in \mathbb{Z}$, $a_1 \leq 0$ then τ_{a_1, a_2} has a residue at $x = 0$ and in these cases cannot be exact. $\square$

22.2 Analytic Theory

The analytic formulation of the preceding theory is obtained by replacing $\mathbf{L}_0$ by L_0, defined as in Chapter 3 with $\lambda = 0$. Using Adolphson's comparison theorem we obtain under the hypotheses of Lemma 22.1, $\mathbf{W}_f$ is isomorphic to $L_0/D_f L_0$ and the latter space is spanned by (the representative of) $1/(1-x)$.

We now assume that (a_1, a_2) and (b_1, b_2) both satisfy the hypothesis of Lemma 22.1. We assume

$$(22.2.1) \qquad \begin{aligned} p(b_1, b_2) &\equiv (a_1, a_2) \quad \bmod \mathbb{Z}^2, \\ a_2 \not\equiv 0 &\equiv b_2 \qquad\quad \bmod \mathbb{Z}, \\ a_1, a_2, b_1, b_2 &\in \mathbb{Q} \cap \mathbb{Z}_p. \end{aligned}$$

Let

$$h_0(x) = x^{b_1 - 1}(1-x)^{b_2}$$

$$\Lambda\left(\frac{1}{1-x}\right) = \frac{(1-x)^p}{1-x^p} = \frac{T_1^p}{T_1^\Phi},$$

$$\mu_i = pb_i - a_i,$$

$$(22.2.2) \qquad G_{f_0, h_0} = \frac{xf(x)}{x^p h_0(x^p)} = x^{-\mu_1}(1-x)^{-\mu_2}\Lambda^{b_2} \in L_0.$$

We define

$$(22.2.3) \qquad \alpha_{f_0, h_0} = \psi \circ G_{f_0, h_0}$$

a mapping of L_0 into itself which maps $D_{f_0} L_0$ into $D_{h_0} L_0$. We may now define the p-adic beta function by setting

$$(22.2.4) \qquad \alpha_{f_0, h_0} \frac{1}{1-x} \equiv B_p(a_1, a_2; b_1, b_2) \frac{1}{1-x} \quad \bmod D_{h_0} L_0.$$

Our object is to compute $B_p(a_1, a_2; b_1, b_2)$ in terms of gamma functions. We shall do more; we write for fixed $s \in \mathbb{N}$

$$(22.2.5) \qquad \alpha_{f_0, h_0} \frac{1}{1-x} = \sum_{j=0}^{s-1} \gamma_j D_{h_0}^j \frac{1}{1-x} + D_{h_0}^s \xi,$$

where $\xi \in L_0$ and compute $\gamma_0, \gamma_1, \ldots, \gamma_{s-1}$.

Theorem. *Let $\mu_1, \mu_2 \in [0, p-1]$. For z close to zero let*

(22.2.6)

$$H(z) = -\frac{\Gamma_p(1 - a_2 - a_1 - pz)}{\Gamma_p(1 - a_2)\Gamma_p(1 - a_1 + pz)} \chi_{\mu_1 + \mu_2, p-1}(p(1 - b_1 - b_2 + z)),$$

an analytic function of z. Then the expansion of H,

(22.2.7) $$H(z) = \gamma_0 + \gamma_1 z + \gamma_2 z^2 + \cdots$$

has coefficients coinciding with the coefficients in (22.2.5).

Proof. We compute $\alpha_{f_0, h_0} 1/(1 - x)$ by applying ψ to a representation in $\Omega((x))$ of

(22.2.8) $$\frac{1}{1 - x} G_{f_0, h_0} = x^{-\mu_1} \frac{(1 - x)^{a_2 - 1}}{(1 - x^p)^{b_2}} = \frac{1}{(1 - x^p)^{b_2}} x^{-\mu_1} \sum_{m=0}^{\infty} A_{1-a_2}(m) x^m$$

in the notation of (21.5.1). Thus

(22.2.9) $$\alpha_{f_0, h_0} \frac{1}{1 - x} = \frac{1}{(1 - x)^{b_2}} \sum_m A_{1-a_2}(\mu_1 + pm) x^m.$$

We may write

$$D_{h_0}^j \frac{1}{1 - x} = \frac{1}{x^{b_1}(1 - x)^{b_2}} E^j(x^{b_1}(1 - x)^{b_2 - 1}) = \frac{1}{x^{b_1}(1 - x)^{b_2}} \sum_{m=0}^{\infty} A_{1-b_2}(m)(m + b_1)^j x^{b_1 + m}.$$

We observe that ξ in the error term of (22.2.5) lies in L_0 and hence represents a bounded analytic function on an annulus

$$1 - \varepsilon < |x| < 1$$

for some $\varepsilon \in (0, 1)$. It follows that we may represent

$$(1 - x)^{b_2} \xi = \sum_{m=-\infty}^{\infty} c_m x^m$$

on the same annulus and so c_m is bounded independently of m as $m \to \infty$.

Substituting (22.2.9) in (22.2.5),

$$(22.2.10)\quad A_{1-a_2}(\mu_1 + pm) = A_{1-b_2}(m) \sum_{j=0}^{s-1} \gamma_j(m + b_1)^j + (m + b_1)^s c_m.$$

We apply (21.5.2) and obtain

$$(22.2.11)\quad -\frac{\Gamma_p(1 - a_2 + \mu_1 + pm)}{\Gamma_p(1 + \mu_1 + pm)} \frac{1}{\Gamma_p(1 - a_2)} \chi_{\mu_1, p-1-\mu_2}(p(1 - b_2 + m))$$
$$= \sum_{j=0}^{s-1} \gamma_j(m + b_1)^j + c_m(m + b_1)^s.$$

Let $m + b_1 = z$, so $\mu_1 + pm = -a_1 + pz$. The theorem follows from (22.2.11). □

22.2.12 **Corollary.** *If $b_1 + b_2, b_1, b_2$ are all outside of $\mathbb{Z}$ then*

$$\mathrm{B}_p(a_1, a_2; b_1, b_2) = \frac{\gamma_p(a_1, b_1)\gamma_p(a_2, b_2)}{\gamma_p(a_1 + a_2, b_1 + b_2)}.$$

Remark. We eliminate here the hypothesis that μ_1 and μ_2 lie in $[0, p - 1]$.

Proof. If $\mu_1, \mu_2 \in [0, p - 1]$ then

$$\gamma_p(a_i, b_i) = \pi^{\mu_i}\Gamma_p(a_i);$$

and if $\mu_1 + \mu_2 < p$ then

$$\gamma_p(a_1 + a_2, b_1 + b_2) = \pi^{\mu_1+\mu_2}\Gamma_p(a_1 + a_2).$$

If $\mu_1 + \mu_2 \geq p$ then $\mathrm{Rep}(-a_1 - a_2) = \mu_1 + \mu_2 - p$ and so

$$\gamma_p(a_1 + a_2, b_1 + b_2 - 1) = \pi^{\mu_1+\mu_2-p}\Gamma_p(a_1 + a_2).$$

If μ_1, μ_2 lie in $[0, p - 1]$ then the assertion follows from the preceding theorem with the aid of (21.4.3). By showing that the two sides of the asserted equation transform in the same way under substitutions $a_i \to a_i + m_i$, $b_i \to b_i + n_i$; $m_i, n_i \in \mathbb{Z}$, we complete the proof of the corollary. □

22.3

For our application we will be interested in two cases in which (a_1, a_2) (resp. (b_1, b_2)) do not satisfy the conditions of Lemma 22.1. Clearly we could transform this situation into that one. It seems simpler to proceed directly. We will be brief since the method is the same as in the preceding lemma.

Lemma. (i) *Let* $a, a' \in \mathbb{Q} \cap \mathbb{Z}_p - \mathbb{Z}$, $pa' - a = \mu_a \in [0, p-1]$,

$$f_0 = (1-x)^{-a},$$

$$h_0 = (1-x)^{-a'}.$$

Then $1/x$ *represents a basis of* $\mathbf{W}_{f_0}$ *and*

$$\alpha_{f_0, h_0} \frac{1}{x} = \gamma_0 \frac{1}{x} + \gamma_1 D_{h_0} \frac{1}{x} + \cdots,$$

where the γ_i *are the coefficients of the local expansion of*

$$-\frac{\Gamma_p(a+pz)}{\Gamma_p(a)} \frac{1}{\Gamma_p(1+pz)} = \gamma_0 + \gamma_1 z + \cdots. \tag{22.3.1}$$

(ii) *Let* $b', b \in \mathbb{Q} \cap \mathbb{Z}_p - \mathbb{Z}$, $pb' - b = \mu_b \in [0, p-1]$,

$$f_0 = x^{b-1}(1-x)^{-b},$$

$$h_0 = x^{b'-1}(1-x)^{-b'}.$$

Then 1 *represents a basis of* $\mathbf{W}_{f_0}$ *and*

$$\alpha_{f_0, h_0} 1 = \gamma_0 \cdot 1 + \gamma_1 D_{h_0} 1 + \cdots,$$

where the γ_i *are the coefficients of the local expansion of*

$$-\frac{\Gamma_p(pz)}{\Gamma_p(b)\Gamma_p(1-b+pz)} = \gamma_0 + \gamma_1 z + \cdots. \tag{22.3.2}$$

Proof. (i) We compute $\alpha_{f_0, h_0}(1/x)$ by representing

$$\frac{1}{x} G_{f_0, h_0} = \frac{1}{x^p} \frac{(1-x)^{-a}}{(1-x^p)^{-a'}} \tag{22.3.1.1}$$

as an element of $\Omega((1/x))$. Thus

$$\alpha_{f_0, h_0} \frac{1}{x} = \frac{1}{x(1-x)^{-a'}} \sum_{m=0}^{\infty} A_a(pm) x^m. \tag{22.3.1.2}$$

We can write this in the form

$$\gamma_0 \frac{1}{x} + \gamma_1 \frac{1}{xh_0} E\left(xh_0 \frac{1}{x}\right) + \gamma_2 \frac{1}{xh_0} E^2\left(xh_0 \frac{1}{x}\right) + \cdots$$
$$= \frac{1}{xh_0} \{\gamma_0(1 - x)^{-a'} + \gamma_1 E(1 - x)^{-a'} + \gamma_2 E^2(1 - x)^{-a'} + \cdots\}.$$

Comparing sides we obtain

(22.3.1.3) $$A_a(pm) = A_{a'}(m)[\gamma_0 + \gamma_1 m + \cdots].$$

Assertion (22.3.1) now follows from Lemma 21.5,

$$\frac{A_a(pm)}{A_{a'}(m)} = -\frac{\Gamma_p(a + pm)}{\Gamma_p(1 + pm)} \frac{1}{\Gamma_p(a)}$$

by considering the right side as analytic function of m for m close to zero.

(ii) We compute $\alpha_{f_0, h_0}(1)$ by representing

(22.3.2.1) $$G_{f_0, h_0} = x^{-\mu_b} \frac{(1 - x)^{-b}}{(1 - x^p)^{-b'}}$$

as an element of $\Omega((x))$. We thus obtain

(22.3.2.2) $$\alpha_{f_0, h_0}(1) = \frac{1}{(1 - x)^{-b'}} \sum_{m=0}^{\infty} A_b(pm + \mu_b)x^m$$

and we know that this may be represented by a sum

$$\gamma_0 1 + \gamma_1 \frac{1}{x^{b'}(1 - x)^{-b'}} E(x^{b'}(1 - x)^{-b'}) + \cdots,$$

i.e.,

$$\sum_{m=0}^{\infty} A_b(pm + \mu_b)x^{m+b'} = \gamma_0 x^{b'}(1 - x)^{-b'} + \gamma_1 E(x^{b'}(1 - x)^{-b'}) + \cdots$$

and so

(22.3.2.3) $$A_b(pm + \mu_b) = A_{b'}(m)(\gamma_0 + \gamma_1(m + b') + \cdots).$$

The assertion (22.3.2) in this case again follows by means of Lemma 21.5,

$$\frac{A_b(\mu_b + pm)}{A_{b'}(m)} = -\frac{\Gamma_p(b + \mu_b + pm)}{\Gamma_p(1 + \mu_b + pm)} \frac{1}{\Gamma_p(b)},$$

putting $m = -b' + z$ and considering the right side as analytic function of z for z close to zero. □

Chapter 23

Beta Function as Residues

We use the notation of Chapter 22. In all cases the dimension of $\mathbf{W}_{f_0}$ is 1 but $1/(1-x)$ is not a representative of the basis unless condition (22.1.1) is satisfied. This means that the formulae for the dual space will degenerate if (22.1.1) is not satisfied. For this reason we will first consider the case in which (22.1.1) is satisfied and then consider the two degenerate cases of Lemma 22.3.

23.1

We suppose here that condition (22.2.1) is satisfied by (a_1, a_2) and by (b_1, b_2). The dual theory involves the construction of

$$\mathbf{K}_{f_0} \subset R = R_0 \oplus R_1 \oplus R_\infty$$

annihilated by

(23.1.1) $$-D^*_{f_0} = \gamma_-^{(0)} \circ D_{1/xf_0},$$

where $\gamma_-^{(0)}$ is the projection of R' into R.

Writing

$$\left(\frac{1}{1-x}\right)^* = (\xi_0, \xi_1, \xi_\infty)$$

we find

(23.1.2) $$D_{1/xf_0}\xi_\nu = -\frac{a_2}{T_1}, \qquad \nu = 0, 1, \infty.$$

We compute

(23.1.3)

$$\xi_0(a_1, a_2, x) = -a_2 \sum_{m=0}^{\infty} (-T_0)^m \sum_{i+k=m} \binom{a_2}{k}\binom{-a_2-1}{i} \frac{1}{i-a_1+1},$$

$$\xi_1(a_1, a_2, x) = a_2 \sum_{m=0}^{\infty} (-T_1)^m \sum_{i+k=m} \binom{a_1-1}{k}\binom{-a_1}{i} \frac{1}{i-a_2},$$

$$\xi_\infty(a_1, a_2, x) = -a_2 T_\infty \sum_{m=0}^{\infty} (-T_\infty)^m \sum_{i+k=m} \binom{a_2}{k}\binom{-a_2-1}{i} \frac{1}{i+a_1+a_2}.$$

The homogeneous equation corresponding to (23.1.2) has solution f_0 which is never uniform at $x = 1$ since $a_2 \notin \mathbb{Z}$. It is uniform at infinity if $a_1 + a_2 \in \mathbb{Z}$ but cannot lie in $T_\infty \Omega[[T_\infty]]$ if $a_1 + a_2 \geq 1$. It is uniform at $x = 0$ if $a_1 \in \mathbb{Z}$ but cannot lie in $\Omega[[x]]$ if $a_1 \leq 0$.

Lemma. *Subject to hypothesis 22.2.1 for a_1, a_2, b_1, b_2; let*

(23.1.4) $\quad p \geq 1 + \mu_1,$

(23.1.5) $\quad p \geq -(\mu_1 + \mu_2) + 1,$

$$M_\nu = x^{p-1-\mu_1}(1-x)^{-\mu_2-1}\Lambda_{b_2}\left(\frac{1}{1-x}\right)\xi_\nu(b_1, b_2, x^p),$$

$$\Lambda_{b_2} = \Lambda^{b_2}.$$

Let

$$\theta = B_p(a_1, a_2; b_1 b_2).$$

Then

(23.1.6)
$$\theta = \operatorname{Res}_1 M_1 + (-1)^{\mu_1} \frac{b_2}{b_1+b_2}\binom{a_2-1}{-1-\mu_1-\mu_2},$$

(23.1.7)
$$\frac{a_2}{a_1+a_2}\theta = \operatorname{Res}_1 (1-x) M_1 + (-1)^{\mu_1} \frac{b_2}{b_1+b_2}\binom{a_2}{-\mu_1-\mu_2}.$$

Proof. By the definition of $\alpha^*_{f_0, h_0}$ and of B_p,

$$\theta = \left\langle \alpha^*_{f_0, h_0}\left(\frac{1}{1-x}\right)^*, \frac{1}{1-x}\right\rangle = \sum_\nu \operatorname{Res}_\nu M_\nu.$$

Since

$$D_{f_0} 1 = a_1 + a_2 - \frac{a_2}{1-x},$$

we obtain

$$\frac{a_2}{a_1 + a_2}\theta = \left\langle \alpha^*_{f_0, h_0}\left(\frac{1}{1-x}\right)^*, 1\right\rangle = \sum_\nu \mathrm{Res}_\nu M_\nu \cdot (1-x).$$

By (23.1.4), $\mathrm{Res}_0 M_0 = 0 = \mathrm{Res}_0(1-x)M_0$. We now write

$$M_\infty = (-1)^{\mu_2+1}x^{-2-\mu_1-\mu_2}\left(1 - \frac{1}{x}\right)^{-\mu_2-1}\Lambda_{b_2}\left(\frac{1}{1-x}\right)x^p\xi_\infty(b_1, b_2; x^p),$$

$$\Lambda_{b_2}\left(\frac{1}{1-x}\right) = (1 - T_\infty)^{pb_2}(1 - T_\infty^p)^{-b_2}.$$

Hence

$$(-1)^{\mu_2+1}M_\infty = x^{-2-\mu_1-\mu_2}Y,$$

where

$$Y = (1 - T_\infty)^{a_2-1}(1 - T_\infty^p)^{-b_2}x^p\xi_\infty(b_1, b_2, x^p).$$

Thus

$$-(-1)^{\mu_2+1}\mathrm{Res}_\infty M_\infty = \text{coefficient of } T_\infty^{-1-\mu_1-\mu_2} \text{ in } Y.$$

Likewise

$$(-1)^{\mu_2+1}\mathrm{Res}_\infty(1-x)\cdot M_\infty = \text{coefficient of } T_\infty^{-\mu_1-\mu_2} \text{ in } Y\cdot(1 - T_\infty).$$

We observe

$$Y \equiv (1 - T_\infty)^{a_2-1}\frac{-b_2}{b_1 + b_2} \mod T_\infty^p\Omega[[T_\infty]]$$

and so by (23.1.5)

$$(-1)^{\mu_1}\mathrm{Res}_\infty M_\infty = \frac{b_2}{b_1 + b_2}\binom{a_1 - 1}{-1 - \mu_1 - \mu_2},$$

$$(-1)^{\mu_1}\mathrm{Res}_\infty(1-x)M_\infty = \frac{b_2}{b_1 + b_2}\binom{a_2}{-\mu_1 - \mu_2}.$$

This completes the proof. □

23.2 $a_1 = 1 = b_1$.

Here we are concerned with $f_0 = (1-x)^{-a}$, $h_0 = (1-x)^{-a'}$.

$$D_{f_0} = E + (1-a) + \frac{a}{1-x} = \frac{1}{x(1-x)^{-a}} \circ E \circ x(1-x)^{-a}. \tag{23.2.1}$$

We easily verify that $1/x$ represents a basis of $\mathbf{W}_{f,0} = \mathbf{L}_0/D_{f_0}\mathbf{L}_0$. It then follows that $\{1/x,\ a/(1-x) = D_{f_0}(1/x)\}$ represents a basis of the factor space $\mathbf{W}_{f_0}^{(2)} = \mathbf{L}_0/D_{f_0}^2\mathbf{L}_0$. The dual space here involves the construction of $K_{f_0}^{(2)} \subset R$ annihilated by $D_{f_0}^{*2}$. Here

$$-D_{f_0}^* = \gamma_-^{(0)}(1-x)^{-a} \circ E \circ \frac{1}{(1-x)^{-a}} \tag{23.2.2}$$

and so the unique element in the kernel of $D_{f_0}^*$ is

$$\left(\frac{1}{x}\right)^* = ((1-x)^{-a}, 0, 0), \tag{23.2.3}$$

while an independent element of $K_{f_0}^{(2)}$ may be found by solving the equation

$$-D_{f_0}^*\eta = \left(\frac{1}{x}\right)^* \cdot k, \tag{23.2.4}$$

where k is a constant. This leads to the equation

$$\frac{1}{(1-x)^a} E(1-x)^a(\eta_0, \eta_1, \eta_\infty) = \frac{1}{1-x} + k((1-x)^{-a}, 0, 0). \tag{23.2.5}$$

Now the point is that the image of η_0 on the left side must lie in $x\Omega[[x]]$ and so we must choose $k = -1$ to put the 0-component on the right side in this ideal. We obtain

$$\left(\frac{a}{1-x}\right)^* = (\eta_0, \eta_1, \eta_\infty),$$

$$\eta_0(a, x) = (1-x)^{-a} \sum_{m=1}^{\infty} \frac{(1-a)_m}{m!} \frac{x^m}{m},$$

$$\eta_1(a, x) = -\sum_{m=0}^{\infty} \frac{(1-x)^m}{a+m}, \tag{23.2.6}$$

$$\eta_\infty(a, x) = (1-T_\infty)^{-a} \sum_{m=0}^{\infty} \frac{T_\infty^{m+1}}{m-a+1} \frac{(1-a)_m}{m!}.$$

(Note that

$$\left\langle\left(\frac{a}{1-x}\right)^*, \frac{a}{1-x}\right\rangle = \operatorname{Res}_1 \eta_1 \cdot \frac{a}{1-x} = 1,$$

$$\left\langle\left(\frac{a}{1-x}\right)^*, \frac{1}{x}\right\rangle = \operatorname{Res}_0 \eta_0 \cdot \frac{1}{x} = 0,$$

and it is to achieve this last identity that η_0 is chosen as indicated rather than replaced by its sum with a constant multiple of $(1-x)^{-a}$.)

Lemma. *Let* $\hat{\hat{M}}_v = T_1^{\mu_a}\Lambda^{-a} \cdot (1/x)\eta_v(a', x^p)$ $(v = 0, 1, \infty)$ *(where* Λ *denotes* $\Lambda(1/(1-x)))$. *Then under the hypotheses of Lemma* 22.3(i)

$$\left\langle \alpha^*_{f_0, h_0}\left(\frac{1}{1-x}\right)^*, \frac{1}{x}\right\rangle = \operatorname{Res}_1 \hat{\hat{M}}_1. \tag{23.2.7}$$

Proof. It follows from the definitions since $\chi^{p^{-1}}G_{f_0, h_0} = T_1^{\mu_a}\hat{\hat{\Lambda}}^{-a'}$ that the indicated pairing is just $\sum_v \operatorname{Res}_v M_v$. It is clear, however, that $\hat{\hat{M}}_0 \in x^{p-1}\Omega[[x]]$ and that $\hat{\hat{M}}_\infty \in T^{p+1-\mu_a}\Omega[[T_\infty]]$. These have zero residues and this completes the proof of the lemma. ☐

We now compute the above pairing.

23.2.8 **Corollary.** *Let* $a, a' \in \mathbb{Z}_p \cap \mathbb{Q} - \mathbb{Z}$, $pa' - a = \mu_a \in [0, p-1]$. *Then*

$$\alpha_{f_0, h_0}\begin{pmatrix} 1/x \\ D_{f_0}(1/x)\end{pmatrix} \equiv \begin{pmatrix} 1 & \theta_2 \\ 0 & p\end{pmatrix}\begin{pmatrix} 1/x \\ D_{h_0}(1/x)\end{pmatrix} + D^2_{h_0} L_0,$$

where $\operatorname{Res}_1 \hat{\hat{M}}_1 = \theta_2 = p[(\Gamma'_p/\Gamma_p)(a) - (\Gamma'_p/\Gamma_p)(1)]$.

Proof. Let us write the matrix of the mapping of $\mathbf{W}^{(2)}_{f_0}$ into $\mathbf{W}^{(2)}_{h_0}$ in the form

$$\begin{pmatrix} \theta_1 & \theta_2 \\ \theta_3 & \theta_4\end{pmatrix}.$$

Applying D_{f_0} to the relation

$$\alpha_{f_0, h_0}\frac{1}{x} = \theta_1\frac{1}{x} + \theta_2 D_{h_0}\frac{1}{x} \bmod D^2_{h_0} L_0,$$

we deduce from (3.8.2) that

$$\alpha_{f_0, h_0} \frac{a}{1-x} = p\theta_1 \frac{a'}{1-x} \bmod D_{h_0}^2 L_0.$$

Thus $\theta_3 = 0$, $\theta_4 = p\theta_1$. The evaluation of θ_2 may be deduced from equation (22.3.1) since it may be identified with γ_2 in that equation. □

23.3 $a_1 + a_2 = 0$

We are concerned here with $f_0 = x^{b-1}(1-x)^{-b}$, $h_0 = x^{b'-1}(1-x)^{-b'}$.

We verify that $1 \notin D_{f_0}\mathbf{L}_0$ and so $\{1, D_{f_0}1\}$ is a basis of $\mathbf{W}_{f_0}^{(2)} = \mathbf{L}_0/D_{f_0}^2\mathbf{L}_0$. The dual space, $\mathbf{K}_f^{(2)}$ in R has basis dual to $\{1, D_{f_0}1 = b/(1-x)\}$

(23.3.1)
$$1^* = \left\{0, 0, -\frac{1}{x}\left(1-\frac{1}{x}\right)^{-b}\right\},$$

(23.3.2)
$$\left(\frac{b}{1-x}\right)^* = \{\xi_0, \xi_1, \xi_\infty\},$$

which satisfies

$$x^{b-1}(1-x)^{-b}E\frac{1}{x^{b-1}(1-x)^{-b}}\xi_\nu = \frac{1}{1-x} + \begin{cases} 0 & \text{if } \nu \neq \infty, \\ k\dfrac{1}{x}\left(1-\dfrac{1}{x}\right)^{-b} & \text{if } \nu = \infty. \end{cases}$$

We find $k = 1$

$$\xi_0(b, x) = (1-x)^{-b}\sum_{m=0}^{\infty} A_{1-b}(m)\frac{x^m}{m+1-b},$$

(23.3.3)
$$\xi_1(b, x) = -\sum_{m=0}^{\infty} T_1^m \sum_{i+j=m} A_{1-b}(i)A_b(j)\frac{1}{j+b},$$

$$\xi_\infty(b, x) = T_\infty(1-T_\infty)^{-b}\sum_{m=1}^{\infty}\frac{T_\infty^m A_{1-b}(m)}{m+b}.$$

In the following lemma we again use Λ to denote $\Lambda(1/(1-x))$.

Lemma. *Let* $\hat{M}_\nu = x^{p-1-\mu_b}(1-x)^{\mu_b}\Lambda^{-b'}\xi_\nu(b', x^p)$ $(\nu = 0, 1, \infty)$. *Then under the hypotheses of Lemma* 2.2.3(*ii*)

$$\left\langle \alpha_{f_0, h_0}^*\left(\frac{b'}{1-x}\right)^*, 1\right\rangle = \operatorname{Res}_1 \hat{M}_1.$$

Proof. The left side coincides with $\sum_v \mathrm{Res}_v\, \hat{M}_v$. We note that $\hat{M}_0$ is analytic at zero while $\hat{M}_\infty$ lies in $T_\infty^{2p-(p-1-\mu_b)}\Omega[[T_\infty]]$ since ξ_∞ is divisible by T_∞^2. This completes the proof. □

Corollary.

(23.3.4)
$$\mathrm{Res}_1 \hat{M}_1 = p(-1)^{\mu_b}\left(\frac{\Gamma_p'}{\Gamma_p}(0) - \frac{\Gamma_p'}{\Gamma_p}(1-b)\right).$$

Proof. The point here is that

$$\mathrm{Res}_1\, \hat{M}_1 = \left\langle \alpha_{f_0, h_0}^* \left(\frac{b'}{1-x}\right)^*, 1 \right\rangle = \left\langle \left(\frac{b'}{1-x}\right)^*, \alpha_{f_0, h_0} 1 \right\rangle$$
$$= \langle (D_{h_0} 1)^*, \alpha_{f_0, h_0} 1 \rangle = \gamma_1,$$

where

$$\alpha_{f_0, h_0}(1) \equiv \gamma_0 1 + \gamma_1 D_{h_0} 1 \mod D_{h_0}^2 L_0.$$

The coefficient γ_1 may be computed from (22.3.2). □

Chapter 24

Singular Disks, Part I

The object of this chapter is to examine the matrix B (of equation (4.5.1)) for λ close to zero. We know that this matrix has analytic continuation in all cases and so the relation of the local behavior of B near 1 with its local behavior near $\lambda = 0$ makes sense without any restriction on a, b, c other than (1.1). We will assume conditions (9.0.1)–(9.0.3).

This we believe is the correct way to formulate the problem of Kobbitz [22] and Diamond [8] concerning "the value at $\lambda = 1$ of the analytic extension of $F(a, b, c, x)/F(a', b', c', x^p)$," the difficulty being precisely that the ratio has no analytic continuation unless a, b, c satisfy rather unpleasant hypotheses.

While these questions are of interest, we are more particularly interested in the question of the existence of a deformation type mapping of $\mathbf{K}_{f_0}$ into $\mathbf{K}_{f_\lambda}$, for λ close to 0 in such a way that the diagram

$$\begin{array}{ccc} \mathbf{K}_{f_0} & \xrightarrow{T_{0\,\lambda,f}} & \mathbf{K}_{f_\lambda} \\ \uparrow{\scriptstyle \alpha^*_{f_0,h_0}} & ? & \uparrow{\scriptstyle \alpha^*_{f,h}} \\ \mathbf{K}_{h_0} & \xrightarrow[T_{0,\lambda^p,h}]{} & \mathbf{K}_{h_{\lambda^p}} \end{array}$$

commutes. Unfortunately condition (4.6.1.1) is violated here and so the construction has not been established. We hope that the considerations of this section will lead to a resolution of this question.

We shall use the classical representation

$$F(a, b, c, x) = \sum_{s=0}^{\infty} \frac{(a)_s(b)_s}{(c)_s s!} x^s. \tag{24.0.1}$$

24.1 **Lemma.** *For $c \in (0, 1)$, a solution matrix of (2.4.5.1) at the origin is given by*

$$\begin{pmatrix} 1 & 0 \\ 0 & \lambda^{-c} \end{pmatrix} U(\lambda),$$

where

$$U(\lambda) = \begin{pmatrix} (c-b)F(a,b,1+c,\lambda) & cF(a,b,c,\lambda) \\ (1-c)F(a-c,b-c,1-c,\lambda) & (c-a)\lambda F(a+1-c,b+1-c,2-c,\lambda) \end{pmatrix}.$$

Proof. This result is well known but a discussion may be useful. We noted that if (u_1, u_2) is a solution of (2.4.5.1) then u_1 is a branch of $F(a, b, c+1, \lambda)$ and u_2 is a branch of $F(a, b, c, \lambda)$. If we choose the unique branch holomorphic at $\lambda = 0$ we obtain the first row of the matrix U. The Riemann data of (2.4.5.2) is

$$\begin{pmatrix} 0 & 1 & \infty \\ 0 & 0 & a \\ -c & 1-c-a-b & b \end{pmatrix}$$

(the first row lists the singularities, the columns give the corresponding exponents). Multiplication by λ^c changes the data to

$$\begin{pmatrix} 0 & 1 & \infty \\ c & 0 & a-c \\ 0 & 1+c-a-b & b-c \end{pmatrix},$$

i.e., to a branch of $F(a-c, b-c, 1-c, \lambda)$. This explains the first coefficient in the second row of U. The second coefficient may be explained similarly. This completes the proof. □

24.1.1 **Notation.** By the indicated transformations we obtain solution matrices of the corresponding differential equations: Also we define algebraic functions, $\tilde{f}, \tilde{h}$. In the notation of (3.7.0) let μ_a, μ_b, μ_c all lie in $[0, p-1]$.

$(a, b, c) \mapsto$	(a', b', c')	$1-a, 1-b, 1-c$	$1-a', 1-b', 1-c'$
$\begin{pmatrix} 1 & 0 \\ 0 & \lambda^{-c} \end{pmatrix} U$	$\begin{pmatrix} 1 & 0 \\ 0 & \lambda^{-c} \end{pmatrix} V$	$\begin{pmatrix} 1 & 0 \\ 0 & \lambda^{c-1} \end{pmatrix} \tilde{U}$	$\begin{pmatrix} 1 & 0 \\ 0 & \lambda^{c'-1} \end{pmatrix} \tilde{V}$
f	h	$\tilde{f}$	$\tilde{h}$

24.2 **Lemma** ($c \neq 0, 1$).

$$c(1-c)\begin{pmatrix} b-c & 0 \\ 0 & (a-c)\lambda \end{pmatrix} = U\begin{pmatrix} (a-c)\lambda & 0 \\ 0 & (b-c)(1-\lambda) \end{pmatrix}\tilde{U}^t,$$

$$c'(1-c')\begin{pmatrix} b'-c' & 0 \\ 0 & (a'-c')\lambda \end{pmatrix} = V\begin{pmatrix} (a'-c')\lambda & 0 \\ 0 & (b'-c')(1-\lambda) \end{pmatrix}\tilde{V}^t.$$

Proof. For z, satisfying (4.6.1.1) we have the commutative diagram,

$$\begin{array}{ccc} \mathbf{K}_{f,z} & \xrightarrow{\hat{D}_{f,z}} & \mathbf{W}_{\tilde{f},z} \\ \Big\downarrow T_{z,\lambda} & & \Big\uparrow \tilde{T}^t_{z,\lambda} \\ \mathbf{K}_{f,\lambda} & \xrightarrow[\hat{D}_{f,\lambda}]{} & \mathbf{W}_{\tilde{f},\lambda} \end{array}$$

This gives the matrix relation

$$C_f(z, \lambda)(\hat{D}_{f,\lambda})C_{\tilde{f}}(z, \lambda)^t = (\hat{D}_{f,z}) \tag{24.2.1}$$

so the right side is independent of λ. From this we deduce a constant 2×2 matrix, X, such that

$$X = \begin{pmatrix} 1 & 0 \\ 0 & \lambda^{-c} \end{pmatrix} U \begin{pmatrix} (a-c)\lambda & 0 \\ 0 & (b-c)(1-\lambda) \end{pmatrix} \tilde{U}^t \begin{pmatrix} 1 & 0 \\ 0 & \lambda^{c-1} \end{pmatrix}.$$

thus

$$\begin{pmatrix} 1 & 0 \\ 0 & \lambda^{c} \end{pmatrix} X \begin{pmatrix} 1 & 0 \\ 0 & \lambda^{1-c} \end{pmatrix}$$

is uniform near zero and hence X is a diagonal matrix. We thus obtain constants x_1, x_4 such that

$$\begin{pmatrix} x_1 & 0 \\ 0 & x_4\lambda \end{pmatrix} = U \begin{pmatrix} (a-c)\lambda & 0 \\ 0 & (b-c)(1-\lambda) \end{pmatrix} \tilde{U}^t. \tag{24.2.2}$$

We compute x_1 by setting $\lambda = 0$ and using the formula for the matrix, U. We compute x_4 by taking determinants and observing that $x_1 x_4/(a-c)(b-c) = \det U(0)\tilde{U}(0)$. The assertion of the lemma for U follows from these calculations. The assertion for V is then clear. □

24.3 **Lemma** ($c \neq 0, 1$) *There exist constants θ_1, θ_4, $\tilde{\theta}_1$, $\tilde{\theta}_4$ such that with B representing $\alpha^*_{f,h}$, B representing $\alpha^*_{\tilde{f},\tilde{h}}$ relative to the bases $\{1^*, (1/(1-x))^*\}$, we have*

$$V(\lambda^p)B = \begin{pmatrix} \theta_1 & 0 \\ 0 & \theta_4\lambda^{\mu_c} \end{pmatrix} U(\lambda), \tag{24.3.1}$$

$$\tilde{V}(\lambda^p)\tilde{B} = \begin{pmatrix} \tilde{\theta}_1 & 0 \\ 0 & \tilde{\theta}_4\lambda^{p-1-\mu_c} \end{pmatrix} \tilde{U}(\lambda), \tag{24.3.2}$$

$$\theta_1\tilde{\theta}_1 = p\,\frac{c'(1-c')}{c(1-c)}\,\frac{b'-c'}{b-c}, \tag{24.3.3}$$

$$\theta_4\tilde{\theta}_4 = p\,\frac{c'(1-c')}{c(1-c)}\,\frac{a'-c'}{a-c}. \tag{24.3.4}$$

Proof. It follows from equation (7.1.2) that there exists constant matrices Θ, $\tilde{\Theta}$ such that

$$\begin{pmatrix} 1 & 0 \\ 0 & \lambda^{-pc'} \end{pmatrix} V(\lambda^p) B = \Theta \begin{pmatrix} 1 & 0 \\ 0 & \lambda^{-c} \end{pmatrix} U(\lambda),$$

$$\begin{pmatrix} 1 & 0 \\ 0 & \lambda^{-p(1-c')} \end{pmatrix} \tilde{V}(\lambda^p) \tilde{B} = \tilde{\Theta} \begin{pmatrix} 1 & 0 \\ 0 & \lambda^{-(1-c)} \end{pmatrix} \tilde{U}(\lambda).$$

By considerations of uniformity at $\lambda = 0$, we find that Θ and Θ are diagonal matrices. We deduce equations (24.3.1) and (24.3.2) directly. We now apply Theorem 4.7 to obtain the matrix relation

$$B(\hat{D}_{f,\lambda}) = p(\hat{D}_{h,\lambda^p})(\tilde{B}^{-1})^t. \tag{24.3.5}$$

Writing B (resp. $\tilde{B}$) in terms of U and V (resp. $\tilde{U}$ and $\tilde{V}$) and substituting in (24.3.5) gives

$$\begin{pmatrix} \theta_1 & 0 \\ 0 & \theta_4 \lambda^{\mu_c} \end{pmatrix} U(\lambda)(\hat{D}_{f,\lambda}) U(\lambda)^t = pV(\lambda^p)(\hat{D}_{h,\lambda^p})\tilde{V}(\lambda^p)^t \begin{pmatrix} \tilde{\theta}_1 & 0 \\ 0 & \theta_4 \lambda^{p-1-\mu_c} \end{pmatrix}^{-1}.$$

By Lemma 24.2, $U(\hat{D}_{f,\lambda})\tilde{U}^t$ and $V\hat{D}_h\tilde{V}$ are explicitly determined diagonal matrices. Equations (24.3.3) and (24.3.4) are easily verified. □

Since V and U are explicitly determined, we conclude that B is determined near zero if θ_1 and θ_4 are computed. In the sense of analytic continuation, this would specify B for all λ satisfying (3.5.0). Our object is to compute the θ_i in terms of the p-adic beta function.

24.4 **Lemma** ($c \neq 0, 1$). *The matrix B is analytic at $\lambda = 0$ (i.e., there is no pole). In the notation of* (4.5.1),

$$B_1(\lambda) \equiv \theta_4 \frac{1-c}{1-c'} \lambda^{\mu_c} \qquad \text{mod } \lambda^{1+\mu_c},$$

$$B_2(\lambda) \equiv \theta_4 \frac{c-a}{1-c'} \lambda^{1+\mu_c} \quad \text{mod } \lambda^{2+\mu_c},$$

(24.4.1)

$$B_3(\lambda) \equiv \theta_1 \frac{c-b}{c'} \qquad \text{mod}(\lambda, \lambda^{\mu_c})$$

$$B_4(\lambda) \equiv \theta_1 \frac{c}{c'} \qquad \text{mod } \lambda.$$

In particular

$$\frac{c}{c'}\theta_1 = B_4(0),$$

(24.4.2)

$$\frac{1-c}{1-c'}\theta_4 = \lim_{\lambda \to 0} \frac{B_1(\lambda)}{\lambda^{\mu_c}}.$$

Proof. The first set of relations follows from (23.3.1) and the fact that

$$V(\lambda^p)^{-1} \equiv -\frac{1}{c'(1-c')}\begin{pmatrix} 0 & -c' \\ -(1-c') & c'-b' \end{pmatrix} \mod \lambda^p.$$

Equation (24.4.2) is now immediate. □

24.5 The Case $c = 1$ (Logarithmic Case)

We briefly indicate the situation is this case, the proofs are as before. The solution matrices are the product of

$$\begin{pmatrix} 1 & 0 \\ \log \lambda & 1 \end{pmatrix}$$

with uniform matrices, again designated by U, V, $\tilde{U}$, $\tilde{V}$.

$$U = \begin{pmatrix} (1-b)F(a, b, 2, \lambda) & F(a, b, 1, \lambda) \\ (1-b)H(a, b, 2, \lambda) & H(a, b, 1, \lambda) \end{pmatrix}, \tag{24.5.1}$$

where in general for $n \in \mathbb{N}$, $n \geq 1$ we define $H(a, b, n, x)$ by the conditions that it lie in $Q((x))$, that

$$y = H(a, b, n, x) + F(a, b, n, x) \log x \tag{24.5.2}$$

be a formal solution of the equation ($\delta = x(d/dx)$)

$$(x(\delta + a)(\delta + b) - \delta(\delta + n - 1))y = 0 \tag{24.5.3}$$

and finally that the coefficient of 1 in the Laurent series representation of H be zero. The differential equation implies that putting $P = (\delta + a)(\delta + b)$, $Q = \delta(\delta + n - 1)$,

$$(xP(\delta) - Q(\delta))H + (xP'(\delta) - Q'(\delta))F = 0 \tag{24.5.4}$$

and so we deduce that

(24.5.5)

$$U(\lambda) \equiv \begin{pmatrix} (1-b)\left(1+\frac{ab}{2}\lambda\right) & 1+ab\lambda \\ \frac{1}{1-a}\frac{1}{\lambda}+(1-b)\left(\frac{a+b}{2}-\frac{3}{4}ab\right)\lambda & (a+b-2ab)\lambda \end{pmatrix} \bmod \lambda^2.$$

For $\tilde{U}$ we need solutions of $(2.4.5.2)_{1-a,1-b,0}$ and of $(2.4.5.3)_{1-a,1-b,0}$. For the latter we need branches of "$F(1-a, 1-b, 0, \lambda)$". As is well known, in general $x^{c-1}F(a, b, c, x)$ and $F(a-c+1, b-c+1, 2-c, x)$ satisfy the same second-order differential equation and so $F(a, b, c, x)$ is "equivalent" to $x^{1-c}F(a-c+1, b-c+1, 2-c, x)$. In particular then, when $c = 0$, we formally replace "$F(a, b, 0, x)$" by $x\,F(a+1, b+1, 2, x)$ and likewise "$F(1-a, 1-b, 0, x)$" by $x\,F(2-a, 2-b, 2, x)$. We thus obtain

(24.5.6)

$$\tilde{U} = \begin{pmatrix} F(1-a, 1-b, 1, \lambda) & -(1-a)\lambda F(2-a, 2-b, 2, \lambda) \\ H(1-a, 1-b, 1, \lambda) & -(1-a)\lambda H(2-a, 2-b, 2, \lambda) \end{pmatrix}$$

$$\equiv \begin{pmatrix} 1+(1-a)(1-b)\lambda & -(1-a)\lambda \\ (a+b-2ab)\lambda & -1/(1-b) \end{pmatrix} \bmod \lambda^2.$$

24.5.7 **Lemma** ($c = 1$).

(24.5.7.1) $$\begin{pmatrix} 0 & -1 \\ 1 & 0 \end{pmatrix} = U\begin{pmatrix} (a-1)\lambda & 0 \\ 0 & (b-1)(1-\lambda) \end{pmatrix}\tilde{U}^t.$$

Proof. By the argument of Lemma 24.2, there exists a constant matrix X such that the right side of (24.5.7.1) coincides with

$$\begin{pmatrix} 1 & 0 \\ -\log\lambda & 1 \end{pmatrix} X \begin{pmatrix} 1 & -\log\lambda \\ 0 & 1 \end{pmatrix},$$

which therefore by reason of uniformity at $\lambda = 0$ must be of the form

$$x_0\begin{pmatrix} 0 & -1 \\ 1 & x_4 \end{pmatrix}.$$

It suffices to compute the right side of (24.5.7.1) mod λ. Since the second and third factors have no pole at zero, it is enough to use U mod λ. The product of the first two factors has no pole at zero and hence we need $\tilde{U}^t$ modulo λ. The verification follows from (24.5.5) and (24.5.6). □

24.5.8 **Lemma** ($c = 1$). *There exist constants γ, θ, such that*

$$V(\lambda^p)B = \theta\begin{pmatrix} 1 & 0 \\ \gamma & p \end{pmatrix} U,$$

$$\tilde{V}(\lambda^p)\tilde{B} = \theta^{-1}\begin{pmatrix} 1 & 0 \\ \gamma & p \end{pmatrix}\tilde{U}.$$

Note. The explicit value of γ is given by Theorem 26.1.

Proof. Following the proof of Lemma 24.3, there exists a constant matrix Θ such that

$$V(\lambda^p)BU(\lambda)^{-1} = \begin{pmatrix} 1 & 0 \\ -p\log\lambda & 1 \end{pmatrix}\Theta\begin{pmatrix} 1 & 0 \\ \log\lambda & 1 \end{pmatrix}.$$

This imposes conditions on Θ from which the first assertion follows. It is clear that there exists $\tilde{\theta}$, $\tilde{\gamma}$ such that

$$\tilde{V}(\lambda^p)\tilde{B}\tilde{U}^{-1} = \tilde{\theta}\begin{pmatrix} 1 & 0 \\ \tilde{\gamma} & p \end{pmatrix}.$$

Thus as in the proof of Lemma 24.3,

$$\theta\begin{pmatrix} 1 & 0 \\ \gamma & p \end{pmatrix}\begin{pmatrix} 0 & -1 \\ 1 & 0 \end{pmatrix} = p\begin{pmatrix} 0 & -1 \\ 1 & 0 \end{pmatrix}\tilde{\theta}^{-1}\begin{pmatrix} 1 & \tilde{\gamma} \\ 0 & p \end{pmatrix}^{-1}$$

and so $\theta\tilde{\theta} = 1$, $\gamma = \tilde{\gamma}$. This completes the proof. □

24.5.9 **Lemma.** *B is analytic at $\lambda = 0$ and*

$$\theta = B_4(0),$$

$$\theta\gamma(1 - a') = \left.\frac{B_2}{\lambda^p}\right|_{\lambda=0}.$$

Proof.

$$\det V(\lambda) \equiv -\frac{1}{1-a'}\frac{1}{\lambda^p} - \frac{a'b'}{1-a'} \quad \mathrm{mod}(\lambda^p)$$

and hence

$$V(\lambda^p)^{-1} = -((1-a')\lambda^p(1-a'b'\lambda^p + (\lambda^{2p}))$$

$$\times \begin{pmatrix} (\lambda^p) & -1 + (\lambda^p) \\ -\dfrac{1}{1-a'}\dfrac{1}{\lambda^p} + (\lambda^p) & (1-b') + (\lambda^p) \end{pmatrix}$$

$$= \begin{pmatrix} 0 & (1-a')\lambda^p \\ 1 - a'b'\lambda^p & -(1-a')(1-b')\lambda^p \end{pmatrix} + (\lambda^{2p}).$$

Thus

$$V(\lambda^p)^{-1}\begin{pmatrix} 1 & 0 \\ \gamma & p \end{pmatrix} =$$

$$\begin{pmatrix} (1-a')\gamma\lambda^p & p(1-a')\lambda^p \\ 1 - \lambda^p(a'b' + \gamma(1-a')(1-b')) & -p(1-a')(1-b')\lambda^p \end{pmatrix} + (\lambda^{2p}).$$

Hence

$$\theta^{-1}B = \begin{pmatrix} p\dfrac{1-a'}{1-a}\lambda^{p-1} + (\lambda^p) & (1-a')\gamma\lambda^p + (\lambda^{p+1}) \\ (1-b) + (\lambda) & 1 + (\lambda) \end{pmatrix}.$$

The lemma is now clear. □

Chapter 25

Singular Disks, Part II. Nonlogarithmic Case

We assume (9.0.1)–(9.0.3), $c \neq 1$; μ_a, μ_b, μ_c all lie in $[0, p-1]$.

Theorem. *In notation of* (24.3.1),

$$\theta_4 = (-1)^{\mu_b - \mu_c} \frac{\gamma_p(c-1, c'-1)\gamma_p(1-a, 1-a')}{\gamma_p(1+c-a, 1+c'-a')},$$

$$\theta_1 = \frac{\gamma_p(b, b')\gamma_p(c-b, c'-b')}{\gamma_p(1+c, 1+c')},$$

$$\det B(\lambda) = \theta_1\theta_4 \frac{c(1-c)}{c'(1-c')} \lambda^{\mu_c}(1-\lambda)^{\mu_a+\mu_b-\mu_c}\Lambda_{c'-a'-b'}\left(\frac{1}{1-\lambda}\right).$$

Proof. It follows from (24.4.2) that it is enough to show

$$B_4(0) = B_p(b, c-b; b', c'-b'),$$

$$\lim_{\lambda\to 0} \frac{B_1}{\lambda^{\mu_c}} = (-1)^{\mu_b-\mu_c} \cdot \frac{a'-c'}{a-c}\frac{a}{a'} B_p(c, -a; c', -a').$$

Part I. Calculation of B_4

We apply the method of proof of Theorem 4.5. Let

(25.1)
$$\left(\frac{1}{1-x}\right)^* = (\xi_0, \xi_1, \xi_{1/\lambda^p}, \xi_\infty),$$

$$B_4(\lambda) = \left\langle \alpha^*_{f,h}\left(\frac{1}{1-x}\right)^*, \frac{1}{1-x}\right\rangle,$$

$$= \sum_{v\in S} J_v(\lambda),$$

where

$$J_\nu = \operatorname{Res}_\nu M_\nu,$$

$$M_\nu = T_0^{k_0} T_1^{k_1} T_{1/\lambda}^{k_2} G_0 \xi_{\nu^p}(a', b', c', \lambda^p, x^p).$$

In the present application

$$\text{(25.2)} \qquad \begin{aligned} k_0 &= p - 1 - \mu_b, \\ k_1 &= \mu_b - \mu_c - 1, \\ k_2 &= \mu_a. \end{aligned}$$

The central point is that in calculating ξ_{ν^p} we use equation (2.3.5.10) with $(a, b, c, \lambda, \nu) \mapsto (a', b', c', \lambda^p, \nu^p)$ and

$$\text{(25.3)} \qquad \begin{aligned} A &= b' - c', \\ A\lambda^p + B &= 0. \end{aligned}$$

Since $k_0 \geqslant 0$, M_0 has no singularity at $x = 0$, i.e.,

$$\text{(25.4)} \qquad J_0(\lambda) = 0.$$

Since G_0 is analytic at infinity and

$$\xi_\infty(a', b', c', \lambda^p, x^p) \in T_\infty^p \Omega[[T_\infty^p]],$$

while

$$T_0^{k_0} T_1^{k_1} T_{1/\lambda}^{k_2} = \pm x^{p-2-\mu_c+\mu_a} \lambda^{\mu_c} (1 - T_\infty)^{\mu_b - \mu_c - 1} \left(1 - \frac{1}{\lambda} T_\infty\right)^{\mu_a},$$

we conclude that $\pm J_\infty(\lambda)$ is the coefficient of $T_\infty^{\mu_a - \mu_c - 1}$ in

$$\lambda^{\mu_c} (1 - T_\infty)^{\mu_b - \mu_c - 1} \left(1 - \frac{1}{\lambda} T_\infty\right)^{\mu_a} \frac{1}{T_\infty^p} \xi_\infty(a', b', c', \lambda^p, x^p).$$

Clearly we need only consider $(1/T_\infty^p)\xi_\infty \bmod T_\infty^p$ which is$(A + (B/\lambda^p)) \times (1/(c' - a')) = 0$. This shows that

$$\text{(25.5)} \qquad J_\infty(\lambda) = 0.$$

For $v = 1/\lambda$ we observe that $\xi_{1/\lambda^p}(a', b', c', \lambda^p, x^p)$ is a sum of two series in $(1 - x^p\lambda^p)$, $1/(1 - \lambda^p)$, one with the factor $\lambda^p A$, the other with factor $B = -A\lambda^p$. Thus ξ_{1/λ^p} introduces a factor λ^p. We compute

(25.5.1)

$$G_0 = \Lambda_{-a'}\left(\frac{1}{T_{1/\lambda}}\right)\Lambda_{c'-b'}\left(\frac{1}{1-\lambda}\right)\frac{((1 + (1/(\lambda - 1))T_{1/\lambda})^p)^{c'-b'}}{(1 - (1/(\lambda^p - 1))((1 - T_{1/\lambda})^p - 1))^{c'-b'}},$$

a Laurent series in $T_{1/\lambda}$ whose coefficients have no pole at $\lambda = 0$. Furthermore

(25.5.2)

$$T_0^{k_0}T_1^{k_1}T_{1/\lambda}^{k_2} = \frac{1}{\lambda^{k_0+k_1}}(\lambda - 1)^{k_1}T_{1/\lambda}^{k_2}(1 - T_{1/\lambda})^{k_0}\left(1 - \frac{1}{1-\lambda}T_{1/\lambda}\right)^{k_1}.$$

Thus aside from a factor $\lambda^p/\lambda^{k_0+k_1}$, $M_{1/\lambda}$ is a Laurent series in $T_{1/\lambda}$ whose coefficients have no singularity at $\lambda = 0$. We observe that in computing $\mathrm{Res}_{1/\lambda} M_{1/\lambda}$, a factor $1/\lambda$ is introduced which shows that $J_{1/\lambda}(\lambda)$ has a factor $\lambda^{p-1-(k_0+k_1)} = \lambda^{1+\mu_c}$. We conclude

$$J_{1/\lambda}(0) = 0. \tag{25.6}$$

We have thus shown that

$$B_4(0) = J_1(0).$$

We write

$$T_0^{k_0}T_1^{k_1}T_{1/\lambda}^{k_2} = (1 - \lambda)^{k_2}T_1^{k_1}(1 - T_1)^{k_0}\left(1 + \frac{\lambda}{1-\lambda}T_1\right)^{k_2}, \tag{25.6.1}$$

(25.6.2)

$$G_0 = \Lambda_{c'-b'}\left(\frac{1}{T_1}\right)\Lambda_{-a'}\left(\frac{1}{1-\lambda}\right)\frac{(1 + (\lambda/(1 - \lambda))T_1)^{pa'}}{(1 - (\lambda^p/(1 - \lambda^p))((1 - T_1)^p - 1))^{a'}},$$

$$-\xi_1(a', b', c', \lambda^p, x^p) = (b' - c')\cdot\text{power series in}\left(\frac{\lambda^p}{\lambda^p - 1},\ (1 - T_1)^p - 1\right) + \lambda^p(b' - c')\cdot\text{similar power series.} \tag{25.6.3}$$

Since all the factors of $M_1(\lambda)$ are analytic at $\lambda = 0$, we conclude that

$$J_1(0) = \mathrm{Res}_{x=1} M_1(0). \tag{25.7}$$

When $\lambda = 0$,

$$T_0^{k_0} T_1^{k_1} T_{1/\lambda}^{k_2} \mapsto T_0^{p-1-\mu_b} T_1^{\mu_b - \mu_c - 1},$$

$$G_0 \mapsto \Lambda_{c'-b'}\left(\frac{1}{T_1}\right),$$

$$\xi_1 \mapsto -(b' - c') \sum_{m=0}^{\infty} (x^p - 1)^m \sum_{i+k=m} \binom{b'-1}{k} \binom{-b'}{i} \frac{1}{i + b' - c'}$$
$$= \xi_1(b', c' - b', x^p).$$

Taking the product we find that $M_1(0)$ coincides with M_1 as used in Lemma 23.1 under the substitution

$$(a_1, a_2, b_1, b_2) \mapsto (b, c - b, b', c' - b').$$

Our assertion concerning $B_4(0)$ now follows from that lemma.

Part II. Calculation of B_1

We now use

(25.8)
$$B_1(\lambda) = \langle \alpha_{j,h}^* 1^*, 1 \rangle = \sum_{v \in S} J_v(\lambda),$$

$$J_v(\lambda) = \mathrm{Res}_v\, M_v.$$

In this application

$$k_0 = p - 1 - \mu_b,$$

$$k_1 = \mu_b - \mu_c,$$

$$k_2 = \mu_a,$$

and the main point is that in using (2.3.5.10), we have

(25.9)
$$A = 0,$$

$$B = \lambda^p (a' - c').$$

Again M_0 is analytic at $x = 0$ so

$$J_0(\lambda) = 0. \tag{25.10}$$

Since T_∞^p divides ξ_∞, we consider

$$(-1)^t T_0^{k_0} T_1^{k_1} T_{1/\lambda}^{k_2} T_\infty^p = x^{\mu_a - \mu_c - 1}(1 - T_\infty)^{k_1}\left(1 - T_\infty \frac{1}{\lambda}\right)^{k_2} \lambda^{\mu_a},$$

where

$$t = \mu_b + \mu_a - \mu_c.$$

Since

$$G_0 \frac{1}{T_\infty^p} \xi_\infty(a', b', c', \lambda^p, x^p)$$

is analytic at ∞,

$$J_\infty(\lambda) = 0 \quad \text{if } \mu_a < \mu_c. \tag{25.11}$$

We now assume $\mu_a \geq \mu_c$ and conclude that $-J_\infty(\lambda)$ is the coefficient of $T_\infty^{\mu_a - \mu_c}$ in

$$y = (-1)^t \lambda^{\mu_a}(1 - T_\infty)^{k_1}\left(1 - T_\infty \frac{1}{\lambda}\right)^{k_2} G_0 \cdot \frac{1}{T_\infty^p} \xi_\infty(a', \ldots, x^p).$$

For this it is enough to compute y mod T_∞^p. We observe

$$\frac{1}{T_\infty^p} \xi_\infty(a', \ldots, \lambda^p, x^p) \equiv \left(A + \frac{B}{\lambda^p}\right) \frac{1}{c' - a'} \equiv -1 \mod T_\infty^p,$$

$$\begin{aligned} G_0(x) &= \left(\frac{(1 - T_\infty)^p}{1 - T_\infty^p}\right)^{c' - b'} \left(\frac{(1 - (1/\lambda)T_\infty)^p}{1 - (1/\lambda^p)T_\infty^p}\right)^{-a'} \\ &\equiv (1 - T_\infty)^{p(c' - b')}\left(1 - \frac{1}{\lambda} T_\infty\right)^{-pa'} \mod T_\infty^p. \end{aligned}$$

This shows that

$$(1 - T_\infty)^{k_1}\left(1 - T_\infty \frac{1}{\lambda}\right)^{k_2} G_0(x) \equiv (1 - T_\infty)^{c - b}\left(1 - \frac{1}{\lambda} T_\infty\right)^{-a} \mod T_\infty^p.$$

Thus $-J_\infty(\lambda)$ is the coefficient of $T_\infty^{\mu_a-\mu_c}$ in

$$(-1)^{t+1}\lambda^{\mu_a}(1-T_\infty)^{c-b}\left(1-\frac{1}{\lambda}T_\infty\right)^{-a}.$$

Thus

$$(-1)^t J_\infty(\lambda)=\lambda^{\mu_a}\sum_{i+j=\mu_a-\mu_c}(-1)^{\mu_a-\mu_c}\binom{c-b}{i}\binom{-a}{j}\frac{1}{\lambda^j}$$

and so

$$(-1)^{\mu_b}\frac{J_\infty(\lambda)}{\lambda^{\mu_c}}=\lambda^{\mu_a-\mu_c}\sum_{i+j=\mu_a-\mu_c}\binom{c-b}{i}\binom{-a}{j}\frac{1}{\lambda^j}.$$

We conclude

$$\lim_{\lambda\to 0}\frac{J_\infty(\lambda)}{\lambda^{\mu_c}}=(-1)^{\mu_b}\binom{-a}{\mu_a-\mu_c}. \tag{25.12}$$

We assert

$$\lim_{\lambda\to 0}\frac{J_1(\lambda)}{\lambda^{\mu_c}}=0. \tag{25.13}$$

Here we see from (25.6.1) and (25.6.2), and from (25.6.3) (modified for the present situation to read

$$-\xi_1(a',b',c',\lambda^p,x^p)=\frac{\lambda^p}{1-\lambda^p}\cdot\text{power series in }(1-x^p),\frac{\lambda^p}{1-\lambda^p}\Big)$$

that

$$\lim_{\lambda\to 0}\frac{J_1(\lambda)}{\lambda^p}=\operatorname{Res}_1\lim_{\lambda\to 0}\frac{M_1(\lambda)}{\lambda^p}.$$

Since $p>\mu_c$, equation (25.13) is clear.

We now compute $J_{1/\lambda}(\lambda)$. To simplify our calculation we shall replace $T_{1/\lambda}$ by the symbol, y, and use

$$J_{1/\lambda}(\lambda)=-\frac{1}{\lambda}\operatorname{Res}_{y=0}M_{1/\lambda}.$$

We rewrite (25.5.1) as

$$G_0 = \Lambda_{-a'}\left(\frac{1}{y}\right)\Lambda_{c'-b'}\left(\frac{1}{1-\lambda}\right)\frac{(1 + (1/(\lambda - 1))y)^{p(c'-b')}}{(1 - (1/(\lambda^p - 1))((1 - y)^p - 1))^{c'-b'}}$$

and (25.5.2) as

$$T_0^{k_0}T_1^{k_1}T_{1/\lambda}^{k_2} = \frac{(\lambda - 1)^{k_1}}{\lambda^{k_0+k_1}}\, y^{k_2}(1 - y)^{k_0}\left(1 - \frac{1}{1-\lambda}\, y\right)^{k_1}$$

and finally $1 - x^p\lambda^p = 1 - (1 - y)^p$,

$$-\xi_{1/\lambda^p}(a', b', c', \lambda^p, x^p) = \lambda^p(a' - c')\cdot\text{power series in } (1 - y)^p - 1, \frac{1}{1 - \lambda^p}.$$

Thus in $(1/\lambda^{\mu_c})J_{1/\lambda} = (1/\lambda^{1+\mu_c})\operatorname{Res}_{y=0} M_{1/\lambda}$ we have the factor

$$\frac{\lambda^p}{\lambda^{k_0+k_1+1+\mu_c}} = 1.$$

This shows that

$$z \stackrel{\text{def}}{=} \lim_{\lambda\to 0}\frac{1}{\lambda^{\mu_c}}J_{1/\lambda}(\lambda) = \operatorname{Res}_{y=0}\overline{M},$$

where

$$\overline{M} = \lim_{\lambda\to 0} -\frac{1}{\lambda^{1+\mu_c}}M_{1/\lambda}.$$

Now

$$\lim_{\lambda\to 0} G_0 = \Lambda_{-a'}\left(\frac{1}{y}\right),$$

$$\lim_{\lambda\to 0}\lambda^{k_0+k_1}T_0^{k_0}T_1^{k_1}T_{1/\lambda}^{k_2} = (-1)^{k_1}y^{k_2}(1 - y)^{k_0+k_1},$$

$$\lim_{\lambda\to 0}\frac{1}{\lambda^p}\xi_{1/\lambda^p}(a', b', c', \lambda^p, x^p) = \frac{a' - c'}{a'}\,\xi_1(c', -a', (1 - y)^p),$$

in the notation of (22.2), the point being that the coefficient of $((1 - y)^p - 1)^m$ in this last limit is $-(a' - c')\theta_m$ where

$$\theta_m = \sum_{k+l+i+j=m} \binom{b'-1}{k}\binom{c'-b'}{l}\binom{-b'}{i}\binom{b'-c'}{j}\frac{1}{a'+i+j}$$

$$= \sum_{r+s=m} \frac{1}{a'+r} \sum_{i+j=r} \binom{-b'}{i}\binom{b'-c'}{j} \cdot \sum_{k+l=s} \binom{b'-1}{k}\binom{c'-b'}{l}$$

$$= \sum_{r+s=m} \frac{1}{a'+r} \binom{-c'}{r}\binom{c'-1}{s}$$

and so

$$-\frac{a'}{a'-c'}\bar{M} = (-1)^{k_1} y^{k_2} (1-y)^{k_0+k_1} \Lambda_{-a'}\left(\frac{1}{y}\right) \xi_1(c', -a', (1-y)^p).$$

We now write $y = 1 - T$, so that

$$-\frac{a'}{(a'-c')}\bar{M} = (-1)^{k_1} T^{p-1-\mu_c} (1-T)^{\mu_a} \Lambda_{-a'}\left(\frac{1}{1-T}\right) \xi_1(c', -a', T^p).$$

But $\mathrm{Res}_{y=0} = -\mathrm{Res}_{T=1}$. Therefore

$$\frac{a'(-1)^{k_1} z}{a'-c'} = \mathrm{Res}_{T=1}\, T^{p-1-\mu_c}(1-T)^{\mu_a} \Lambda_{-a'}\left(\frac{1}{1-T}\right) \xi_1(c', -a', T^p).$$

Under the substitution

$$(a_1, a_2, b_1, b_2, x) \mapsto (c, -a, c', -a', T)$$

the function $(1 - x)M_1$ in Lemma 23.1 transforms into the function whose residue we are to calculate. It follows that

$$+\frac{a'(-1)^{k_1}}{a'-c'} z = -\frac{a}{c-a} B_p(c, -a; c', -a') - (-1)^{\mu_c} \frac{-a'}{c'-a'} \binom{-a}{\mu_a - \mu_c}$$

and the second assertion of the theorem now follows from

$$\lim_{\lambda \to 0} \frac{B_1(\lambda)}{\lambda^{\mu_c}} = z + (-1)^{\mu_b} \binom{-a}{\mu_a - \mu_c}. \quad \square$$

Chapter 26

Singular Disks, Part III. Logarithmic Case

We assume here that $c = 1$; a, b satisfy (9.0.1)–(9.0.3). We use $G = \Gamma'_p/\Gamma_p$ to denote the logarithmic derivative of the (Morita) p-adic gamma function discussed in Chapter 21. We again assume that μ_a, μ_b, μ_c, lie in $[0, p-1]$.

26.1 **Theorem.** *In the notation of Lemma* 24.5.8,

$$\theta = (-1)^{\mu_b},$$

$$\gamma = p((G(a) - G(1)) + (G(b) - G(1))),$$

$$\det B(\lambda) = p\frac{1-a'}{1-a}\lambda^{p-1}(1-\lambda)^{\mu_a+\mu_b-(p-1)}\Lambda_{1-a'-b'}\left(\frac{1}{1-\lambda}\right).$$

Proof. It follows from Lemma 24.5.9 that it is enough to compute $B_4(0)$ and $\lim_{\lambda\to 0} B_2(\lambda)/\lambda^p$.

The calculation of $B_4(0)$ proceeds as in the nonlogarithmic case. We now have

$$k_0 = p - 1 - \mu_b,$$

$$k_1 = \mu_b - p,$$

$$k_2 = \mu_a.$$

Thus once again $J_0 = 0$, while M_∞ lies in

$$T_\infty^{-(k_0+k_1+k_2)+p}\Omega[[T_\infty]] \subset T_\infty^2\Omega[[T_\infty]],$$

which shows that $J_\infty = 0$. The same argument used previously shows that (25.6) is again valid. We now observe that (25.7) holds with $M_1(0)$ coinciding with M_1 as used in Lemma 23.1 under the substitution

$$(a_1, a_2, b_1, b_2) \mapsto (b, 1-b, b', 1-b').$$

The assertion for $B_4(0)$ now follows from Lemma 23.1. We proceed to the calculation of $B_2(\lambda)$.

$$B_2(\lambda) = \left\langle \alpha^*_{f,h} 1^*, \frac{1}{1-x} \right\rangle = \sum_{v \in S} \operatorname{Res}_v M_v,$$

where

$$M_v = T_0^{k_0} T_1^{k_1} T_{1/\lambda}^{k_2} G_0 \xi_{vp}(a', b', 1, \lambda^p, x^p).$$

In this calculation

$$k_0 = p - 1 - \mu_b,$$

$$k_1 = \mu_b - (p-1) - 1,$$

$$k_2 = \mu_a.$$

Clearly $p - 1 \geq \mu_b$ implies $J_0(\lambda) = 0$. Again as element of $\Omega((T_\infty))$, the ideal of (M_∞) is contained in $(x)^{k_0+k_1+k_2-p} = (x)^{\mu_a - 1 - p} \in (T_\infty^2)$. Hence $J_\infty(\lambda) = 0$.

We recall that in the calculation of ξ_{vp} we must use (2.3.5.10) with $A = 0$, $B = \lambda^p(a' - 1)$. In computing $M_1(\lambda)/\lambda^p$, we have the factors

$$T_0^{k_0} T_1^{k_1} T_{1/\lambda}^{k_2} = (1-\lambda)^{k_2} T_1^{k_1} (1 - T_1)^{k_0} \left(1 + \frac{\lambda}{1-\lambda}\right)^{k_2},$$

$$G_0 = \Lambda_{1-b'}\left(\frac{1}{T_1}\right) \Lambda_{-a'}\left(\frac{1}{1-\lambda}\right) \frac{(1 + (\lambda/(1-\lambda))T_1)^{-pa'}}{(1 - (\lambda^p/(1-\lambda^p))((1-T_1)^p - 1))^{-a'}},$$

$$\frac{1-\lambda^p}{\lambda^p} \xi_1(a', b', c', \lambda^p, x^p) = \text{power series in } (T_1 - 1)^p - 1 \quad \text{and} \quad \frac{\lambda^p}{1-\lambda^p}.$$

We conclude that

$$\lim_{\lambda \to 0} \frac{1}{\lambda^p} J_1(\lambda) = \operatorname{Res}_1 \lim_{\lambda \to 0} \frac{M_1(\lambda)}{\lambda^p}.$$

Letting γ_1 be the limit on the left we find

$$\gamma_1 = \operatorname{Res}_1 \hat{M}_1 \cdot (a' - 1),$$

where

$$\hat{M}_1 = T_0^{p-1-\mu_b} T_1^{\mu_b} \Lambda^{-b'} \xi_1(b', x^p),$$

ξ_1 being given by (23.3.3). The point is that

$$\lim_{\lambda\to 0} T_0^{k_0}T_1^{k_1}T_{1/\lambda}^{k_2} = T_1^{k_1}T_0^{k_0} = T_0^{p-1-\mu_b}T_1^{\mu_b-p},$$

$$\lim_{\lambda\to 0} G_0 = \Lambda^{1-b'},$$

$$\lim_{\lambda\to 0}\frac{1}{\lambda^p}\xi_1(a', b', 1, \lambda^p, x^p) = (1-a')T_1^{\phi}\sum_{m=0}^{\infty}(-T_1^{\phi})^m\sum_{k+i=m}\binom{b'-1}{k}\binom{-b'}{i}$$
$$\times\frac{1}{i+b'}$$
$$= -(1-a')T_1^{\phi}\xi_1(b', x^p).$$

From Corollary 23.3.4 we deduce by (21.4.9)

$$\begin{aligned}\gamma_1 &= p(a'-1)(-1)^{\mu_b}(G(0)-G(1-b))\\ &= p(1-a')(-1)^{\mu_b}(G(b)-G(1)).\end{aligned}$$

To compute $(1/\lambda^p)J_{1/\lambda}(\lambda)$ we use

$$\frac{1}{\lambda^p}J_{1/\lambda}(\lambda) = -\mathrm{Res}_{y=0}\frac{1}{\lambda^{p+1}}M_{1/\lambda}$$

where we write y for $T_{1/\lambda}$. We consider the three factors of $M_{1/\lambda}$,

$$T_0^{k_0}T_1^{k_1}T_{1/\lambda}^{k_2} = \frac{(\lambda-1)^{k_1}}{\lambda^{-1}}y^{k_2}(1-y)^{k_0}\left(1-\frac{1}{1-\lambda}y\right)^{k_1},$$

$$G_0 = \Lambda_{-a'}\left(\frac{1}{y}\right)\Lambda_{1-b'}\left(\frac{1}{1-\lambda}\right)\frac{(1+(1/(\lambda-1))y)^{p(1-b')}}{(1-(1/(\lambda^p-1))((1-y)^p-1))^{1-b'}}$$

and

$$-\xi_{1/\lambda^p}(a', b', 1, \lambda^p, x^p) = \lambda^p\cdot\text{power series in }((1-y)^p-1)\text{ and }\frac{1}{1-\lambda^p}.$$

Thus $(1/\lambda^{p+1})M_{1/\lambda}$ is a product of factors which as power series in y or in $1/y$ have coefficients which are holomorphic at $\lambda = 0$. We conclude that

$$\gamma_2 \overset{\text{def}}{=} \lim_{\lambda\to 0}\frac{1}{\lambda^p}J_{1/\lambda}(\lambda) = -\mathrm{Res}_{y=0}\lim_{\lambda\to 0}\frac{1}{\lambda^{p+1}}M_{1/\lambda}(\lambda).$$

We observe

$$\lim_{\lambda=0} \frac{1}{\lambda} T_0^{k_0} T_1^{k_1} T_{1/\lambda}^{k_2} = (-1)^{k_1} y^{k_2} (1-y)^{k_0+k_1},$$

$$\lim_{\lambda=0} G_0 = \Lambda_{-a'}\left(\frac{1}{y}\right),$$

$$\lim_{\lambda\to 0} \frac{1}{\lambda^p} \xi_{1/\lambda^p}(a', b', 1, \lambda^p, x^p) = (1-a') \sum_{m=0}^{\infty} (1-(1-y)^p)^m \frac{1}{m+a'}.$$

Thus

$$\frac{1}{1-a'} \gamma_2 = -\mathrm{Res}_{y=0}(-1)^{k_1} y^{k_2} (1-y)^{k_1+k_0} \Lambda_{-a'}\left(\frac{1}{y}\right) \times \sum_{m=0}^{\infty} \frac{1}{m+a'} (1-(1-y)^p)^m,$$

$$\frac{(-1)^{k_1}}{1-a'} \gamma_2 = \mathrm{Res}_{x=1}\, T_1^{k_2} T_0^{-1} \Lambda_{-a'}\left(\frac{1}{T_1}\right) \sum_{m=0}^{\infty} \frac{1}{m+a'} (1-x^p)^m$$

$$= -\mathrm{Res}_{x=1}\, T_1^{\mu_a} \Lambda^{-a'} \cdot \left(\frac{1}{x}\right) \eta_1(a', x^p),$$

with η_1 as defined by (23.2.6). Thus by Corollary 23.2.8,

$$\gamma_2 = (-1)^{k_1+1}(1-a')p(G(a)-G(1))$$
$$= (-1)^{\mu_b}(1-a')p(G(a)-G(1)).$$

We now obtain the asserted formula for $\lim_{\lambda\to 0} B_2(\lambda)/\lambda^p$ by adding γ_1 and γ_2. The formula for γ is now obtained from Lemma 24.5.9. □

26.2 *Remark*. For $n = 2$, the Gauss multiplication formula gives

$$\Gamma_p\left(\frac{x}{2}\right)\Gamma_p\left(\frac{x+1}{2}\right) = \Gamma_p(x) \cdot (\tfrac{1}{2})^{-t+(p-1)y}\, \Gamma_p(\tfrac{1}{2}),$$

with t a local constant, $x = -t + py$. Thus

$$\frac{1}{2}\left(G\left(\frac{x}{2}\right) + G\left(\frac{x+1}{2}\right)\right) = G(x) + \frac{1}{p} \log(\tfrac{1}{2})^{p-1}.$$

Substituting $x = 1$

$$\frac{p}{2}(G(\tfrac{1}{2}) - G(1)) = \log(\tfrac{1}{2})^{p-1}$$

and so for $a = b = \frac{1}{2}$, $c = 1$, the constant γ is found to be $4\log(\frac{1}{2})^{p-1}$, agreeing with the value originally found in Dwork [11, equation (8.17)] by entirely different methods.

26.2.1 But is the Frobenius structure of our differential equation unique? Let E_{sad} be the field of analytic elements whose supports are superadmissible sets. Suppose we could choose $\mathbf{A}_\lambda$, $\tilde{\mathbf{A}}_\lambda \in \mathrm{GL}(2, E_{sad})$ such that equation (7.1.2) holds with $A_{\bar{z}}$, $\tilde{A}_{\bar{z}} \in \mathrm{GL}(2, \Omega)$, not necessarily obtained by specialization of λ. Putting $B = \tilde{\mathbf{A}}_\lambda^{-1} A_\lambda$, $G = \tilde{A}_z^{-1} A_z \in \mathrm{GL}(2, \Omega)$, we find

$$C_f(z, \lambda)B = GC_f(z, \lambda).$$

If (u_1, u_2) is the unique solution of (2.4.5.1) bounded on $D(z, 1^-)$ then $(u_1, u_2)B = k(u_1, u_2)$ where $k \in \Omega$, $k \neq 0$. It follows that either $B = kI$ or $u_2/u_1 \in E_{sad}$ which contradicts Theorem 16.11 if $l = 1$ and we are in Cases 1 or 2. This justifies the comparison of the two evaluations of γ for $(\frac{1}{2}, \frac{1}{2}, 1)$.

26.3

For $a \in \mathbb{Q} \cap \mathbb{Z}_p$, the calculation of $\Gamma_p'/\Gamma_p(a)$ may be performed by means of the formula

$$\log \Gamma_p(x) = \sum_{\substack{i=0 \\ |x+i|=1}}^{p=1} G_p\left(\frac{x+i}{p}\right),$$

where G_p is Diamond's p-adic log gamma function which is locally holomorphic on the complement of $\mathbb{Z}_p$ in Ω. This permits us to reduce the calculation to that of $G_p'(=\psi_p)$ on $\mathbb{Q} - \mathbb{Z}_p$. Extending to the p-adics the classical theorem of Gauss, Diamond gives for $r, f \in \mathbb{N}$, $r < f$, $\mathrm{ord}(r/f) < 0$,

$$G_p'\left(\frac{r}{f}\right) = -\log f - \gamma_p + \sum_{s=1}^{f-1} \rho^{-sr} \log(1 - \rho^s),$$

where ρ is a primitive fth root of unity, and γ_p is the p-adic Euler constant which need not concern us here since according to Diamond,

$$\frac{\Gamma_p'}{\Gamma_p}(1) = \frac{1-p}{p}\gamma_p$$

and the γ_p disappears in the calculation of $(\Gamma_p'/\Gamma_p)(a) - (\Gamma_p'/\Gamma_p)(1)$.

For further details see Diamond [7]. By this method, Diamond has obtained the formula for $a \in \mathbb{Q} \cap \mathbb{Z}_p$, $0 \leq a < 1$, $d = \text{denom } a$,

$$\frac{\Gamma_p'}{\Gamma_p}(a) - \frac{\Gamma_p'}{\Gamma_p}(1) = \sum_{\substack{z^d = 1 \\ z \neq 1}} \left[(z^{da} - 1) - \frac{1}{p}(z^{da'} - 1) \right] \log(1 - z).$$

Appendix (by Alan Adolphson)

L-Functions

In this appendix we explain the connection between character sums, L-functions, and the map $\bar{\alpha}_{f,h}: \mathbf{W}_{f,\lambda} \to \mathbf{W}_{h,\lambda^p}$.

Let p be a prime, $q = p^f$, $\mathbb{F}_q$ = finite field with q elements, $f \in \mathbb{F}_q[x_1, \ldots, x_n]$ a nonconstant polynomial. Let $\overline{\mathbb{Q}}_p$ denote the algebraic closure of the p-adic numbers, $\mathbb{Q}_p$, and let

$$\chi: \mathbb{F}_q^\times \to \overline{\mathbb{Q}}_p^\times$$

be a nontrivial multiplicative character. (One could replace $\overline{\mathbb{Q}}_p^\times$ by any field of characteristic 0 containing the $(q-1)$st roots of unity.) Let

$$\mathbb{N}_r: \mathbb{F}_{q^r} \to \mathbb{F}_q$$

be the norm map. Composition defines a multiplicative character on $\mathbb{F}_{q^r}^\times$:

$$\chi_r = \chi \circ \mathbb{N}_r: \mathbb{F}_{q^r}^\times \to \overline{\mathbb{Q}}_p^\times.$$

Define a character sum

$$S_r(f, \chi) = \sum_{\substack{x_1, \ldots, x_n \in \mathbb{F}_{q^r} \\ f(x_1, \ldots, x_n) \neq 0}} \chi_r(f(x_1, \ldots, x_n)) \tag{1}$$

and an associated L-function

$$L(f, \chi, T) = \exp\left(\sum_{r=1}^{\infty} S_r(f, \chi)\frac{T^r}{r}\right). \tag{2}$$

(Equivalently, one could extend χ_r by defining $\chi_r(0) = 0$ and then sum over *all* $x_1, \ldots, x_n \in \mathbb{F}_{q^r}$.)

We wish to identify $L(f, \chi, T)$ with an L-function associated to a Kummer covering of affine n-space. Recall the basic definitions: If X is a variety over $\mathbb{F}_q$, the *Frobenius map* $F_q: X \to X$ is defined in terms of "local coordinates" by

$$F_q(x_1, \ldots, x_m) = (x_1^q, \ldots, x_m^q).$$

Let G be a finite group of automorphisms of X, defined over $\mathbb{F}_q$, and let $\pi: G \to \mathrm{GL}(V)$ be a representation of G in a finite-dimensional vector space V over a field of characteristic 0. The associated L-function is defined by

(3)

$$L(X/\mathbb{F}_q, G, \pi, T) = \exp\left(\sum_{r=1}^{\infty} \frac{T^r}{r}\left(\frac{1}{\mathrm{card}(G)} \sum_{g \in G} (\mathrm{tr}\ \pi(g)) \cdot \mathrm{card}(\mathrm{Fix}(F_q^r \circ g^{-1}))\right)\right)$$

where tr denotes the trace and $\mathrm{Fix}(F_q^r \circ g^{-1})$ denotes the set of points of X fixed under the endomorphism $F_q^r \circ g^{-1}$.

Apply this general definition of L-function to the variety in affine $(n + 1)$-space:

$$X: y^{q-1} = f(x_1, \ldots, x_n),$$

where f is as above. The group $\mathbb{F}_q^\times$ acts on X: For $a \in \mathbb{F}_q^\times$, define $g_a: X \to X$ by

$$g_a(x_1, \ldots, x_n, y) = (x_1, \ldots, x_n, ay).$$

The character χ defined above gives a one-dimensional representation of $\mathbb{F}_q^\times$, hence we have an L-function $L(X/\mathbb{F}_q, \mathbb{F}_q^\times, \chi, T)$.

Lemma 1. *For χ nontrivial,*

$$L(X/\mathbb{F}_q, \mathbb{F}_q^\times, \chi, T) = L(f, \chi, T).$$

Proof. We must check that

$$S_r(f, \chi) = \frac{1}{q-1} \sum_{a \in \mathbb{F}_q^\times} \chi(a) \cdot \mathrm{card}(\mathrm{Fix}(F_q^r \circ g_a^{-1})).$$

Now

$$\begin{aligned}\mathrm{Fix}(F_q^r \circ g_a^{-1}) &= \{(x_1, \ldots, x_n, y) \in X \mid x_1^{q^r} = x_1, \ldots, x_n^{q^r} = x_n, y^{q^r} = ay\} \\ &= X_r(a) \cup Y_r \text{ (a disjoint union)},\end{aligned}$$

where

$$X_r(a) = \{(x_1, \ldots, x_n, y) \in X \mid x_i \in \mathbb{F}_{q^r} \text{ for all } i \text{ and } y^{q^r - 1} = a\},$$

$$Y_r = \{(x_1, \ldots, x_n, y) \in X \mid x_i \in \mathbb{F}_{q^r} \text{ for all } i \text{ and } f(x_1, \ldots, x_n) = 0\}.$$

If $y^{q^r-1} = a \in \mathbb{F}_q^\times$, then $(y^{q-1})^{q^r-1} = a^{q-1} = 1$, hence $y^{q-1} \in \mathbb{F}_{q^r}^\times$ and

$$\begin{aligned}\mathbb{N}_r y^{q-1} &= y^{(q-1)(1+q+q^2+\cdots+q^{r-1})} = y^{q^r-1}\\ &= a.\end{aligned}$$

But for $(x_1, \ldots, x_n, y) \in X$, one has $y^{q-1} = f(x_1, \ldots, x_n)$, hence

$$X_r(a) = \{(x_1, \ldots, x_n, y) \in X \mid x_i \in \mathbb{F}_{q^r} \text{ for all } i, \mathbb{N}_r(f(x_1, \ldots, x_n))) = a\}.$$

Therefore

$$\begin{aligned}&\frac{1}{q-1} \sum_{a \in \mathbb{F}_q^\times} \chi(a) \cdot \operatorname{card}(X_r(a))\\ &= \frac{1}{q-1} \sum_{a \in \mathbb{F}_q^\times} \sum_{(x_1, \ldots, x_n, y) \in X_r(a)} \chi(\mathbb{N}_r(f(x_1, \ldots, x_n))).\\ &= \sum_{\substack{x_1, \ldots, x_n \in \mathbb{F}_{q^r}\\ f(x_1, \ldots, x_n) \neq 0}} \sum_{y^{q-1} = f(x_1, \ldots, x_n)} \frac{1}{q-1} \chi_r(f(x_1, \ldots, x_n))\\ &= \sum_{\substack{x_1, \ldots, x_n \in \mathbb{F}_{q^r}\\ f(x_1, \ldots, x_n) \neq 0}} \chi_r(f(x_1, \ldots, x_n)),\end{aligned}$$

since $y^{q-1} = f(x_1, \ldots, x_n)$ has $q - 1$ solutions when $f(x_1, \ldots, x_n) \neq 0$. On the other hand

$$\begin{aligned}\frac{1}{q-1} \sum_{a \in \mathbb{F}_q^\times} \chi(a) \cdot \operatorname{card}(Y_r) &= \frac{\operatorname{card}(Y_r)}{q-1} \sum_{a \in \mathbb{F}_q^\times} \chi(a),\\ &= 0\end{aligned}$$

since χ is nontrivial. QED □

The connection between p-adic analysis and these L-functions is based on two results, the first elementary and the second much deeper:

(1) p-adic analytic lifting of the character
(2) Reich Trace Formula [29]

First recall the Teichmüller character $\omega\colon \mathbb{F}_q^\times \to \overline{\mathbb{Q}}_p^\times$, defined by:

$\omega(a) =$ the unique root of unity of order prime to p in $\overline{\mathbb{Q}}_p^\times$ whose reduction mod p is a.

We extend ω by making the convention $\omega(0) = 0$. The group of characters of $\mathbb{F}_q^\times$ is cyclic of order $q - 1$, generated by ω. Note that $S_r(f, \omega^d) = S_r(f^d, \omega)$, so all the L-functions discussed above are of the form $L(f^d, \omega, T)$ for some d, $0 < d < q - 1$. For $f \in \mathbb{F}_q[x_1, \ldots, x_n]$, let $f^\omega \in \overline{\mathbb{Q}}_p[x_1, \ldots, x_n]$ be the polynomial whose coefficients are obtained by applying ω to the coefficients of f. For convenience we write x for the n-tuple $(x_1, \ldots, x_n)$. Let "ord" denote the valuation on $\overline{\mathbb{Q}}_p$ with ord $p = 1$. Let $\mathbf{0} = \{x \in (\overline{\mathbb{Q}}_p)^n \mid \operatorname{ord} x_i \geq 0$ for all $i\}$.

There is an analytic function F defined on a set containing $\mathbf{0} - \{x \in \mathbf{0} \mid \operatorname{ord} f^\omega(x) > 0\}$ such that the following diagram commutes:

$$\begin{array}{ccc} \mathbf{0} - \{\operatorname{ord} f^\omega(x) > 0\} & \xrightarrow{F} & \mathbb{Q}_p^\times \\ \uparrow{\scriptstyle\omega} & & \uparrow{\scriptstyle\omega} \\ \mathbb{F}_q - \{f(x) = 0\} & \xrightarrow[f]{} & \mathbb{F}_q^\times \end{array}$$

namely,

$$F(x) = \left[\frac{f^\omega(x^q)}{f^\omega(x)}\right]^{1/(q-1)} \tag{4}$$

$$= f^\omega(x)\left[\frac{f^\omega(x^q)}{f^\omega(x)^q}\right]^{1/(q-1)},$$

where $x = (x_1^q, \ldots, x_n^q)$. How do we make sense out of taking the $(q - 1)$st root of this rational function? Note that

$$f^\omega(x^q) = f^\omega(x)^q + p \cdot g(x),$$

where the coefficients of $g \in \overline{\mathbb{Q}}_p[x_1, \ldots, x_n]$ have ord ≥ 0. Hence

$$\left[\frac{f^\omega(x^q)}{f^\omega(x)^q}\right]^{1/(q-1)} = \left[1 + p \cdot \frac{g(x)}{f^\omega(x)^q}\right]^{1/(q-1)}$$

converges when expanded according to the binomial formula provided $\operatorname{ord}(pg(x)/f^\omega(x)^q) > 0$. It is easy to check that for all $t \geq 1$, this estimate holds for x in the set

$$\mathbf{D}_t = \left\{(x_1, \ldots, x_n) \in (\overline{\mathbb{Q}}_p)^n \mid \operatorname{ord} x_i \geq \frac{1-t}{qdt} \text{ for all } i, \operatorname{ord} f^\omega(x) < \frac{1}{qt}\right\},$$

where $d = \deg f$ (if $n = 1$, one needs only $\operatorname{ord} f^\omega(x) < 1/q$). Hence

$$F(x) = f^\omega(x)\left[1 + p\,\frac{g(x)}{f^\omega(x)^q}\right]^{(q-1)} \tag{5}$$

is an analytic function on the set $\mathbf{D}_t$, and this formula shows that $F(x)$ mod p coincides with $f(\bar{x})$, where $\bar{x} = x$ mod p. Furthermore, from the definition of F,

$$F(x)^{q-1} = \frac{f^{\omega}(x^q)}{f^{\omega}(x)},$$

so if $x \in \mathbf{0}$ satisfies $x^q = x$, ord $f^{\omega}(x) < 1/qt$, then $F(x)^{q-1} = 1$, i.e., $F(x)$ is a $(q-1)$st root of unity. Thus when x satisfies these conditions, $F(x)$ is the Teichmüller lifting of $f(\bar{x})$:

$$F(x) = \omega(f(\bar{x})). \tag{6}$$

It follows that for $x \in \mathbf{D}_t$, $x^{q^r} = x$,

$$F(x)F(x^q)F(x^{q^2}) \cdot \ldots \cdot F(x^{q^{r-1}}) = \omega(f(\bar{x})f(\bar{x}^q)f(\bar{x}^{q^2}) \cdot \ldots \cdot f(\bar{x}^{q^{r-1}})), \tag{7}$$

hence

$$S_r(f, \omega) = \sum_{\substack{x \in \mathbf{0} \\ x^{q^r} = x \\ \text{ord } f^{\omega}(x) = 0}} F(x)F(x^q) \cdot \ldots \cdot F(x^{q^{r-1}}). \tag{8}$$

The Reich Trace Formula, described below, relates the above L-functions to certain completely continuous operators on a p-adic Banach space (see, for example, Serre [34] for a general discussion of such operators). When doing p-adic analysis, it is convenient to work over a field which is both complete and algebraically closed. Since $\overline{\mathbb{Q}}_p$ is not complete we work over $\mathbb{C}_p$ = completion of $\overline{\mathbb{Q}}_p$, a field which is complete and algebraically closed.

For Reich's theory we need a polynomial whose reduction mod p is a product of distinct irreducible factors. Factor $f(x_1, \ldots, x_n)$ in $\mathbb{F}_q[x_1, \ldots, x_n]$: $f = p_1^{\alpha_1} \cdot \ldots \cdot p_s^{\alpha_s}$, where the $p_i \in \mathbb{F}_q[x_1, \ldots, x_n]$ are irreducible. Set

$$P(x_1, \ldots, x_n) = \prod_{i=1}^{s} p_i^{\omega}(x_1, \ldots, x_n) \in \overline{\mathbb{Q}}[x_1, \ldots, x_n]. \tag{9}$$

For $n > 1$ to assure applicability of Reich's theorem we assume that the homogeneous summand of highest degree of P mod p is square free. This is only a minor inconvenience since this condition is certainly satisfied if f is homogeneous and the calculation of (1) may be reduced to the case in which f is homogeneous of degree divisible by $q - 1$.

For ε, Δ non-negative real numbers, define

$$\mathbf{D}(\varepsilon, \Delta, P) = \{x = (x_1, \ldots, x_n) \in (\mathbb{C}_p)^n \,|\, \text{ord } P(x) \le \varepsilon, \text{ ord } x_i \ge -\Delta \text{ for all } i\}. \tag{10}$$

A short calculation shows that $F \in \mathscr{L}(\varepsilon, \Delta, P)$ for $\varepsilon < 1/qdt$, $\Delta < (t-1)/qd^2t$, any $t \geq 1$ where $\mathscr{L}(\varepsilon, \Delta, P)$ denotes the space of p-adic analytic functions on $\mathbf{D}(\varepsilon, \Delta, P)$, i.e., uniform limits on $\mathbf{D}(\varepsilon, \Delta, P)$ of rational functions with poles outside $\mathbf{D}(\varepsilon, \Delta, P)$. For $\xi \in \mathscr{L}(\varepsilon, \Delta, P)$, define

$$(\psi_q \xi)(x) = \frac{1}{q^n} \sum_{y^q = x} \xi(y). \tag{11}$$

Then by [29], ψ_q maps $\mathscr{L}(\varepsilon/q, \Delta/q, P)$ into $\mathscr{L}(\varepsilon, \Delta, P)$. For $\varepsilon < 1/dt$, $\Delta < (t-1)/d^2t$, any $t \geq 1$, multiplication by F maps $\mathscr{L}(\varepsilon, \Delta, P)$ into $\mathscr{L}(\varepsilon/q, \Delta/q, P)$. Taking the composition gives a map defined for $\varepsilon < 1/dt$, $\Delta < (t-1)/d^2t$, any $t \geq 1$:

$$\psi_q \circ F \colon \mathscr{L}(\varepsilon, \Delta, P) \to \mathscr{L}(\varepsilon, \Delta, P), \tag{12}$$

which by [29] is completely continuous. The point of [34] is that the trace and Fredholm determinant of $\psi_q \circ F$ are well defined and have the usual properties. The main result of [29] is the Reich Trace Formula:

Theorem 1. $(q^r - 1)^n \operatorname{tr}(\psi_q \circ F)^r = \sum F(x)F(x^q) \cdot \ldots \cdot F(x^{q^{r-1}})$, *where the sum is over those* $x \in (\mathbb{C}_p^\times)^n$ *with* $x^{q^r} = x$, $\operatorname{ord} f^{\omega}(x) = 0$.

Note that, unlike the sum in equation (8), this sum excludes those x with some coordinate $x_i = 0$. Thus from the point of view of the p-adic theory, a more natural character sum is

$$S_r^*(f, \omega) = \sum_{\substack{x_1, \ldots, x_n \in \mathbb{F}_{q^r}^\times \\ f(x_1, \ldots, x_n) \neq 0}} \omega(\mathbb{N}_r(f(x_1, \ldots, x_n))) \tag{13}$$

and the Reich Trace Formula reads

$$(q^r - 1)^n \operatorname{tr}(\psi_q \circ F)^r = S_r^*(f, \omega). \tag{14}$$

Of course, $S_r(f, \omega)$ and $S_r^*(f, \omega)$ are related by elementary combinatorics: let A denote any subset (possibly the empty set) of $\{1, 2, \ldots, n\}$, and let f_A be the polynomial in n-card(A) variables obtained from f by setting $x_i = 0$ for $i \in A$. Then

$$\begin{aligned} S_r(f, \omega) &= \sum_A S_r^*(f_A), \\ S_r^*(f, \omega) &= \sum_A (-1)^{\operatorname{card}(A)} S_r(f_A, \omega) \end{aligned} \tag{15}$$

If we put $L^*(f, \omega, T) = \exp(\sum_{r=1}^{\infty} S_r^*(f, \omega)^{(T^r/r)})$, there are corresponding relations between $L^*(f, \omega, T)$ and $L(f, \omega, T)$:

$$L(f, \omega, T) = \prod_A L^*(f_A, \omega, T),$$

(16)

$$L(f^*, \omega, T) = \prod_A L(f_A, \omega, T)^{(-1)^{\mathrm{card}(A)}}.$$

One has ([34, Corollary 3 to Proposition 7])

$$\det(I - T(\psi_q \circ F) = \exp\left(-\sum_{r=1}^{\infty} \mathrm{tr}(\psi_q \circ F)^r \frac{T^r}{r}\right)$$

and furthermore, $\det(I - T(\psi_q \circ F)) \in 1 + T\mathbb{C}_p[[T]]$ converges for all $T \in \mathbb{C}_p$, i.e., is a p-adic entire function. If we define an operator δ on power series $g(T) \in 1 + T\mathbb{C}_p[[T]]$ by

$$g(T)^\delta = \frac{g(T)}{g(qT)},$$

then equation (14) is equivalent to

$$\text{(17)} \qquad L^*(f, \omega, T)^{(-1)^{n-1}} = \det(I - T(\psi_q \circ F))^{\delta^n}.$$

Theorem 2. *$L(f, \omega, T)$ is a rational function.*

Proof. It follows from equations (16) and (17) that $L(f, \omega, T)$ is a quotient of p-adic entire functions. The theorem is then a consequence of Dwork's rationality criterion [8a, Theorem 3]. □

Reich used his trace formula to show the rationality of (3) subject to the following hypothesis concerning the field of definition:

(R) *For each prime power m which occurs as the order of a cyclic factor group of a subgroup of G, there exists an integer l prime to m such that the equation $X^m - 1$ splits in $\mathbb{F}_{q^l}$.*

A p-adic proof of rationality of (3) was given by Monsky [27] by means of a trace formula which may be viewed as a generalization of that of Reich. This question was also discussed by Dwork [10] who observed that the restriction R could be removed by extending the Reich trace formula to treat functions on hypersurfaces. This extension is indeed provided by the work of Monsky.

Our discussion up to this point has on the whole been simply an expansion of §e of [29]. We now consider $L(f, \omega, T)$ in the special case $n = 1$, i.e., f a polynomial in one variable. We make the assumption that $X: y^{q-1} = f(x)$ is absolutely irreducible. A reference for the following information is Hasse [19a].

As before, factor $f(x)$ in $\mathbb{F}_q[x]$:

$$f(x) = p_1(x)^{\alpha_1} \cdot \ldots \cdot p_s(x)^{\alpha_s}, \tag{18}$$

where the $p_i(x) \in \mathbb{F}_q[x]$ are irreducible. We assume that $0 < \alpha_i < q - 1$ for all i. (This can always be accomplished by a change of variable that leaves the function field of X unchanged and affects $L(f, \omega, T)$ only in a trivial way.) We regard X as a cover of the affine line with possible ramification at $p_1(x), \ldots, p_s(x)$. Let $\tilde{X}$ be the complete nonsingular model of the function field of X, which may also be ramified at infinity. Let χ be a nontrivial multiplicative character of $\mathbb{F}_q^\times$. The difference between $L(X/\mathbb{F}_q, \mathbb{F}_q^\times, \chi, T)$ and $L(\tilde{X}/\mathbb{F}_q, \mathbb{F}_q^\times, \chi, T)$ is that the former contains no local factors for $p_1(x), \ldots, p_s(x)$ or infinity. Let $I_1, \ldots, I_s, I_\infty$ be the inertia groups for these places and let $F_1, \ldots, F_s, F_\infty$ be representatives of the Frobenius classes. Then

$$L(\tilde{X}/\mathbb{F}_q, \mathbb{F}_q^\times, \chi, T) = \tag{19}$$

$$L(X/\mathbb{F}_q, \mathbb{F}_q^\times, \chi, T) \cdot \prod_{i=1}^{s, \infty} \begin{cases} (1 - T^{\deg p_i}\chi(F_i))^{-1} & \text{if } \chi(I_i) = 1, \\ 1 & \text{if } \chi(I_i) \neq 1. \end{cases}$$

Now I_i is the group of $[(q-1)/(\alpha_i, q-1)]$th roots of unity in $\mathbb{F}_q^\times$, so $\chi(I_i) = 1$ if and only if $\chi^{(\alpha_i, q-1)} = 1$ (if $i = \infty$, replace $(\alpha_i, q-1)$ by $(\deg f, q-1)$).

It is well known (see [19a]) that $L(\tilde{X}/\mathbb{F}_q, \mathbb{F}_q^\times, \chi, T)$ is a polynomial of degree

$$-2 + \sum_{i=1}^{s, \infty} \begin{cases} \deg p_i & \text{if } \chi(I_i) \neq 1, \\ 0 & \text{if } \chi(I_i) = 1, \end{cases} \tag{20}$$

(where deg "p_∞" $= 1$) hence by (19), $L(X/\mathbb{F}_q, \mathbb{F}_q^\times, \chi, T)$ is a polynomial of degree

$$\left(\sum_{i=1}^{s} \deg p_i\right) - 1, \tag{21}$$

a number dependent only on the irreducible factors of f, not on the exponents α_i or the character χ. Furthermore, by a theorem of Weil, the reciprocal roots of $L(\tilde{X}/\mathbb{F}_q, F_q^\times, \chi, T)$ have absolute value $q^{1/2}$. Thus the reciprocal roots of

$L(X/\mathbb{F}_q, \mathbb{F}_q^\times, \chi, T)$ are either roots of unity or have absolute value $q^{1/2}$. Note that for the Teichmuller character ω, $\omega^d = 1$ if and only if $(q-1)|d$. Hence by (19) and Lemma 1,

(22)

$$L(\tilde{X}/\mathbb{F}_q, \mathbb{F}_q^\times, \omega, T) = L(f, \omega, T)\cdot\begin{cases}(1 - T\omega(F_\infty))^{-1} & \text{if } (q-1)|\deg f,\\ 1 & \text{if } (q-1)\nmid \deg f.\end{cases}$$

When $n = 1$, we may work with $\mathscr{L}(\varepsilon, \Delta, P)$ for any $\Delta > 0$, $\varepsilon < 1/d$. Equation (17) reads

$$L^*(f, \omega, T) = \det(I - T(\psi_q \circ F))/\det(I - qT(\psi_q \circ F)). \tag{23}$$

We will realize this polynomial as $\det(I - T\bar{\alpha})$, where $\bar{\alpha}$ is a linear operator on a finite-dimensional p-adic vector space. This is done by finding a differential operator which commutes with $\psi_q \circ F$ up to a factor of q.

An elementary chain rule calculation shows

$$\psi_q \circ \left(x\frac{d}{dx}\right) = q\left(x\frac{d}{dx}\right) \circ \psi_q. \tag{24}$$

Since $\psi_q(G(x^q)) = G(x)$, we have formally

$$\psi_q \circ F(x) = f^\omega(x)^{1/(q-1)} \circ \psi_q \circ f^\omega(x)^{-1/(q-1)}.$$

Hence, formally, $\psi_q \circ F$ commutes up to a factor of q with

$$D = f^\omega(x)^{1/(q-1)} \circ \left(x\frac{d}{dx}\right) \circ f^\omega(x)^{-1/(q-1)} \tag{25}$$

$$= x\frac{d}{dx} - \frac{1}{q-1}\frac{x(f^\omega(x))'}{f^\omega(x)}.$$

This latter differential operator has rational function coefficients with poles only at the zeros of $f^\omega(x)$, hence operates on $\mathscr{L}(\varepsilon, \Delta, P)$. As operators on $\mathscr{L}(\varepsilon, \Delta, P)$,

$$(\psi_q \circ F) \circ D = qD \circ (\psi_q \circ F). \tag{26}$$

We replace $\mathscr{L}(\varepsilon, \Delta, P)$ by

$$\mathscr{L} = \bigcup_{\substack{\Delta > 0\\ 0<\varepsilon<1/d}} \mathscr{L}(\varepsilon, \Delta, P).$$

This does not affect $\det(I - T(\psi_q \circ F))$, which is independent of ε, Δ, and allows us to apply [2] to obtain information about D. It is not hard to check that D has no kernel on $\mathscr{L}$, and by (26) we have a commutative diagram with exact rows

(27)
$$\begin{array}{ccccccccc}
0 & \longrightarrow & \mathscr{L} & \xrightarrow{D} & \mathscr{L} & \longrightarrow & \mathscr{L}/D\mathscr{L} & \longrightarrow & 0 \\
 & & \Big\downarrow{\scriptstyle q(\psi_q \circ F)} & & \Big\downarrow{\scriptstyle \psi_q \circ F} & & \Big\downarrow{\scriptstyle \bar{\alpha}} & & \\
0 & \longrightarrow & \mathscr{L} & \xrightarrow{D} & \mathscr{L} & \longrightarrow & \mathscr{L}/D\mathscr{L} & \longrightarrow & 0
\end{array}$$

where $\bar{\alpha}$ is the endomorphism of $\mathscr{L}/D\mathscr{L}$ induced by $\psi_q \circ F$. Thus by (23) and a variation of [34, Proposition 9]

(28)
$$L^*(f, \omega, T) = \det(I - T\bar{\alpha}).$$

By [2, Remark 1 after Corollary 2], $\mathscr{L}/D\mathscr{L}$ is finite dimensional and

(29)
$$\dim \mathscr{L}/D\mathscr{L} = \deg P - 1 + \begin{cases} 0 & \text{if } f(0) = 0, \\ 1 & \text{if } f(0) \neq 0. \end{cases}$$

One can construct a one-sided inverse to $\psi_q \circ F$ on $\mathscr{L}$, which induces an inverse to $\bar{\alpha}$ since $\mathscr{L}/D\mathscr{L}$ is finite dimensional. Thus $\bar{\alpha}$ is an isomorphism, so by (28) and (29),

(30)
$$\deg L^*(f, \omega, T) = \deg P - 1 + \begin{cases} 0 & \text{if } f(0) = 0, \\ 1 & \text{if } f(0) \neq 0, \end{cases}$$

which (by Lemma 1) reproves (21), since $L^*(f, \omega, T)$ and $L(f, \omega, T)$ are identical except when 0 is unramified, in which case $L^*(f, \omega, T)$ is missing the corresponding linear Euler factor.

Instead of fixing the polynomial $f(x) \in \mathbb{F}_q[x]$, one may allow it to vary in a family, i.e., consider $f_\lambda(x) \in \overline{\mathbb{F}}_p[x]$ where $\lambda \in \overline{\mathbb{F}}_p$ is a parameter. The challenge then is to describe how the $\mathscr{L}_\lambda/D_\lambda\mathscr{L}_\lambda$ and $\bar{\alpha}_\lambda$ vary with λ. In this book, Dwork treats this topic for a certain family of polynomials, namely,

(31)
$$f_\lambda(x) = x^{\alpha_1}(1 - x)^{\alpha_2}(1 - x\lambda)^{\alpha_3} \in \mathbb{F}_p(\lambda)[x].$$

(In fact one may treat rational functions this way: $\alpha_1, \alpha_2, \alpha_3$ need not be positive. The associated function (cf. (4)

$$F_\lambda(x) = \left[\frac{f_\lambda^\omega(x^q)}{f_\lambda^\omega(x)}\right]^{1/(q-1)}$$

is holomorphic except for x near 0, 1, $1/\lambda$ and satisfies (where $q = p^{\deg \lambda}$)

$$\sum_{\substack{x \in \mathbb{F}_{q^r} \\ x \neq 0, 1, 1/\lambda}} \chi_r(x^{\alpha_1}(1-x)^{\alpha_2}(1-x\lambda)^{\alpha_3}) = \sum_{\substack{x \in \mathbb{C}_p \\ x \not\equiv 0, 1, 1/\lambda (\mathrm{mod}\, p) \\ x = x^{q^r}}} F(x)F(x^q) \cdot \ldots \cdot F(x^{q^{r-1}}).)$$

This choice of f_λ is motivated by the classical Euler integral formula for the hypergeometric function ${}_2F_1(a, b; c; \lambda)$ (see [28a, §27]). Indeed, Dwork shows how the variation of cohomology with λ is described by the classical hypergeometric differential equation, regarded p-adically.

One further point: for $f \in \mathbb{F}_q[x]$, the operator related to $L(f, \omega, T)$ was was $\psi_q \circ F$, both ψ_q and F depending on q. When considering $f_\lambda \in \mathbb{F}_p(\lambda)[x]$, q is not fixed and will depend on λ. To handle this problem, one defines an operator "over $\mathbb{F}_p$," and recovers the operator related to $L(f_\lambda, \omega, T)$ by taking iterates of it. We will try to explain this in discussing the operators considered here by Dwork.

Let $a, b, c \in \mathbb{Z}_p \cap \mathbb{Q}$, and consider the algebraic function (we think of x and λ as p-adic variables here)

$$f_0(x, \lambda) = x^b(1-x)^{c-b}(1-x\lambda)^{-a}. \tag{32}$$

(This is the function denoted by Dwork in Chapter 1 as $xf(x, y)$.) Choose $a', b', c' \in \mathbb{Q} \cap \mathbb{Z}_p$ such that $pa' - a$, $pb' - b$, $pc' - c$ all lie in $\mathbb{Z}$. We define a'', b'', c'', etc. recursively by requiring $pa^{(j+1)} - a^{(j)}$, $pb^{(j+1)} - b^{(j)}$, $pc^{(j+1)} - c^{(j)}$ all lie in $\mathbb{Z}$. Note that there exists a positive integer s such that $q = p^s$ satisfies: $(q-1)(a, b, c) \in \mathbb{Z}^3$. Hence $a^{(s)} - a$, $b^{(s)} - b$, $c^{(s)} - c$ all lie in $\mathbb{Z}$. Therefore we can and do assume that $a^{(s)}$, $b^{(s)}$, $c^{(s)}$ are chosen to be a, b, c, respectively. Put

$$f_j(x, \lambda) = x^{b(j)}(1-x)^{c(j)-b(j)}(1-x\lambda)^{-a(j)}. \tag{33}$$

(The function $f_1(x, \lambda)$ was denoted by Dwork in §3.7 as $xh(x, \lambda)$.)

Consider the expression

$$\begin{aligned} G_0(x, \lambda) &= \frac{f_0(x, \lambda)}{f_1(x^p, \lambda^p)} \\ &= x^{b-pb'} \frac{(1-x)^{c-b}}{(1-x^p)^{c'-b'}} \frac{(1-x\lambda)^{-a}}{(1-x^p\lambda^p)^{-a'}}. \end{aligned} \tag{34}$$

(This is denoted $G_{f,h}$ by Dwork in equation (3.7.1).) Since $b - pb' \in \mathbb{Z}$, the first factor on the right is well defined. By the lemma in §3.7, the other two factors are well defined and in fact $G_0(x, \lambda)$ is holomorphic except near $x = 0, 1, 1/\lambda$. If we put $P_\lambda(x) = x(1-x)(1-x\lambda)$, then in our earlier notation, $G_0 \in \mathscr{L}(\varepsilon, \Delta, P_\lambda)$ for some $\varepsilon, \Delta > 0$. If we set

$$G_j(x, \lambda) = \frac{f_j(x, \lambda)}{f_{j+1}(x^p, \lambda^p)}, \tag{35}$$

then similarly

$$G_j \in \mathscr{L}(\varepsilon, \Delta, P_\lambda) \tag{36}$$

for some $\varepsilon, \Delta > 0$.

Define (as in (3.5.6))

$$(\psi_p \xi)(x) = \sum_{y^p = x} \xi(y). \tag{37}$$

Note that

$$\psi_p\left(\frac{1}{1 - x\lambda}\right) = \frac{1}{1 - x\lambda^p},$$

i.e., ψ_p is *not* an endomorphism of $\mathscr{L}(\varepsilon, \Delta, P_\lambda)$. Instead,

$$\psi_p \colon \mathscr{L}(\varepsilon, \Delta, P_\lambda) \to \mathscr{L}(\varepsilon, \Delta, P_{\lambda^p}). \tag{38}$$

Define for $j = 0, 1, \ldots, s - 1$

$$\alpha_{j,\lambda} = \psi_p \circ G_j(x, \lambda) \colon \mathscr{L}(\varepsilon, \Delta, P_\lambda) \to \mathscr{L}(\varepsilon, \Delta, P_{\lambda^p}). \tag{39}$$

(In the notation of §3.8, $\alpha_{0,\lambda} = \alpha_{f,h}$.) Consider the composition

$$\begin{aligned} \text{(40)}\quad & \alpha_{s-1,\lambda^{p^{s-1}}} \circ \alpha_{s-2,\lambda^{p^{s-2}}} \circ \cdots \circ \alpha_{0,\lambda} \\ &= (\psi_p \circ G_{s-1}(x, \lambda^{p^{s-1}})) \circ (\psi_p \circ G_{s-2}(x, \lambda^{p^{s-2}})) \circ \cdots \circ (\psi_p \circ G_0(x, \lambda)) \\ &= (\psi_p)^s \circ G_{s-1}(x^{p^{s-1}}, \lambda^{p^{s-1}}) G_{s-2}(x^{p^{s-2}}, \lambda^{p^{s-2}}) \cdot \ldots \cdot G_0(x, \lambda) \\ &= \psi_q \circ \left[\frac{f_0(x, \lambda)}{f_0(x^q, \lambda^q)}\right], \end{aligned}$$

mapping $\mathscr{L}(\varepsilon, \Delta, P_\lambda)$ into $\mathscr{L}(\varepsilon, \Delta, P_{\lambda^q})$. Suppose $\lambda \in \overline{\mathbb{Q}}_p$ satisfies $\lambda^q = \lambda$. Then this composition is an endomorphism of $\mathscr{L}(\varepsilon, \Delta, P_\lambda)$. The function

$$F(x, \lambda) = \frac{f_0(x, \lambda)}{f_0(x^q, \lambda^q)} = \frac{f_0(x, \lambda)}{f_0(x^q, \lambda)} \tag{41}$$

satisfies: If $x \in \overline{\mathbb{Q}}_p$, $x^q = x$, $x \neq 0, 1, 1/\lambda$, then $F(x, \lambda)$ is the Teichmuller lifting of $g_{\bar\lambda}(\bar x)$, where $g_{\bar\lambda}$ is the rational function

$$g_{\bar\lambda}(\bar x) = \bar x^{b(1-q)}(1 - \bar x)^{(c-b)(1-q)}(1 - \bar x\bar\lambda)^{-a(1-q)} \tag{42}$$

and $\bar{x}$, $\bar{\lambda}$ denote the reductions mod p of x, λ respectively. By the Reich Trace Formula

$$(q^r - 1)\mathrm{tr}(\psi_q \circ F(x, \lambda))^r = \sum_{\substack{\bar{x} \in \mathbb{F}_{q^r} \\ \bar{x} \neq 0, 1, 1/\bar{\lambda}}} \omega(\mathbb{N}_r(g_{\bar{\lambda}}(\bar{x}))), \tag{43}$$

hence

$$L(g_{\bar{\lambda}}, \omega, T) = \frac{\det(I - T(\psi_q \circ F(x, \lambda)))}{\det(I - qT(\psi_q \circ F(x, \lambda)))}. \tag{44}$$

There are differential operators commuting with the individual factors in the composition (40) (see the Lemma of §3.8):

$$\alpha_{j,\lambda} \circ D_{j,\lambda} = pD_{j+1,\lambda^p} \circ \alpha_{j,\lambda}, \tag{45}$$

where

$$\begin{aligned} D_{j,\lambda} &= \frac{1}{f_j(x, \lambda)} \circ x \frac{d}{dx} \circ f_j(x, \lambda) \\ &= x\frac{d}{dx} + \frac{xf'_j}{f_j}, \end{aligned} \tag{46}$$

a differential operator whose coefficients are rational functions with poles only at 0, 1, $1/\lambda$. Thus putting

$$\begin{aligned} \mathscr{L}_\lambda &= \bigcup_{\varepsilon, \Delta} \mathscr{L}(\varepsilon, \Delta, P_\lambda), \\ \mathbf{W}_{j,\lambda} &= \mathscr{L}_\lambda / D_{j,\lambda} \mathscr{L}_\lambda, \end{aligned} \tag{47}$$

it follows that $\alpha_{j,\lambda}$ induces a map

$$\bar{\alpha}_{j,\lambda}\colon \mathbf{W}_{j,\lambda} \to \mathbf{W}_{j+1,\lambda^p}. \tag{48}$$

Hence if $\lambda^q = \lambda$, the composition (40) induces a map

$$\bar{\alpha}\colon \mathbf{W}_{0,\lambda} \to \mathbf{W}_{0,\lambda} \tag{49}$$

and (44) and (45) imply

$$\det(I - T\bar{\alpha}) = L(g_{\bar{\lambda}}, \omega, T). \tag{50}$$

This identifies the L-function related to the operators $\alpha_{j,\lambda}$ and spaces $\mathbf{W}_{j,\lambda}$ studied by Dwork, who generally restricts his attention to the case $j = 0$, the other cases being similar.

Bibliography

[1] Adolphson, A. A p-adic theory of Hecke polynomials. *Duke Math. J.*, **43** (1976), 115–145.

[2] Adolphson, A. An index theorem for p-adic differential operators. *Trans. Amer. Math. Soc.*, **216** (1976), 279–293.

[3] Adolphson, A., Dwork, B., and Sperber, S. Growth of solutions of linear differential equations at a logarithmic singularity. *Trans. Amer. Math. Soc.* (to appear).

[4] Amice, Y. *Les Nombres p-adiques*. Presses Univ. de France, 1975.

[4a] Baldassarri, F. Differential modules and singular points of p-adic differential equations. *Adv. Math.*, **44** (1982), 155–179.

[4b] Barsky, D. On Morita's p-adic gamma function. *Math. Proc. Comb. Phil. Soc.*, **89** (1981), 23–27.

[5] Boyarsky, M. p-adic gamma function and Dwork cohomology. *Trans. Amer. Math. Soc.*, **257** (1980), 359–369.

[6] Clark, D. A note on the p-adic convergence of solutions of linear differential equations. *Proc. Amer. Math. Soc.*, **17** (1966), 262–269.

[7] Diamond, J. The p-adic log gamma function and p-adic Euler constants. *Trans. Amer. Math. Soc.*, **233** (1977), 321–337.

[8] Diamond, J. Hypergeometric series with p-adic variable. *Pacific J. Math.*, **94** (1981), 265–276.

[8a] Dwork, B. On the rationality of the zeta function of an algebraic variety, *Amer. J. Math.*, **82** (1960), 631–648.

[9] Dwork, B. A deformation theory for the zeta function of a hypersurface. *Proc. Int. Cong. Math.*, (1962), 247–259.

[9a] Dwork, B. On the zeta function of a hypersurface. *Pub. Math. IHES*, **12** (1962), 5–68.

[10] Dwork, B. On the rationality of zeta functions and L-series. *Proc. Conf. Local Fields, Driebergen 1966*. Springer-Verlag, 1967, pp. 40–55.

[11] Dwork, B. *P*-adic cycles. *Pub. Math. IHES*, **37** (1969), 327–415.

[12] Dwork, B. Normalized period matrices, I. *Ann. Math.*, **94** (1971), 337–388.

[13] Dwork, B. Normalized period matrices, II. *Ann. Math.*, **98** (1973), 1–57.

[14] Dwork, B. p-adic differential equations, IV. *Ann. Sci. Ecole Normale Sup.*, **6** (1973), 295–315.

[15] Dwork, B. Arithmetic theory of differential equations. *Symposia Matematica Vol. XXIV, 1981*. Istituto Nazionale de Alta Matematica Francesco Severi. 225–243.

[16] Dwork, B. and Robba, P. On ordinary linear p-adic differential equations. *Trans. Amer. Math. Soc.*, **231** (1977), 1–46.

[17] Dwork, B. and Robba, P. Effective p-adic bounds for solutions of homogeneous linear differential equations. *Trans. Amer. Math. Soc.*, **259** (1980), 559–577.

[18] Goursat, E. *Lecons Sur Les Séries Hypergéometriques*. Hermann & Co., Paris, 1936.

[19] Gross, B. and Koblitz, N. Gauss sums and the p-adic Γ-function. *Ann. Math.*, **109** (1979), 569–581.

[19a] Hasse, H. Theorie der Relativ-Zyklischen Algebraischen Funktionenkörper, Inbesondere bei Endlichem Konstantenkörper, *J. Reine Angew. Math.*, **172** (1934), 37–54.

[20] Honda, T. Formal groups obtained from generalized hypergeometric functions. *Osaka J. Math.*, **9** (1972), 447–462.

[21] Katz, N. Crystaline cohomology, Dieudonné modules, and Jacobi sums. *Tata Inst. Bombay 1979* (Automorphic Forms, Representation Theory and Arithmetic). 165–246.

[22] Koblitz, N. The hypergeomtric function with p-adic parameters. *Proc. Queen's (Ontario) Number Theory Conference (1979)*, pp. 319–328.
[23] Lang, S. *Cyclotomic Fields, II.* Springer-Verlag, 1980.
[23a] Lazand, M. Les zéros d'une fonction analytique d'une variable sur un corps valué complet. *Pub. Math. IHES*, **14** (1962), 47–75.
[24] Lubin, J. Canonical subgroups of formal groups. *Trans. Amer. Math. Soc.*, **251** (1979), 103–127.
[25] Lutz, E. Sur l'equation $y^2 = x^3 - Ax - B$ dans les corps p-adiques. *J. Reine Angew. Math.*, **177** (1937), 238–247.
[26] Messing, W. On the nilpotence of the hypergeometric equation. *J. Math. Kyoto Univ.*, **12** (1972), 369–383.
[27] Monsky, P. Formal cohomology, III. *Ann. Math.*, **93** (1971), 315–343.
[28] Morita, Y. A p-adic analogue of the Γ-function. *J. Fac. Sci. Univ. Tokyo Sect. IA Math.*, **22** (1975), 255–266.
[28a] Poole, E. G. C. *Introduction to the Theory of Linear Differential Equations*, Oxford University Press, Oxford, 1936 (Dover reprint, 1960).
[29] Reich, D. A p-adic fixed point formula. *Amer. J. Math.*, **91** (1969), 835–850.
[30] Robba, P. On the index of p-adic differential operators, I. *Ann. Math.*, **101** (1975), 280–316.
[31] Robba, P. Solutions bornées des systems differentiels linéaires, Applications aux fonctions hypergéometriques. GEAU 1975/76, Exp. 5.
[32] Robba, P. Disque genérique et équations differentielles. GEAU 1976/77, Exp. 8.
[33] Robba, P. Prolongement analytique. *Asterisque*, **10** (1973).
[34] Serre, J-P. Endomorphismes complètement continus des espaces de banach p-adiques, *Pub. Math. IHES.*, **12** (1962), 69–85.
[35] Sibuya, Y. and Sperber, S. Arithmetic properties of power series solutions of algebraic differential equations. *Ann. Math.*, **113** (1981), 111–157.
[36] Watson, G. *Theory of Bessel Functions.* Cambridge University Press, Cambridge 1966.

Index of Symbols

(Chapters 10, 11, and 12 are more or less individually self-contained with independent notation.)

Index

Grundlehren der mathematischen Wissenschaften

A Series of Comprehensive Studies in Mathematics

A Selection

161. Bhatia/Szegö: Stability of Dynamical Systems
162. Nevanlinna: Analytic Functions
163. Stoer/Witzgall: Convexity and Optimization in Finite Dimensions I
164. Sario/Nakai: Classification Theory of Riemann Surfaces
165. Mitrinovic/Vasic: Analytic Inequalities
166. Grothendieck/Dieudonné: Eléments de Géometrie Algébrique I
167. Chandrasekharan: Arithmetical Functions
168. Palamodov: Linear Differential Operators with Constant Coefficients
169. Rademacher: Topics in Analytic Number Theory
170. Lions: Optimal Control of Systems Governed by Partial Differential Equations
171. Singer: Best Approximation in Normed Linear Spaces by Elements of Linear Subspaces
172. Bühlmann: Mathematical Methods in Risk Theory
173. Maeda/Maeda: Theory of Symmetric Lattices
174. Stiefel/Scheifele: Linear and Regular Celestial Mechanic. Perturbed Two-body Motion – Numerical Methods – Canonical Theory
175. Larsen: An Introduction to the Theory of Multipliers
176. Grauert/Remmert: Analytische Stellenalgebren
177. Flügge: Practical Quantum Mechanics I
178. Flügge: Practical Quantum Mechanics II
179. Giraud: Cohomologie non abélienne
180. Landkof: Foundations of Modern Potential Theory
181. Lions/Magenes: Non-Homogeneous Boundary Value Problems and Applications I
182. Lions/Magenes: Non-Homogeneous Boundary Value Problems and Applications II
183. Lions/Magenes: Non-Homogeneous Boundary Value Problems and Applications III
184. Rosenblatt: Markov Processes. Structure and Asymptotic Behavior
185. Rubinowicz: Sommerfeldsche Polynommethode
186. Handbook for Automatic Computation. Vol. 2. Wilkinson/Reinsch: Linear Algebra
187. Siegel/Moser: Lectures on Celestial Mechanics
188. Warner: Harmonic Analysis on Semi-Simple Lie Groups I
189. Warner: Harmonic Analysis on Semi-Simple Lie Groups II
190. Faith: Algebra: Rings, Modules, and Categories I
191. Faith: Algebra II, Ring Theory
192. Mallcev: Algebraic Systems
193. Pólya/Szegö: Problems and Theorems in Analysis I
194. Igusa: Theta Functions
195. Berberian: Baer*-Rings
196. Athreya/Ney: Branching Processes
197. Benz: Vorlesungen über Geometric der Algebren
198. Gaal: Linear Analysis and Representation Theory
199. Nitsche: Vorlesungen über Minimalflächen
200. Dold: Lectures on Algebraic Topology
201. Beck: Continuous Flows in the Plane
202. Schmetterer: Introduction to Mathematical Statistics
203. Schoeneberg: Elliptic Modular Functions
204. Popov: Hyperstability of Control Systems

205. Nikollskii: Approximation of Functions of Several Variables and Imbedding Theorems
206. André: Homologie des Algébres Commutatives
207. Donoghue: Monotone Matrix Functions and Analytic Continuation
208. Lacey: The Isometric Theory of Classical Banach Spaces
209. Ringel: Map Color Theorem
210. Gihman/Skorohod: The Theory of Stochastic Processes I
211. Comfort/Negrepontis: The Theory of Ultrafilters
212. Switzer: Algebraic Topology—Homotopy and Homology
213. Shafarevich: Basic Algebraic Geometry
214. van der Waerden: Group Theory and Quantum Mechanics
215. Schaefer: Banach Lattices and Positive Operators
216. Pólya/Szegö: Problems and Theorems in Analysis II
217. Stenström: Rings of Quotients
218. Gihman/Skorohod: The Theory of Stochastic Processes II
219. Duvant/Lions: Inequalities in Mechanics and Physics
220. Kirillov: Elements of the Theory of Representations
221. Mumford: Algebraic Geometry I: Complex Projective Varieties
222. Lang: Introduction to Modular Forms
223. Bergh/Löfström: Interpolation Spaces. An Introduction
224. Gilbarg/Trudinger: Elliptic Partial Differential Equations of Second Order
225. Schütte: Proof Theory
226. Karoubi: K-Theory. An Introduction
227. Grauert/Remmert: Theorie der Steinschen Räume
228. Segal/Kunze: Integrals and Operators
229. Hasse: Number Theory
230. Klingenberg: Lectures on Closed Geodesics
231. Lang: Elliptic Curves: Diophantine Analysis
232. Gihman/Skorohod: The Theory of Stochastic Processes III
233. Stroock/Varadhan: Multi-dimensional Diffusion Processes
234. Aigner: Combinatorial Theory
235. Dynkin/Yushkevich: Markov Control Processes and Their Applications
236. Grauert/Remmert: Theory of Stein Spaces
237. Köthe: Topological Vector-Spaces II
238. Graham/McGehee: Essays in Commutative Harmonic Analysis
239. Elliott: Probabilistic Number Theory I
240. Elliott: Probabilistic Number Theory II
241. Rudin: Function Theory in the Unit Ball of $\mathbb{C}^n$
242. Blackburn/Huppert: Finite Groups I
243. Blackburn/Huppert: Finite Groups II
244. Kubert/Lang: Modular Units
245. Cornfeld/Fomin/Sinai: Ergodic Theory
246. Naimark: Theory of Group Representations
247. Suzuki: Group Theory I
248. Suzuki: Group Theory II
249. Chung: Lectures from Markov Processes to Brownian Motion
250. Arnold: Geometrical Methods in the Theory of Ordinary Differential Equations
251. Chow/Hale: Methods of Bifurcation Theory
252. Aubin: Nonlinear Analysis on Manifolds. Monge—Ampère Equations

GPSR Compliance

The European Union's (EU) General Product Safety Regulation (GPSR) is a set of rules that requires consumer products to be safe and our obligations to ensure this.

If you have any concerns about our products, you can contact us on ProductSafety@springernature.com

In case Publisher is established outside the EU, the EU authorized representative is:

Springer Nature Customer Service Center GmbH
Europaplatz 3
69115 Heidelberg, Germany

Batch number: 10620143

Printed by Printforce, the Netherlands